Advanced Instrumentation, Data Interpretation, and Control of Biotechnological Processes

Advanced Instrumentation, Data Interpretation, and Control of Biotechnological Processes

Edited by

Jan F.M. Van Impe
Katholieke Universiteit Leuven,
Department of Food and Microbial Technology,
Leuven, Belgium

Peter A. Vanrolleghem
Universiteit Gent,
Department of Applied Mathematics, Biometrics and Process Control,
Gent, Belgium

and

Dirk M. Iserentant
Katholieke Universiteit Leuven,
Department of Food and Microbial Technology,
Leuven, Belgium

Springer-Science+Business Media, B.V.

A C.I.P. Catalogue record for this book is available from the Library of Congress.

ISBN 978-90-481-4954-4 ISBN 978-94-015-9111-9 (eBook)
DOI 10.1007/978-94-015-9111-9

Printed on acid-free paper

Originally published by Kluwer Academic Publishers in 1998.
Softcover reprint of the hardcover 1st edition 1998

To my parents
J.V.I.

Contents

Preface

This book is the outgrowth of the COMETT II Course on Advanced Instrumentation, Data Interpretation, and Control of Biotechnological Processes organized by the Katholieke Universiteit Leuven and the Universiteit Gent, and held at Gent, Belgium, October 1994.

The editors of the present volume were very fortunate to find all invited speakers prepared to write state-of-the-art expositions based on their lectures. Special thanks are due to all of them. The result is an account of recent advances in instrumentation, data interpretation, and model based optimization and control of bioprocesses. For anyone interested in this emerging field, this text is of value and provides comprehensive reviews as well as new and important trends and directions for the future, motivated and illustrated by a wealth of applications.

The typesetting of all this material represented a tremendous amount of work. I am most grateful to my wife, Myriam Uyttendaele, and to Kurt Gheys, who did most of the proof-reading. Their efforts have increased a lot the uniformity in style and presentation of the different manuscripts. Many thanks also to the co-editors, for their continued support.

Kluwer Academic Publishers is gratefully acknowledged for publishing this book, thus contributing to the transfer of the latest research results into large scale industrial applications.

Leuven, august 1997
Jan F.M. Van Impe

Preface

Introduction

The scope of the field of biotechnological processes is very wide, covering such processes as fermentations for production of high-valued specialist chemicals (e.g., pharmaceuticals), high-volume production of foods and feeds (yoghurt, cheese, beer, ...) as well as biological waste treatment, handling solid (composting), liquid (activated sludge) and gaseous wastes (biofilters). Compared to other engineering disciplines, the introduction of modern optimization and control strategies is lagging behind. Two main reasons can be identified. First, the living organisms (or part thereof) that are central to these processes make the mathematical modeling of the processes a difficult task, and, since models are central to the development of control systems, therefore also the on-line control problem is complex. The other difficulty stems from the absence, in most cases, of cheap and reliable instrumentation suited to real-time monitoring.

In this book a number of advanced techniques are introduced to deal with these problems. In the first part modern on-line hardware sensors *are discussed in detail (FIA, viable biomass measurement, membrane inlet mass spectrometry, flow cytometry, microcalorimetry). In the second part several aspects of* model based optimization and control *are dealt with, starting from on-line data acquisition, processing, error diagnosis and data reconciliation, over bioprocess modeling and identification, and development of combined hardware-software sensors, up to on-line (optimal) adaptive and model based predictive control algorithms.*

The book is directed to engineers, researchers, and students in the field of process control and systems theory as applied to industrial biotechnological processes, as well as to bioengineers who have some background in control engineering and would like to apprehend better how advanced control theory applies to biological processes.

The Editors

Part I

Hardware Sensors

Chapter 1

Flow Injection Analysis for On-line Monitoring of a Wastewater Treatment Plant

S. ISAACS and H. SØEBERG
Technical University of Denmark
Department of Chemical Engineering
Building 229 DK-2800 Lyngby (DENMARK)

ABSTRACT – *Flow injection analysis (FIA) is a technique by which an analytical procedure can be performed in an automatic fashion, with low reagent consumption, high reproducibility, high accuracy and high measurement frequency. Coupled to an automatic system for sample withdrawal and preparation, FIA lends itself well to the on-line monitoring of chemical and biotechnological processes. In this chapter the basic principles of FIA are explained, based on the example of a FIA system developed for the monitoring and control of an activated sludge wastewater treatment plant.*

KEYWORDS – *Activated Sludge, Ammonia Monitoring, Flow Injection Analysis, Nitrate Monitoring, Phosphate Monitoring, Process Monitoring, Wastewater Treatment*

J.F.M. Van Impe et al. (eds.), Advanced Instrumentation, Data Interpretation, and Control of Biotechnological Processes, 3-40.

1.1 Introduction

Much effort is being expended today towards the development of improved operational and control strategies for wastewater treatment processes based on the activated sludge principle. The ability to automatically monitor concentrations of key constituents in these processes is essential for many of these strategies, and paves the way towards an enhanced understanding of process behavior. With regard to the latter, if we can see what is occurring in the process by sitting in front of a computer screen and observing the course of measured data collected sufficiently often, day in day out, then we can learn more about the process than in the case when only manually collected data, obtained from occasional grab samples or from intensive but short term experimental efforts, is available. With regard to the former, process control, whether it be an automatic PID or model based strategy, or simply operator induced manual manipulations, requires measurements on which decisions can be based.

The number of process variables which we would like to automatically monitor are many, and include both component concentrations (e.g., biomass, organic substrates, nutrient pollutants) as well as reaction rates (e.g., nitrification, denitrification, phosphate uptake). The former tell us the state of the process and can be used directly in feedback control strategies. The latter indicate how the process is doing, and find use in model based and predictive control strategies. For activated sludge processes designed for nutrient removal, the key components to be measured include the nutrients themselves, namely ammonia, nitrate and phosphate. If the frequency at which these components are measured is high enough, reaction rates can be derived from their time course. It is the automatic measurement of these three components using Flow Injection Analysis (FIA) which makes up the scope of this chapter.

Ammonia and nitrate can be measured using ion selective electrodes, and several papers have appeared dealing with the monitoring and control of activated sludge processes with this measurement technique, e.g., [1]. Here the focus is on FIA technology, and how it has been applied to the measurement of ammonia, nitrate and phosphate at a pilot plant scale activated sludge process located in Lundtofte, Denmark, at a research facility belonging to the Department of Environmental Engineering (IMT) at the Technical University of Denmark (DTU). The measurement system has been developed over a five year period, and has already been in service for more than 5 years for the development and testing of new control strategies.

1.2 What is Flow Injection Analysis?

1.2.1 Basic principles

Flow Injection Analysis (FIA) is a technique by which an entire analytical procedure, whose steps may include (among others) sample preparation, separation, chemical reaction and detection, can be performed automatically, reproducibly, and at a relatively high sample rate. The method consists basically of injecting a precise volume of sample fluid containing the analyte into a moving carrier stream, which may or may not contain reagents with which the analyte reacts. The injected sample forms a zone which is then transported through the system. During this transport, the zone undergoes whatever steps are required to form a detectable component, and a signal is recorded as this component passes through a flow-through detector. The analyte concentration of the original sample is then determined by comparing some measure of the detected signal, e.g., peak height, with a calibration curve. It is out of the scope of this chapter to provide a comprehensive tutorial on FIA technology. Nor will any of the numerous ways in which the technology can be applied be discussed, other than that which is implemented at the activated sludge pilot plant. An excellent book dealing with FIA operation and design principles exists [2], and the reader is referred to this and the over 1400 references cited within, particularly if insight into other methods of sample injection, detection and data analysis, as well as other techniques such as gradient, stop flow and merging zone, is desired. The brief discussion which ensues is meant only to provide enough basic information to the complete novice, so that the application to wastewater treatment described in the remainder of this chapter can be understood.

A very simple FIA setup is shown schematically in Figure 1.1a, consisting of a single carrier/reagent line and a coiled length of tubing as reactor module. The pump (P) continuously propels the carrier stream through the injection valve (IV), reaction module (R) and the detector (D). For the majority of the time the injection valve is set in a position which causes the carrier stream to bypass the sample. The sample volume is determined by the dimensions of either an external loop of tubing or the volume of a cavity within the valve, depending on the type of injection valve employed. At the time of injection, which is considered the start of a FIA cycle, the position of the injection valve is switched so that the sample volume becomes a part of the carrier stream. In this manner, precisely the same volume of sample is injected each FIA cycle.

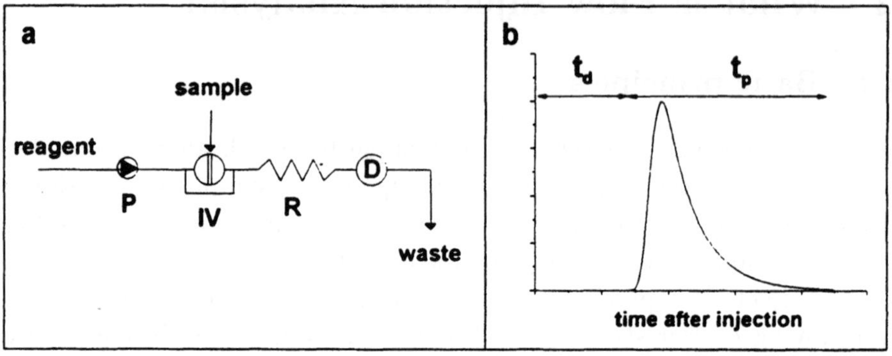

Figure 1.1: (a) Schematic of a simple FIA analyzer and (b) a typical detected FIA signal (the abscissa shows units of time after injection and the ordinate shows the detected quantity, e.g., absorbance units).

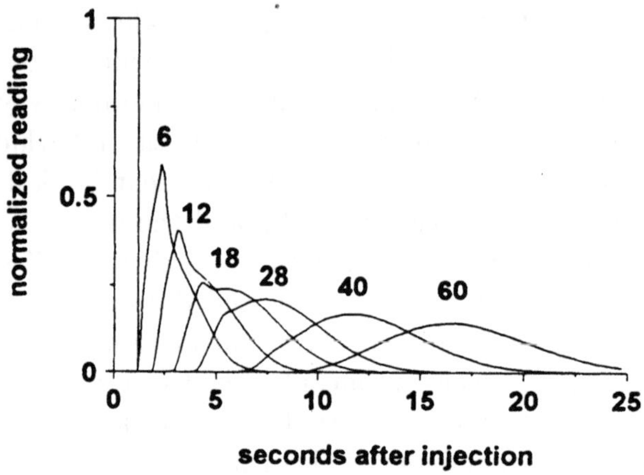

Figure 1.2: An initially plug-like sample zone becomes more rounded due to dispersion as it travels through the FIA system.

As the sample volume travels towards the detector, dispersion into the surrounding carrier fluid occurs, and the once plug-like volume of sample spreads and becomes more rounded axially (Figure 1.2). At the same time, any reagents contained in the carrier fluid disperse into the zone occupied by the sample fluid, and reaction occurs. This reaction may, for example, be the formation of a color, the intensity of which should be linearly related to the concentration of the analyte. In this case, a spectrophotometer equipped with a flow-through cell can be employed as detector, and a signal which is continuous in time and looks something like that shown in Figure 1.1b would be obtained. Since each FIA measurement is performed in a reproducible manner, the extent of reaction (provided that reagents are present in excess) and the amount of dispersion are the same from measurement to measurement. If the FIA system is properly designed, certain characteristics of the detected signal curve will be linearly related to the analyte concentration and, hence, can be used as a measure of analyte concentration in an arbitrary sample volume. These include peak height, peak area, or peak height at a specific location (in time) on the signal curve. Most commonly employed is the peak height, as its value is very simple to determine.

The curves in Figure 1.2 were produced by simulation using the method based on orthogonal collocation described in [3], for the case of transport along a straight tube. The parameters employed were: inner tube radius 0.25 mm; injected volume 10 μL; flow rate 0.5 mL/min; diffusion coefficient 3.6 10^{-5} cm/s. Each curve in the figure represents the mixing-cup average concentration as a function of elapsed time observed at different distances from the point of injection. The distance, or tubing length, in centimeters is given by the numbers above the curves. It is interesting to note that the signal curve becomes gaussian very shortly after injection. The tailing which is very typical of FIA curves (see Figure 1.1b and Figure 1.13 below) is caused by volumes which act as mixing chambers such as the internal volume of a flow-through spectrophotometric detection cell.

1.2.2 Advantages of FIA

The particular characteristics of FIA, as well as some of the advantages, become apparent when the technology is compared to manually performed analyses. With the latter, reactions are carried out to completion or until an equilibrium is attained. With FIA, on the other hand, reactions do not need to go to completion, since the technology allows each sample measurement to be performed in a reproducible manner, where each sample volume experiences the same degree of dispersion and the same reaction time. This reproducibility means that analyte concentrations can be found by comparison with calibration curves prepared from standards of known concentrations which have been treated in exactly the same manner. Obviously, this means that the development of a calibration curve is essential for FIA, and this calibration should be performed at regular intervals, both as a validity check and as a means to compensate for minor changes in system properties.

Clearly, the most significant advantage of FIA over manually performed analyses is that FIA is automatic, and hence human errors and labor effort can be spared. Additionally, reactions are performed with FIA in tubes (and similar flow-through modules) where sample segregation occurs by means of a volume of carrier fluid between each injected sample. This lends nicely to miniaturization, where low pumping rates and small diameter tubes allow for small sample volume requirements and low reagent consumption. Another advantage is the relatively high sample rate afforded by a well designed FIA system, which can easily approach 60 to 120 measurements per hour. Referring to Figure 1.1b, the time required for the front edge of the signal curve to reach the detector after injection is given as t_d, and the width (in time) of the peak is given as t_p. The total time elapsed after injection for completion of a measurement is thus $T = t_d + t_p$. However, the minimum time between measurement cycles is t_p, since the detected signals will first begin to overlap, and hence interfere with each other, when the time between injections becomes less than the signal width t_p.

1.2.3 Some design principles

To obtain high sensitivity and good linearity in a FIA system at least five conditions must be satisfied: proper sample preparation, proper injection, proper mixing, proper reaction and proper detection. The combination can be optimized by using proper dimensions, proper flow rate, proper timing and proper geometry. Due to the desired high frequency of analysis the flow is never stopped. If the flow is stopped the chances of clogging increase due to bacterial growth and crystallisation on the walls of the tubing. Additionally, the elastic *memory* of the tubing can produce undesired variance unless special care is taken. If more than one pump is used it is essential that pulsations are synchronized, partly to avoid *backflow* in parts of the (elastic) system, and partly to avoid any differences in pumping frequency (aliasing effects) which will increase the measurement variance.

If photometric detection is employed, and to reduce clogging, sample filtration is needed. The combination of a *fast* recirculation loop and a spiral cross flow filter allows for a sufficient filtration rate while at the same time maintaining the filter surface clear of particles due to a strong secondary flow induced by centrifugal forces. By designing the filter to employ a relatively low recirculation loop flow rate the stress induced can be minimized so that bacterial flocks (important in biological systems) are not affected.

An important source of error in FIA is variation in the injected volume of sample. This can be reduced by increasing the injected volume but an upper limit is found if proper mixing would be obtained (this is important for the linearity at high analyte concentrations) and proper flushing of the system is also essential between samples to obtain a steady baseline. The injection volume should be selected to give a proper balance between sensitivity, linearity and variance. The interaction of the axial and radial concentration profiles during transportation through the FIA system can be influenced by proper geometry. At the injection point the primary function is mixing aided by convection and diffusion. In the reaction zone the dispersion should be minimum. Relatively large cross sections can be used for mixing, but for the reaction the situation is reversed. The cross section of the tubing is fixed from above by the *diffusion time* and from below by the chances of clogging. However, an artificial reduction of the cross section can be induced by coiling the tubes. The secondary flow across a coiled tube caused by centrifugal forces makes the flow pattern look like an array of parallel tubes (i.e., close to plug flow). The *number* of tubes depends on the curvature and the flow rate. The Meander coil (Figure 1.3) is found to be very efficient at low flow rates whereas a regular coil needs ten times the flow rate to produce the same effect. This means a factor ten in the amount of reactant used. The flow rate to use is decided from the geometry to select the desired number of tubes. The height of the peak at the detector is proportional to this number (5 for the Meander coil, 2 to 5 for a regular coil).

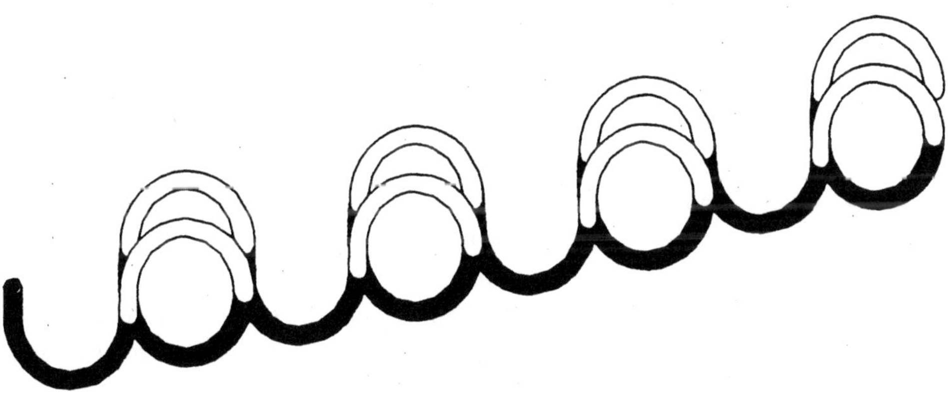

Figure 1.3: In the Meander coil, the flowing liquid encounters a continual series of bends in different directions. The coil is easily prepared by weaving the tubing through a perforated metal plate. Here, the shaded and unshaded tubing segments are respectively under and over the plate.

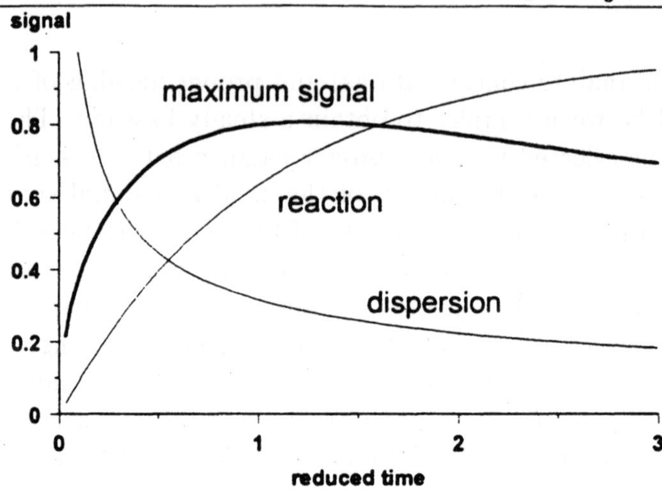

Figure 1.4: The detected signal is maximum when the residence time is close to the time constant of the reaction due to the combined effect of reaction (increasing curve) and dispersion (decreasing curve).

The residence time in the reactor (now selected by the length of the reactor tube) should produce a conversion between 50 % and 90 %. This conversion is found to give maximum sensitivity to the analyte and minimum sensitivity to variations in the flow rate. The reason for this is the complex interaction between reaction and dispersion as illustrated in Figure 1.4.

The decrease in signal level due to dispersion is approximately proportional to the inverse of the square root of time. The increase in signal level due to reaction is proportional to one minus the exponential of negative time (assuming first order kinetics, but the result is not sensitive to details regarding geometry and order of reaction). The signal level which is detected is the product of these two effects.

The detection device (and the chemistry selected) should be designed for maximum sensitivity to the reaction product and minimum sensitivity to the other constituents of the sample. In particular particles and gas bubbles should be avoided or at least easily released, but diffraction effects due to varying ionic strengths should also be considered. These problems are resolved in some industry standard spectrophotometers.

1.3 FIA for Activated Sludge Process Monitoring

The remainder of this chapter describes a FIA system designed for the on-line monitoring of activated sludge wastewater treatment processes. The FIA system is implemented on a pilot plant scale alternating type nutrient removal process (trademark BIODENIPHO developed by I. Krüger A/S, Søborg, Denmark; for a process description see, e.g., [4][5]). A schematic of the pilot plant is shown in Figure 1.5. The FIA system measures the concentrations of ammonia (NH_4-N), nitrite plus nitrate (NO_x-N) and phosphate (PO_4-P) at four locations in the process: the inlet water (SP1), at the exit of the anaerobic zone (SP2), in the aeration tank T2 (SP3) and in the effluent water (SP4). For a description of the pilot plant as well as application examples of the FIA system for process control, see [6][7].

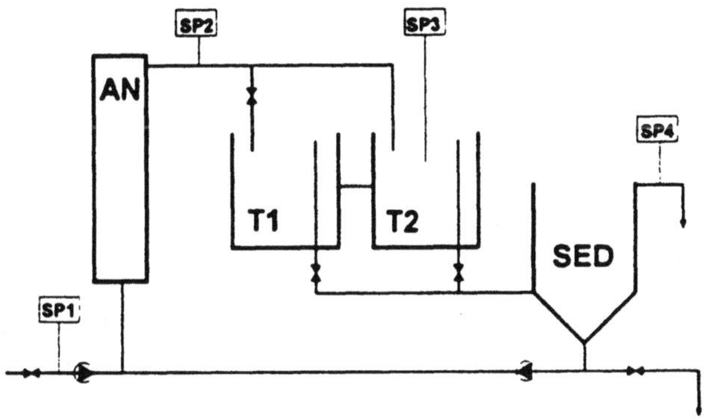

Figure 1.5: Schematic of the BIODENIPHO pilot plant. AN: anaerobic zone; T1, T2: anoxic/aerobic tanks; SED: final settler; SP1-SP4 are the FIA sampling points.

For convenience, the following terminology will be employed:

- *sample point* denotes one of the sample locations SP1 to SP4.
- *pilot plant* denotes the alternating activated sludge process on which the FIA system is implemented;
- *FIA cycle* denotes the measurement of all three components NH_4-N, NO_x-N and PO_4-P in either the standard solution or one of the sample points.
- *measurement cycle* denotes a series of FIA cycles starting with the measurement of the standard solution followed by a fixed sequence of the four sample points;
- *analyzer* refers to the part of the FIA system hardware responsible for carrying out one of the three analyses, for example the ammonia analyzer is shown below in Figure 1.14.

Before beginning with the detailed description of the pilot plant FIA system, some characteristics which set it apart from FIA in general should be pointed out.

- The measured species are simple aqueous ions, and hence relatively simple FIA manifolds can be employed, where detection is performed spectrophotometrically after a few chemical conversion steps.
- The system is intended to always be in operation, 24 hours a day and each day of the year. At a rate of one FIA cycle each 1.5 minutes, this translates to over 350,000 FIA cycles per year.
- Reagent flow rates are in the range of 200 to 300 μL/min which is about 5 to 10 times lower than the flow rate commonly employed in FIA. The reagent pumps employed are highly pulsating, whereas the usual design tendency is to eliminate flow pulsations as much as possible.
- The system has been designed to achieve as large a linear response range as possible in order to measure the entire range of concentrations encountered in the pilot plant without a sample dilution step.

It should also be made clear that the FIA setup described in the following does not necessarily represent the optimal design solution for the automatic monitoring of a wastewater treatment plant. The development was accomplished over a five year period mainly as part of a PhD study, and has been employed since for investigational purposes involving modeling and control of the monitored process. Modifications to the original design ([8][9]) have been made, and are still being performed as the need arises. A number of components are rather expensive, and more routine maintenance is required than originally intended. Many operational problems still arise and the most significant of these will be pointed out. The FIA system can be subdivided into four main units, these being a common sample collection and preparation stage, and three separate analyzers for NH_4-N, NO_x-N and PO_4-P. The system can also be subdivided according to the following tasks.

- Sample acquisition, selection and transport
- Sample injection
- Reagent transport
- Reaction

- Detection
- Standard calibration
- Detected signal analysis
- Automatic and manual trouble prevention

Each of these tasks will be described in turn below, followed by a description of the three analyzers. This will then be followed by a discussion of the problem of carry-over, which is the main cause of noisy measurements in this FIA setup. Finally, the operational implementation will be described along with some example measurements of the pilot plant process.

1.3.1 Sample acquisition, separation, selection and transport

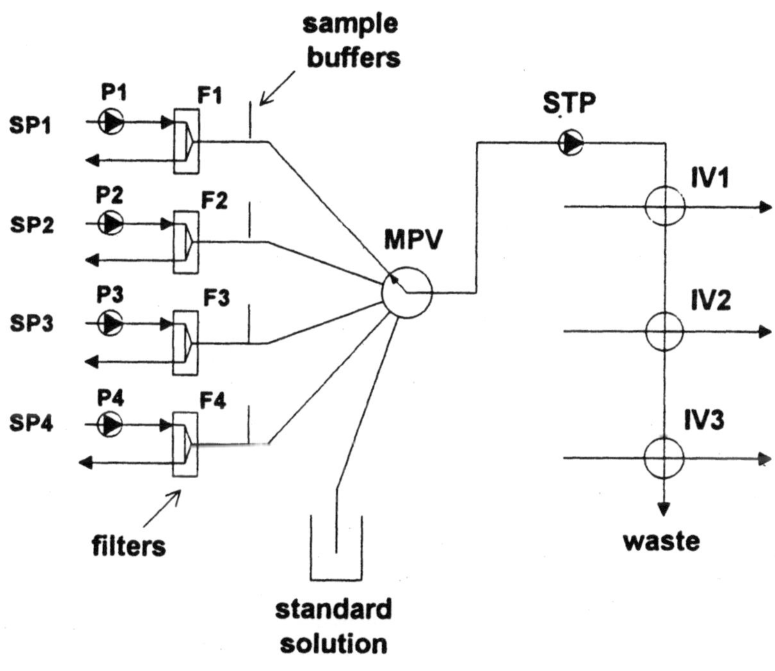

Figure 1.6: Sample preparation and delivery section of the pilot plant FIA system. SP: sample points; P: Mohno pumps; F: cross flow filters; MPV: multiport valve; STP: sample transport pump; IV: injection valves.

The objective here is to provide the injection valves of the three analyzers with a particle free volume of process water from the various sample points. The process water contains activated sludge and other particulate matter, and hence a solids separation is performed (sample acquisition and separation). The particle free process water from each of the four sample points, as well as a standard solution for calibration, is then sent sequentially through a common sample line in order to fill the injection valves of all three analyzers (sample selection and transport). A schematic of the employed setup is shown in Figure 1.6.

Sample acquisition and separation is performed as follows: four pumps of type NE15A from Netch Mohno (P1 to P4) pump the process water from the four sample points (SP1 to SP4) through four cross-flow filter units (F1 to F4) and back to the plant. The speed of the pumps can be adjusted with frequency regulators. This is particularly important for sample points SP2 and SP3 since the water from these two locations contain activated sludge. The shear forces on the bacterial flocs should be minimized, and hence the pumps are manually adjusted to the lowest speed which still provides an adequate filtrate flow rate.

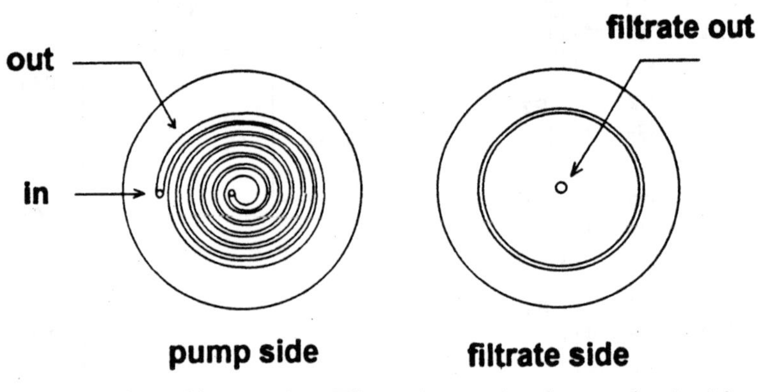

Figure 1.7: Cross-flow filter unit. The schematic shows the inside faces of the two halves of the filter unit housing.

The cross-flow filter unit is shown in Figure 1.7, and consists of two stainless steel discs held together with screws. A semicircular helical pathway is cut in the inner face of the donor side disc, and process water is pumped through the filter in the direction indicated in the figure. The helical flow path ensures that the liquid travels across the filter at a high rate, which aids in maintaining the filter surface clean. The path of flow from the periphery to center of the spiral contributes to establishing a pressure drop across the filter. For filtrate collection, a very shallow circular area opposing the helical pathway of the donor disc has been cut in the acceptor disc. The volume of this circular area is taken up by a 40 micron stainless steel mesh, which serves as filter support and ensures that all of the exposed filter surface is used. The filter material employed is an ultrafiltration membrane with a molecular weight cutoff at 20 10^4 daltons (Danish Sugar Co., Filtration Division, Type ETNA 20A).

Both a large and small version of the cross-flow filter unit are employed. The smaller (larger) unit has the following dimensions in cm: housing diameter: 9.4 (12.4); helical semicircular channel and entrance/exit tube diameter: 0.4 (0.8); outer diameter of helix: 5.5 (9). Originally, only the smaller unit was used for all four sample points. The diameter of the helical pathway within the smaller unit is less than the diameter of the tubes carrying water to and from the filter unit in order to augment the pressure drop across the filter membrane. For sample point SP1, however, the smaller unit was susceptible to plugging due to the occasional presence of particulate matter in the process inlet water. Also, it was feared that the shear stresses occurring within the helix may cause breakup of the bacterial flocs, and hence be detrimental to sludge settleability. Consequently, to reduce shear stress and to avoid abrupt changes in path diameter, the larger version of the cross-flow filtration unit was built, and found to provide satisfactory filtrate flow rates. Sludge settleability is good despite the fact that the sludge from the two sample points SP2 and SP3 are continuously pumped through the larger filter units. Today the smaller unit is only used for the effluent water (SP4), whereas the larger version is employed for the remaining three sample points.

The filtrate leaving each filter unit accumulates in sample buffers, which consist of 2 m of 1.5 mm i.d. PFTE tubing. The sample buffers are open ended, and the filtrate flowing out of the top of the buffers is sent to waste. The purpose of the buffer tubes is that they allow the sample transport pump (STP) flow rate to be higher than the filter flow rates. If this occurs, the sample which has been stored in the buffer since the last measurement is fed to the injection valves in a last-in first-out manner. To ensure robust operation, the buffer volume has been designed to be greater than the volume of liquid pumped by the sample transport pump during a FIA cycle. In this system, the sample transport pump has a flow rate of about 1.2 mL/min, and a FIA cycle lasts 1.5 min, which means the minimum buffer capacity is 1.8 mL. It is essential that the filter flow rates are high enough to completely fill the buffer tubes between measurements.

For the pilot plant, this principle only provides satisfactory results for sample points SP1, SP2 and SP4. These sample points are measured every 12 minutes and exhibit relatively slow concentration dynamics. The concentration dynamics at sample point SP3, on the other hand, are relatively fast (see Figure 1.19) and, for this reason, SP3 is measured every other FIA cycle. Problems therefore occur if the buffer capacity is actually needed. First, if the filtrate flow rate becomes too low the buffer may be emptied, since the time-span of only one FIA cycle is available for buffer filling between each measurement of SP3. This will then cause air to be introduced into the sample line. Second, the composition of the liquid stored in the upper end of the buffer may differ significantly from the composition of the new filtrate due to the faster process dynamics. Hence, for SP3, highest measurement accuracy and robust performance are obtained when the filtrate flow rate always exceeds the sample pump flow rate, meaning that the buffer capacity is not actually employed. With the current cross-flow filter design, no difficulty is encountered in maintaining a filtrate flow rate above 1.2 mL/min for all four sample points, and this is greater than the flow rate of the sample transport pump (in the range of 1 to 1.2 mL/min). The filtrate flow rates are checked regularly, and the ultrafiltration membrane is replaced when the flow rate has noticeably decreased.

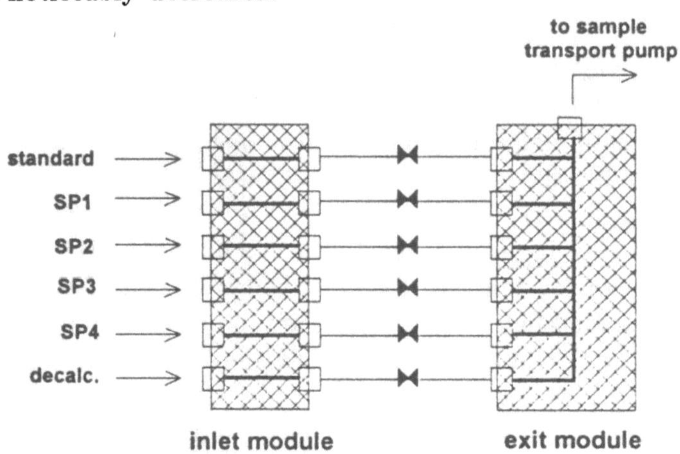

Figure 1.8: Multiport valve.

Sample selection is performed using a home-made multiport valve (MPV). A schematic of the valve is provided in Figure 1.8. With this design, up to six sample sources can be connected to the entrance module consisting of a plexiglass block through which six cylindrical channels have been cut. The exit module serves as a common manifold with six entrances and a common exit. This is where the fluids from the six pathways, or ports, meet. To ensure proper wash-out of a preceding sample, it is important that the common volume within the exit module is as small as possible. 1.0 mm i.d. silicon tubing serves as the connection between the entrance and exit blocks, and these pathways can be individually closed or opened with six air actuated pinch valves (Festo, No. 9336). To select a sample, its pathway is opened by releasing the pressure on its pinch valve, while the remaining pathways are kept closed.

The four sample points SP1 to SP4 and the standard solution occupy five of the sample ports. The remaining port is employed for a decalcifying agent (see later). An advantage of this home-made multiport valve over the commercially available rotating type valves is its random access ability. Any port can be immediately accessed without first stepping through the ports according to sequence. This is of practical use in this application, since the sample location SP3 is selected every second FIA cycle. A difficulty which has been encountered with the current design is that the silicon tubings tend to tear after some time and, hence, need to be periodically replaced. The possibility of extending the lifetime of the tubes by reducing the air pressure controlling the pinch valves has not yet been tried.

Sample transport is performed with the sample transport pump (STP), which is a home-made peristaltic pump using 0.76 mm i.d. Malprene tubing (Watson-Marlow, No. 02-9780076000). The pump pushes the selected sample liquid through the three injection valves (IV1, IV2, IV3) which are connected in series. A peristaltic pump suffices for sample transport since here a precise flow rate is not required. Important only is that the preceding sample has been completely washed out of the three injection valves and replaced by the new sample before injection occurs. Since the sample transport pump is located downstream of the multiport valve, the sample liquid is sucked rather than pushed through this valve. Due to the sucking action, air could conceivably be introduced through leaky connections within or prior to the multiport valve. However, this problem has not been encountered. To maintain a stable and sufficient flow rate through the sample feed line it is important that there is only minor flow resistance prior to the sample transport pump. In this context, a difficulty which has occurred is the plugging of the tubing connections between the cross-flow filter units and the multiport valve due to bacterial growth or calcium precipitation.

To minimize flow resistance, 0.8 mm i.d. PFTE tubing is employed for all connections prior to the multiport valve. After the multiport valve, where the transport pathway is common for all sample liquids, it is important to minimize the holdup volume (see discussion on carry-over in Section 1.3.10), and 0.5 mm i.d. tubing is employed.

1.3.2 Sample injection

Sample injection involves the introduction of a precise volume of sample fluid into the part of the FIA analyzer where reaction and subsequent detection take place. Three external loop injection valves (Valco, No. AC6UWPHC22) are employed, one for each analyzer (IV1, IV2 and IV3 in Figure 1.6). These are six-port air actuated valves manufactured from acid resistant Hastalloy 22 alloy. In the load position the injection loop is part of the pathway through which the sample, pumped by the sample transport pump, is flowing. The injection valve is normally set in the load position, and this serves to fill the injection loop with the next sample to be analyzed. In the inject position the pathway is rearranged so that the injection loop becomes part of the reagent stream. The sample volume is then pushed by the reagent through the reaction part of the analyzer. In the pilot plant, the time period for which the valves are placed in the injection position is 9 seconds. Switching between load and inject positions is computer controlled and precisely timed.

The size of the injected sample is determined by the holdup volume of the external loop plus a volume contribution from within the valve (i.e., connecting volumes and one of the grooves in the valve rotor). The external loop for each analyzer is a 12 cm length of PFTE tubing (0.5 mm i.d. for ammonia and nitrate; 0.3 mm i.d. for phosphate). The smaller inner diameter for phosphate is necessary as otherwise the injected sample volume will be too large to allow linearity within the desired concentration range (see also the discussion regarding flow cell window fouling in the phosphate analyzer description).

1.3.3 Reagent transport

As mentioned previously, precise (i.e., reproducible) reagent flow is the essential characteristic of FIA which enables this analysis method to provide accurate measurement results in spite of the fact that the reactions involved do not necessarily go to completion. Variations in flow rate from sample to sample will translate to variations in reaction holdup times, which will result in decreased accuracy and increased measurement variance. This particular application has been developed to employ extremely low reagent flow rates (from 200 to 300 μL/min). The reason for this is twofold. First, the measurement system has been designed for continual round-the-clock operation. Seven reagents are involved. At the low flow rates employed, only between 2 to 3 liters per week of each reagent is needed, which is manageable from the aspects of chemical costs and labor for reagent preparation. Second, the three analyzers have been designed for good accuracy over a large concentration range without the need for sample dilution. This has been facilitated by the use of low reagent flow rates.

The pumps employed are piston displacement pumps designed for liquid chromatography (Eldex, Type A-60-SF-2). The pumping action is produced by the forward and backward movement of a cylindrical piston within a housing filled with the reagent being pumped. During a forward stroke, the volume of liquid displaced by the extending piston is pushed out of the housing through a one way exit valve. During a backward stroke, the volume made available by the receding piston is filled with new reagent entering through a one way entrance valve. Each pump is driven by a synchronous motor with a pumping frequency of one piston displacement per second. A sapphire rod serves as piston and special materials are employed in the cylinder head and inlet and outlet valves due to the harshness of the reagents. Since the pumped volume is determined each second by the displacement of a rigid piston, a very precise and stable flow rate is provided (if all air is excluded, see comments below).

Since the pumps are equipped with only one piston, the flow rates produced are highly pulsating. However, this is not detrimental to the accuracy of the measurements. Essential for FIA accuracy is that flow rates are reproducible and precise, but not necessarily constant. Reproducibility is assured by the manner of sample injection. The logic which controls the FIA system ensures that sample injection occurs at the start of a full second according to the hardware real time clock of the computer. It is important that the computer's hardware rather than software clock be employed since the accuracy of the latter is insufficient for this purpose. The synchronous motors of the pumps lock the pump speed to the line frequency, which is accurately controlled by the power plant. Since the injection always occurs at the start of a full second, and since the pumps are driven at precisely one piston movement per cycle, each injection occurs at the same point in a piston cycle. Hence, a reproducible flow pattern is assured each FIA cycle. Additionally, the synchronous motors assure that the pump pulsations of all reagents are in phase, i.e., aliasing effects caused by pulsations at slightly different frequencies are avoided.

It is important that the displacement pumps work against a high backpressure as, otherwise, sealing of the piston packing and closing of the the uni-directional inlet valve may not occur properly. Backpressure is provided with 100 psi backpressure regulators (Upchurch Scientific, No. U446) located immediately after the pumps for reagents R3, R4, R5 and R7. Reagents R1, R2 and R6 are strongly acidic, and hence corrosive to this type of backpressure valve. Instead, backpressure for these three reagents is provided using 10 meters of 0.3 mm i.d. PFTE tubing. A difficulty with this is that backpressure is more flow rate dependent than with the backpressure regulators. In both cases, a consistent flow rate is first obtained after an equilibrium with respect to backpressure has been achieved. This means that the system should be operated such that the pumps are always running.

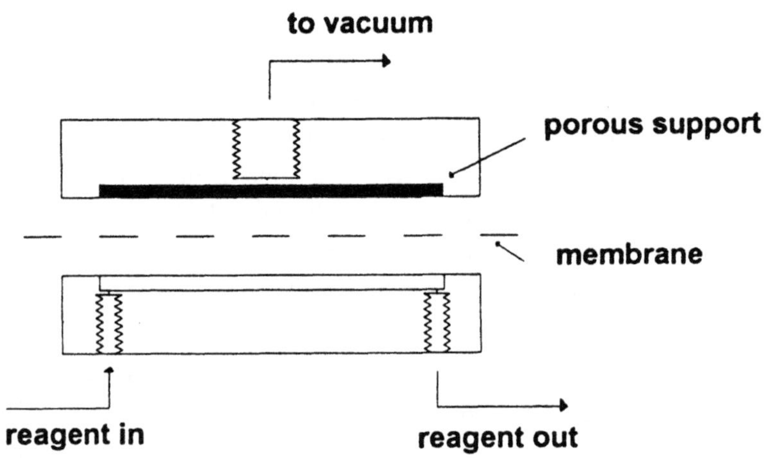

Figure 1.9: Degasser unit employed with reagents R3 to R7.

Experience in excess of five years has shown that the displacement pumps can operate well with minimal maintenance as long as certain precautions are taken. Entrapped gas bubbles within the liquid housing or valves will influence the amount of liquid displaced per piston pulse, and hence affect the flow rate. Entrapped particles or fibers within the entrance and exit valves will cause these valves to not open and close properly. To avoid both problems all reagents are filtered under vacuum through a 0.45 μm pore size cellulose nitrate filter prior to use. Degasser units (Figure 1.9) are also located in the reagent lines before the pumps for reagents R3 through R7. Due to the harsh reagents involved, the degasser units have not been employed for reagents R1 and R2. It is important that the backpressure after the pump be maintained as otherwise the piston seals will begin to leak prematurely.

In this regard many occurances of pump difficulties (seal failures, unsteady flow rate) have been caused by failure of the 100 psi backpressure regulators located after the pumps, either due to failure to open (backpressure too high) or failure to close (backpressure too low).

1.3.4 Reaction

The reaction stages of this application include cadmium catalysed conversion of nitrate to nitrite, the acid/base facilitated transport of ammonia across a gas permeable membrane, and several aqueous phase chemical reactions. The reactions all involve simple acid/base or chemical conversions of at most two steps, and are described in more detail later in Section 1.3.9. All aqueous phase reactions are performed in Meander coil reactors made of PFTE tubing (see Figure 1.3). The reactor sections are supported by a 2 mm thick aluminum plate perforated by a grid of 1.6 mm holes spaced 3.5 mm apart.

1.3.5 Detection

Each of the three analyzers is based on the principle that a color change at a particular wave-length occurs, the extent of which is dependent on the concentration of the substance being measured. This color change is monitored using a spectrophotometer equipped with a flow-through detection cell. Each analyzer employs a Kratos Applied Biosystems No. 757 spectrophotometer with flow cell No. 283 for ammonia and flow cell No. 237 for nitrate and phosphate. These spectrophotometers are relatively expensive, but allow computer controlled adjustment of sensitivity. This is necessary for the ammonia and phosphate analyzers due to the large range of concentrations encountered in the plant (i.e., up to more than 60 mg/L) and since dilution of the measurement samples is not performed. The sensitivity of the ammonia and phosphate analyzers is adjusted before each measurement according to the expected concentration range of the current sample location. For example, ammonia concentrations in the inlet water and after the anaerobic zone (SP1 and SP2) are generally much higher than in the aeration tank (SP3), with the lowest concentrations being encountered in the process outlet (SP4). A sensitivity adjustment is not automatically performed for the nitrate analyzer, due to the smaller concentration range encountered (up to about 10 mg/L).

A problem which often occurred earlier with the ammonia analyzer was the appearance and entrapment of a gas bubble in the detection cell. The pulsating flow of the reagent pumps caused the bubble to move in and out of the light path, producing a very noisy absorbance curve. The occurrence of gas within the flow cell seems to be caused by degassing within the cell, as the problem was resolved by increasing the backpressure downstream of the cell by adding an additional 5 meters of 0.5 i.d. PFTE tubing in the waste line after the cell.

1.3.6 Standard calibration

Common to just about all laboratory analytical procedures is the occasional performance of a standard calibration involving the measurement of one or more standard solutions containing a known amount of the measured analyte. The standard calibration is used to deduce the relationship between the concentration of analyte present in an arbitrary sample and the detected quantity. This is the calibration curve. The analytical procedure is normally designed so that the calibration curve is linear in the concentration range of interest. The calibration curve is then applied to subsequent samples to convert the detected quantity to units of analyte concentration.

The reason for performing a standard calibration is to compensate for physical or chemical changes in the measurement system. Since reactions with FIA do not necessarily go to completion, the actual relationship between detected signal and concentration is dependent on many factors such as reagent properties, catalyst activities and pump flow rates. Changes in these will cause the calibration curve to change with time (measurement drift). In this application, experience has shown that measurement drift can be mainly attributed to the following three factors, although other causes are plausible: 1) Although the displacement pumps are fairly robust in providing a stable flow rate, gradual changes occur due to changes in backpressure (corrosion wear of the backpressure valves), impending valve or seal failure, or the entrapment of gas bubbles in the pump cylinder head and valves; 2) The properties of the two reagents R2 and R7 change slowly with time; 3) The activity of the cadmium catalyst employed for the nitrate analysis decreases with time.

In this application, analyte concentration as a function of the detected quantity (peak height) is assumed to be described by a linear relationship which passes through zero for all three measured species and for all possible concentrations. This is only true over a certain concentration range, but the assumption allows the calibration curve to be uniquely described by one point, and hence the calibration can be performed by measuring only one standard solution. The standard solution employed contains 10 mg/L each of NH_4-N, NO_x-N and PO_4-P (see appendix). The results of a calibration test are shown in Figure 1.10. The abscissa is the analyte concentration of the measured solution (in the range of 0.5 to 50 mg/L) and the ordinate is the resulting calculated concentration (based on a single measurement of the standard solution). Here it can be seen that the calibration curve for each measured analyte is quite linear in this concentration range.

In the pilot plant, the standard calibration is performed by measuring a standard solution once each measurement cycle. A measurement cycle consists of 8 FIA cycles, each of which lasts 1.5 minutes, and so the standard calibration is performed at a rate of once every 12 minutes. The standard solution is connected to the first port of the multiport valve, and the calibration is carried out by selecting this port and performing a normal FIA measurement cycle. A calibration constant is then calculated for each analyzer as one over the respective peak height in absorbance units times the standard solution concentration (10 mg/L). The analyte concentrations in the subsequent samples of the measurement cycle are then calculated by multiplying the sample peak heights by the respective calibration constants.

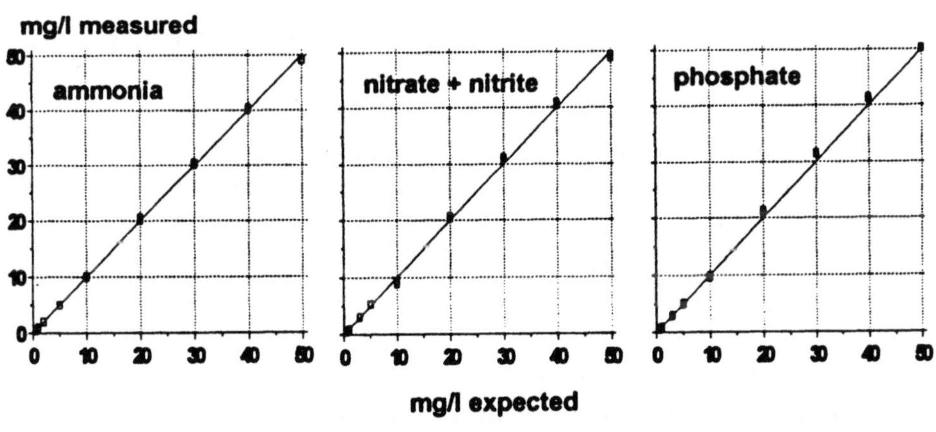

Figure 1.10: Calibration curves covering the range of 0.5 to 50 mg/L NH_4-N, NO_x-N and PO_4-P.

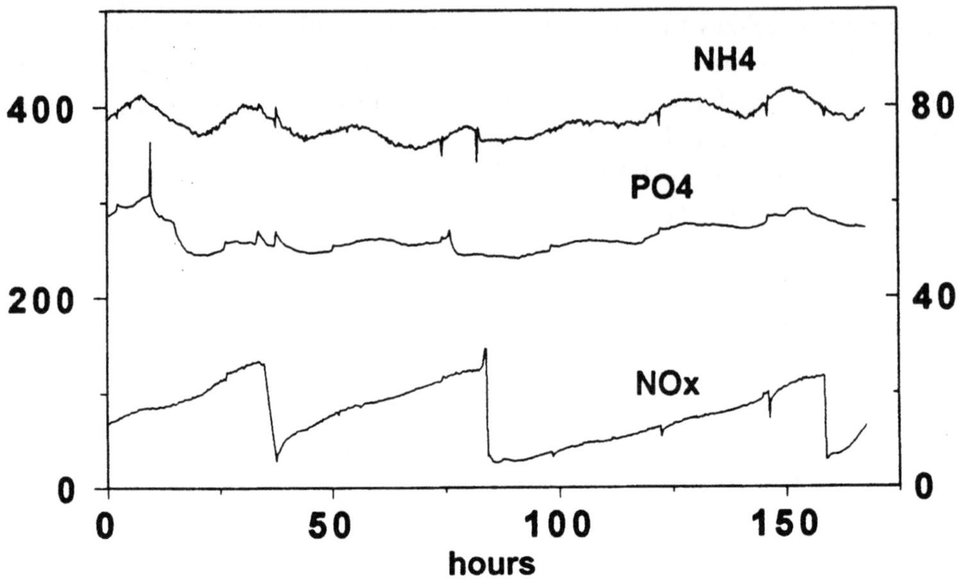

Figure 1.11: Calibration constants over the course of a week. The unsteadiness in the ammonia and phosphate curves is due to a loss of temperature control in the analyzer cabinet.

In addition to providing compensation for gradual measurement drift, the periodic performance of a standard calibration also serves as a validity check. Excessive noise, unusually fast drift, or an unusually high or low calibration constant indicates impending or actual failure of the corresponding analyzer. The course of the calibration constants is illustrated in Figure 1.11 for a week time period. The sudden drops in the nitrate constant correspond to the time instants at which the cadmium column was changed. The gradual rise in the nitrate constant is due to decreasing column efficiency. During this week the rate of loss of column activity was unusually fast, and the column needed to be changed every second or third day. The ammonia and phosphate constants show a distinct diurnal up and down variation. This is unusual, and is due to a loss of temperature control during this time period. The measurements are temperature dependent and, therefore, the three analyzers are contained in a thermostatically controlled cabinet. However, the cabinet is only equipped with the possibility to be heated, but not cooled. The data of Figure 1.11 was collected during an unusually warm summer period when the pilot plant ambient temperature was greater than the setpoint temperature of 23 °C and, hence, the temperature in the cabinet rose and fell according to the daily temperature variation.

1.3.7 Data analysis

As the reagent stream containing the monitored indicator passes through the flow-through detection cell of an analyzer's spectrophotometer, a continuous signal of absorbance versus time is produced. The signals from all three analyzers are sampled by the FIA system's controlling computer at a rate of 5 times a second for a duration of 84 seconds and stored for subsequent analysis. An example of the absorbance curves obtained in the pilot plant is shown in Figure 1.12 (here, for a standard solution measurement). Analyte concentration is determined from the peak height, which is found by subtracting the baseline of the curve from the curve's maximum value. The baseline is taken as the minimum value of the curve preceding the peak. Some rejection of spurious noise and bubble induced false peaks is performed by calculating the derivative of the signal, and rejecting all candidate peak values whose local derivative exceeds a given value.

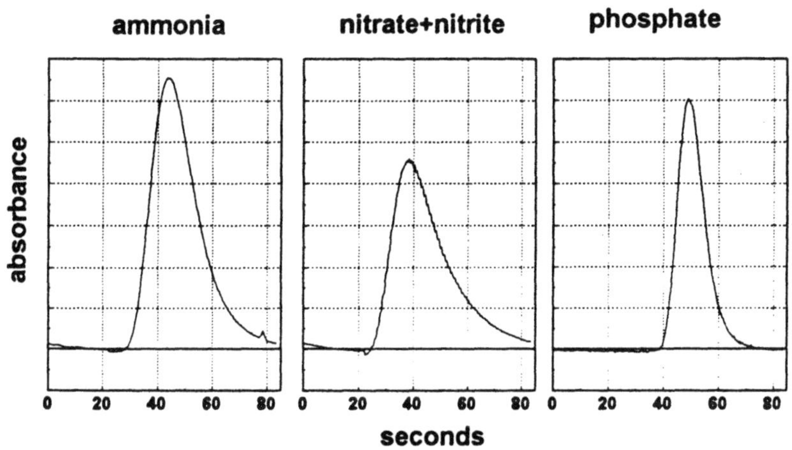

Figure 1.12: Signal curves obtained during one FIA cycle (raw data sampled at a rate of 5 times a second).

This manner of data analysis is rather unsophisticated and could be improved, particularly in the recognition of bubble induced false peaks. For instance, bubbles which are large enough to span several elements in the data vector, and thus provide an acceptable derivative, are not rejected. However, gas bubbles do not appear very often, and if they do it usually is a sign of a failure (e.g., pump failure, broken connection or plugging in a flow path or a large reduction in filtrate flow due to filter plugging). The peak time (i.e., the time of occurrence since injection of the signal peak) of each analyzer is recorded, as this provides a good indication of reagent flow rate precision.

1.3.8 Automatic and manual trouble prevention

The only automatic maintenance step currently implemented is a cleaning cycle performed once each day around midnight. This is to reduce the chance of microbial growth and buildup of precipitated compounds which can cause plugging of the flow pathways. Microbial growth occurs mainly in the tubing between the cross-flow filters and the multiport valve. Precipitation of calcium compounds is particularly a problem in Denmark due to the hardness of the water, and occurs primarily in the sample transport pathway and in the tubing of the sample point SP2. The cleaning cycle consists of stopping the analyses and pumping acid (0.5 M *HCl*) for a period of 2 minutes from a point immediately after the cross-flow filter units (using permanently installed T-joints) through the sample buffers. Simultaneously, the multiport valve is switched to the 6th port. A decalcifying agent (containing citric acid as active ingredient) is connected to this port, and this flushes the sample transport pathway from the multiport valve through all three injection valves with the decalcifying agent.

According to experience, when all equipment is in order the FIA system performs well with only minor maintenance for time periods of up to several weeks. Using 4 liter bottles, reagents need to be replenished only once a week or less. The membranes in the cross-flow filter units need to be changed once a week to once a month (the inlet water position, SP1, exhibits a quicker fouling rate). The reagent pumps are extremely stable and robust when all components (valves, seals, pistons) are clean and in good order. However, satisfactory pump performance can be very difficult to obtain if any of the components are flawed (dirty or damaged) or if the backpressure after the pumps is not stable within the proper working range. A great deal of the routine maintenance effort is devoted to inspecting and changing tubings, fittings and other minor components due to fouling and corrosion wear.

Visual inspection of all tubings is performed routinely, as the automatic cleaning cycle does not cover all flow pathways. Tubings are replaced at any sign of growth or deposit. Other manually performed preventive tasks include daily check of filtrate flow rates, inspection of all fittings and pumps for leakage, and inspection of the recorded times of occurrence of the detected signal peak heights. Drift in these times indicates an unsteady reagent flow rate. The Malprene tubing used with the sample transport peristaltic pump is manually cleaned every 2 or 3 days with 0.5 M *HCl* to remove precipitation products. The linearity of the phosphate measurement is checked by occasionally measuring a solution containing 50 mg/L PO_4-P. Significant departure from linearity at this concentration level indicates that the phosphate analyzer's flow cell windows should be cleaned.

1.3.9 The analyzers

The analyzers are based on standard chemical methods and have been developed in such a manner to allow long term stability, low reagent consumption and low maintenance requirements. The analyzers have been designed to provide a linear response within the range of 0.5 to 50 mg/L for ammonia, 0.5 to 20 mg/L for nitrate and 0.5 to 50 mg/L for phosphate without sample dilution. Each analyzer is described in turn below. All reagent compositions are presented in the appendix.

1.3.9.1 Phosphate

The phosphate analyzer is based on the methods described in [10][11] and employs the following reaction sequence:

$$H_3PO_4 + 12H_2MoO_4 \rightarrow H_3P(Mo_{12}O_{40}) + 12H_2O \tag{1.1}$$

$$2Mo(VI) + Sn(II) \rightarrow 2Mo(V) + Sn(III) \tag{1.2}$$

In the first step, phosphate in the sample forms a complex with the molybdenum(VI) ion. In the second step the ion is reduced to the blue molybdenum(V) ion. The intensity of this color is measured at 610 nm. The analyzer layout is shown in Figure 1.13. Pump flow rates and reactor dimensions are also shown in the figure. Due to the corrosive nature of the reagents, the backpressure *regulators* shown in the figure consist of a 10 meter length of 0.3 mm i.d. PFTE tubing inserted in the reagent line.

Reagent R1 is pumped through the injection valve and hence serves as primary carrier for the injected sample volume. After injection the first reaction step, Equation (1.1), takes place in reactor coil RC1. After RC1, reagent R2 is introduced by means of a T-connector. Then the second reaction step, Equation (1.2), takes place in the second reactor coil RC2. This reactor is physically located within the housing of the spectrophotometer, and detection therefore occurs immediately after reaction. Reagent R2 is acidified with hydrochloric acid, whereas sulphuric acid is employed in reagent R1. The reason for using the more corrosive hydrochloric acid is that precipitation can occur when sulphuric acid is employed, which will lead to plugging of the reagent flow paths. This problem has not been experienced when using hydrochloric acid to acidify reagent R2.

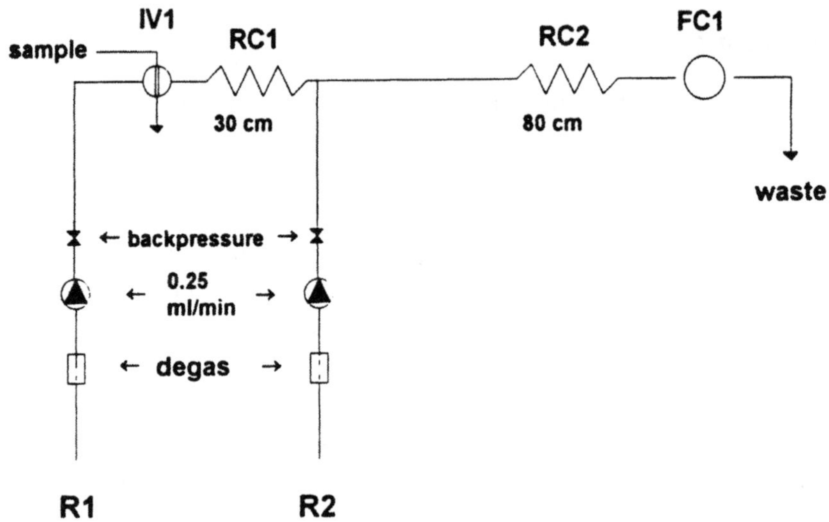

Figure 1.13: Phosphate analyzer. IV: injection valve; RC: reaction coil; FC: flow cell (detector); R: reagent.

The blue color complex formed in this analyzer tends to precipitate on the inner walls of the transport tubes and on the detector's flow cell windows. At some point the flow cell windows need to be cleaned (currently performed manually with 0.5 M *HCl*) as otherwise linearity at high phosphate concentrations will be lost. The frequency at which this must be done can be reduced by minimizing the sample volume, which minimizes the amount of blue complex formed. Earlier, a 2 μL internal volume injection valve was employed for this very reason, and a 1 μL internal volume was also tested. However, difficulties were encountered in maintaining noise-free measurements with the internal volume injection valves. More recently, good results have been obtained by using an external volume injection valve, where 0.3 mm instead of 0.5 mm i.d. tubing is used for the injection loop in order to reduce the injection volume.

1.3.9.2 Ammonia

The complexity of the ammonia analyzer is slightly greater than the phosphate analyzer due to the inclusion of a gas diffusion module. The module allows a separation of the ammonia from other components in the sample liquid, so that a simple acid/base reaction combined with a pH sensitive indicator can be employed for signal detection. Literature references pertaining to this method of ammonia analysis include [12][13][14].

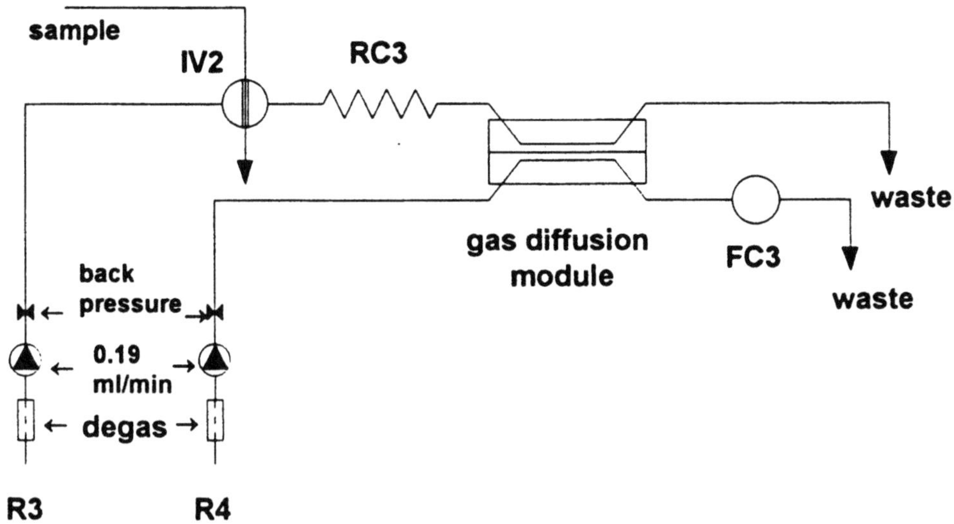

Figure 1.14: Ammonia analyzer. IV: injection valve; RC: reaction coil; FC: flow cell (detector); R: reagent.

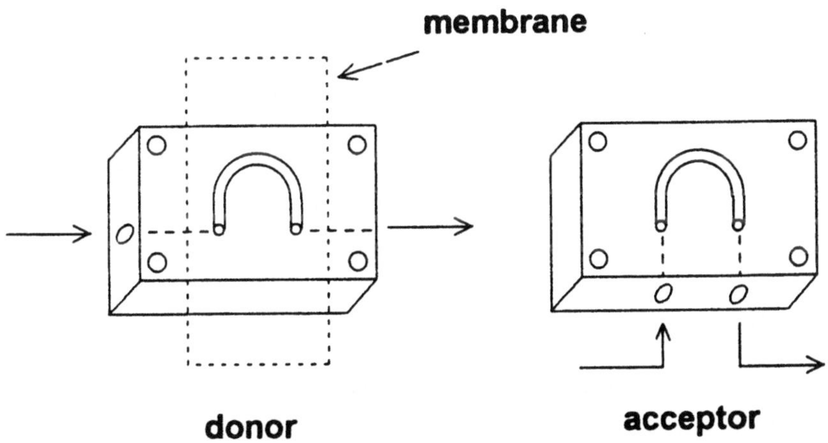

Figure 1.15: Gas diffusion module.

The analyzer layout is shown in Figure 1.14 and a drawing of the gas diffusion module is presented in Figure 1.15. The module consists of two similar stainless steel blocks each with dimensions 2.4 x 3.4 x 0.6 cm. A rectangular channel, 35 mm long, 0.1 mm deep and 2 mm wide, is cut in each block in the form of a U shaped flow path. The U shape is to be preferred over a straight channel, as a curved pathway causes secondary flows to develop during laminar flow, which increases module effectivity. The two channels line up with each other when the two module halves are assembled. A double layer of a hydrophobic membrane is placed between the two block halves to separate the channels. The membrane employed has an effective pore diameter of 0.04 μm and a thickness of 0.025 mm (Hoechst, Type Cellgard 2400). The membrane is impermeable to liquid but allows gas to diffuse through.

The analysis method is founded on ammonia's properties as a base. The pH of the sample volume is lowered to about pH 13 by injection into the basic reagent R3. The pKa value of ammonia is 9.25 so that just about all of the dissolved ammonia ions in the sample are converted to ammonia gas:

$$NH_4^+ + OH^- \leftrightarrow NH_3 + H_2O \tag{1.3}$$

As the sample volume passes through the gas diffusion module, a portion of the ammonia gas diffuses through the Celanese membrane from the donor to the acceptor side. The acceptor side reagent, R4, is a weak pH buffer adjusted to pH 6.8 and contains a pH sensitive indicator. Reversal of Equation (1.3) occurring on the acceptor side causes a change in pH, and a corresponding change in indicator color, which is then monitored at 590 nm.

The gas diffusion module has been found to be very robust. Only seldom it has been found necessary to replace the membrane due to fouling or suspected damage. It is important, however, that large pressure differences between the donor and acceptor side do not occur, as otherwise the membrane can be deformed and possibly tear. In this regard it is important that the pulsations of the two reagent pumps must be synchronized.

1.3.9.3 Nitrate

The nitrate analyzer actually measures both nitrate and nitrite as nitrite. The method is based on passing the sample through a reduction column containing cadmium granules covered with copper. While passing through the column, nitrate is reduced to nitrite, which then reacts with sulphanilamide in reagent R6 to form a diazonium salt. This is mixed with reagent

R7 to form a diazo dye, which is then detected at 540 nm. The layout of the analyzer is shown in Figure 1.16. Relevant literature references include [15][16][17].

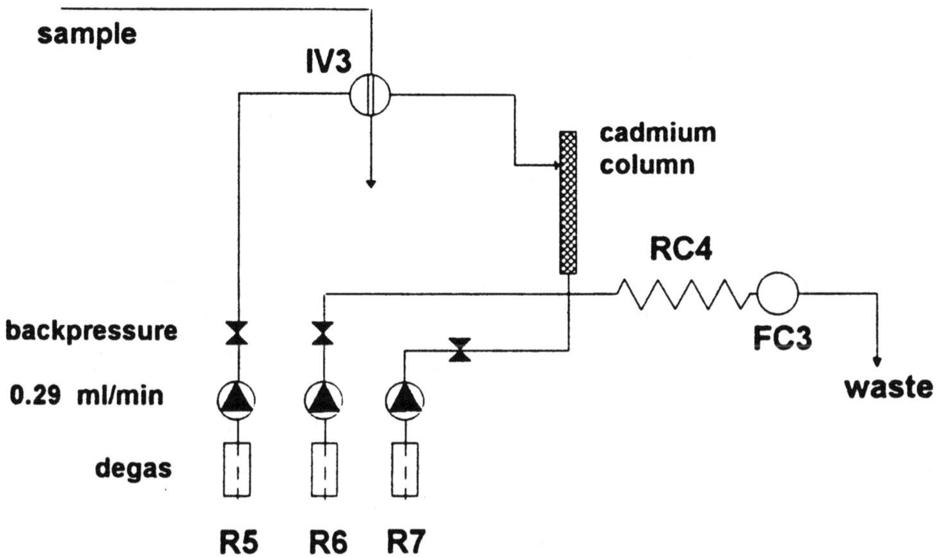

Figure 1.16: Nitrate analyzer. IV: injection valve; RC: reaction coil; FC: flow cell (detector); R: reagent.

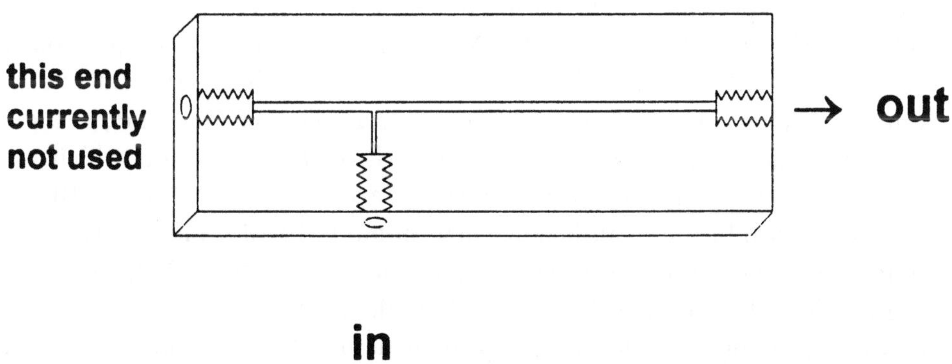

Figure 1.17: Cadmium catalyst column.

The cadmium column is made of a 1 x 3.5 x 7.5 cm plexiglass block and is shown in Figure 1.17. The diameter of the column void which is cut in the block is 0.2 cm. The overall length of the column is 5 cm and the effective length from the side point of reagent entry is 3.5 cm. The column has this particular form since it was originally designed to allow automatic regeneration by periodically pumping a solution containing hydrochloric acid through the back end. However, by adding copper sulfate to reagent R5, which serves as carrier and pushes the sample through the column, the column life could be extended from a few hours (without copper sulfate) to between 3 to 10 days (with copper sulfate), depending on the nitrate loading. Therefore, no automatic regeneration is currently performed, and the back entrance is simply closed off with a plug. The column is replaced when its activity has declined substantially, which is determined by the magnitude of the nitrate calibration constant (the calibration constant is inversely proportional to peak height and hence a higher constant means a lower catalyst efficiency).

Nitrogen present as nitrate (NO_3-N) and nitrite (NO_2-N) could be separately measured by extending the system to allow a precise portion of the sample liquid to bypass the cadmium column. In the bypassed sample, only nitrite would be measured, and the nitrate concentration could be found by difference. This has not been done here in order to minimize analyzer complexity, and since nitrite concentrations are normally quite low in the activated sludge process being monitored. The following caution is advised, however, if a sizeable nitrite fraction is expected to be encountered. The sample concentration is calculated from the calibration constant which is found here by measuring a solution containing 10 mg/L NO_3-N and 0 mg/L NO_2-N. A cadmium column of less than 100 % efficiency means that only partial conversion of nitrate to nitrite occurs. Hence, the NO_x-N reading (NO_x-N refers to NO_3-N plus NO_2-N) which is obtained by comparison to a NO_3-N standard solution will be inaccurate if a significant amount of nitrite is present. This can be illustrated with an example. Assume that, at a given time, β is the fraction of nitrate passing through the column which is reduced to nitrite. This means that the standard solution containing 10 mg/L NO_3-N will produce a peak height corresponding to 10 β mg/L NO_2-N. Assume further that the measured sample contains 10 mg/L NO_x-N with the composition: 10(1-α) mg/L NO_3-N plus 10α mg/L NO_2-N, where α is the fraction of nitrogen in the sample present as nitrite. After passing through the column, the sample will have a NO_2-N content equal to $10(1-\alpha)\beta+10\alpha$ mg/L. The sample NO_x-N concentration is determined by direct comparison with the standard solution peak height:

$$NO_X\text{-}N\ (\mathrm{mg/L}) = \frac{\text{sample peak height}}{\text{standard peak height}} \cdot 10 = \frac{10\,(1-\alpha)\,\beta + 10\alpha}{10\beta} \cdot 10$$
$$= \left(1 - \alpha + \frac{\alpha}{\beta}\right) \cdot 10 \qquad (1.4)$$

If nitrite in the sample is negligible (α small), a measurement result near to the correct value of 10 mg/L NO_x-N will be obtained, regardless of the column efficiency β. As the fraction of nitrite in the sample increases (i.e., as α increases), measurement error will occur unless the column efficiency is high ($\beta \approx 1$). In this example, a sample composed entirely of nitrite ($\alpha = 1$) will give a reading of $10/\beta$ instead of 10 mg/L NO_x-N.

1.3.10 The problem of carry-over

Most of the measurement noise encountered with the pilot plant FIA system can be attributed to the problem of carry-over. This is where the current measurement is affected in some fashion by the previous measurement. What is meant by carry-over here is not the problem by which adjacent signal peaks overlap because the signal peak duration is longer than the FIA cycle duration. The 1.5 minutes employed is more than sufficient for the baseline absorbance level to be regained for each of the three analyzers (see Figure 1.12). Carry-over in the pilot plant system occurs when analyte from the previous sample becomes mixed with the new sample prior to injection. This type of carry-over can occur, for example, when the sample transport pump flow rate is too low to allow the new sample liquid to completely wash out the old before the time of injection, resulting in the measurement of a mixture of the old and new sample.

Whether or not carry-over occurs in a particular FIA system can easily be determined off-line by having the system switch between two solutions of known concentration, where one, for example, may be pure water. If carry-over occurs the measurement result will not be stable at its anticipated value until two or more consecutive FIA cycles are performed on the same solution. More difficult to assess, however, is when carry-over begins to occur during operation, for example due to some change in the FIA system's properties. The history of the pilot plant FIA system shows a good example of this. Originally, the system was implemented using a purely sequential measurement order, where a measurement cycle consisted of five FIA cycles performed in the sequence: standard solution, SP1, SP2, SP3 and SP4. At some point in time carry-over began to occur, partly due to configurational changes which affected flow rates and holdup volumes and partly due to other causes of carry-over described below. However, this went completely unnoticed until the currently employed nonsequential measurement cycle (standard solution, SP3, SP1, SP3, SP2, SP3, SP4, SP3) was implemented.

To understand this, consider the measurement of phosphate. The PO_4-P concentration of the standard solution is 10 mg/L while the sample points SP1, SP2, SP3 and SP4 contain in the order of 6, 25, 5 and 2 mg/L, respectively. SP1, SP2 and SP4 exhibit relatively slow concentration dynamics compared to the duration of a complete measurement cycle. The dynamics at SP3 is faster, but the concentration ranges only between 0 and about 5 mg/L. Therefore, with the sequential sequence, each measurement location was preceded by a measurement of essentially the same concentration level. This means that the error caused by partial (but reproducible!) mixing with the previous sample was approximately constant with time, and hence went unnoticed. As soon as the nonsequential sequence was implemented, however, the problem of carry-over became apparent, as the sample location SP3 was then preceded by a measurement sample containing largely different PO_4-P concentrations.

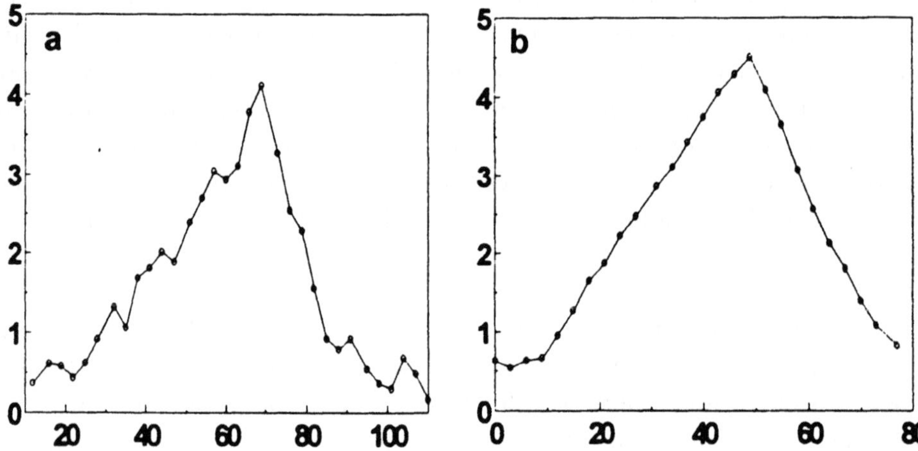

Figure 1.18: Phosphate measurements (a) with and (b) without carry-over noise.

An example of carry-over corrupted measurements of phosphate is shown in Figure 1.18. The leftmost plot shows phosphate measurements at sample point SP3 during a time when carry-over was significant. Every fourth measurement seems to be perturbed in a similar manner. This is a sign that carry-over is occurring, since, with the nonsequential measurement cycle, every fourth measurement of SP3 is preceded by the same sample point. As a comparison, the rightmost plot in Figure 1.18 shows relatively noise-free measurements obtained when little carry-over occurred.

Carry-over in the pilot plant seems to be caused by a variety of factors, not all of which are completely understood. One source of carry-over is the incomplete wash-out of the previous sample, which should be avoidable through proper design. Here it is important to minimize the holdup volume of the sample transport pathway from the point where the pathway becomes common to all sample locations. This can be accomplished by using small diameter transport tubing (0.5 mm i.d. is employed here as narrower tubing will be more subject to plugging), minimizing the volume of the common manifold in the multiport valve, and locating the multiport valve, sample transport pump and all analyzers as close to each other as possible. In choosing a proper flow rate for the sample transport pump it must be remembered that dispersion will widen the transitional front between sample liquids as the front moves along the sample transport pathway.

Carry-over in the pilot plant has also been attributed to damage or wear of an injection valve's inner wall or rotor. Improper sealing of the rotor can cause leakage and retainment of sample within the valve, which then can mix with the subsequent sample. Carry-over also may be caused by adsorption or retainment within the transport pathway, perhaps due to the presence of calcium deposits. There is no conclusive evidence for this. However, carry-over related noise has been observed to disappear after cleaning or replacing the sample transport pump Malprene tubing and after flushing the sample buffer and transport tubes with 0.5 M HCl.

1.3.11 Implementation and an example

The pilot plant FIA system is operated with a PC computer using a program written in the ASYST Software System. ASYST was chosen as a matter of convenience, since the software contains built in routines and commands which ease the programming of communication tasks. The software also allows one high priority task to be run in the background, which is necessary for the proper timing of injections and data sampling.

The controlling program continually performs a loop which carries out one FIA cycle. The sample position which is to be injected is determined by an element in a vector which defines the measurement sequence of each measurement cycle. Currently this vector is composed of 8 elements defining the sample position sequence: standard solution, SP3, SP1, SP3, SP2, SP3, SP4, SP3. The standard solution is measured at the start of each measurement cycle, and the thus calculated calibration constants are used during the remainder of the measurement cycle to calculate the sample analyte concentrations.

For each FIA cycle, the following sequence of actions occurs: the spectrophotometer sensitivities are adjusted according to which sample position is to be measured, the multiport valve is switched to the position which is next to be injected, and all three injection valves are placed in the inject position for 9 seconds. The sample which is then injected is the previous sample, since the sample from the new multiport valve position has not had time to reach the injection loops before the injection. Data is sampled and collected for 84 seconds, and the data analysis is performed during the subsequent FIA cycle. This means that the delay time in the pilot plant system is three minutes from the time of injection until the analyte concentrations are calculated. Added to this is the delay time associated with transport and filtration of the raw sample liquid.

An example of the pilot plant measurements over the course of one day are shown in Figure 1.19. NO_x-N is not shown for SP1 and SP2 since the measurements remained below the detection limit during the entire day. Why a multirate sampling sequence is employed, where SP3 is measured every other FIA cycle, should be clear from this figure. The concentrations at SP3 change rapidly in tact with the one to two hour process operational cycle while the concentrations at the remaining locations change much more gradually. It should be noted that the measurements are clean enough to allow derivation of the nitrification, denitrification and phosphate uptake rates in the measured anoxic/aerobic tank, sample point SP3.

Acknowledgments

Funding for the work associated with preparing this manuscript and for the development and implementation of the pilot plant FIA system has been provided by the Center for Biotechnological Process Research of Denmark. The pilot plant is a part of the research facilities of the Institute for Environmental Engineering (IMT) at the Technical University of Denmark.

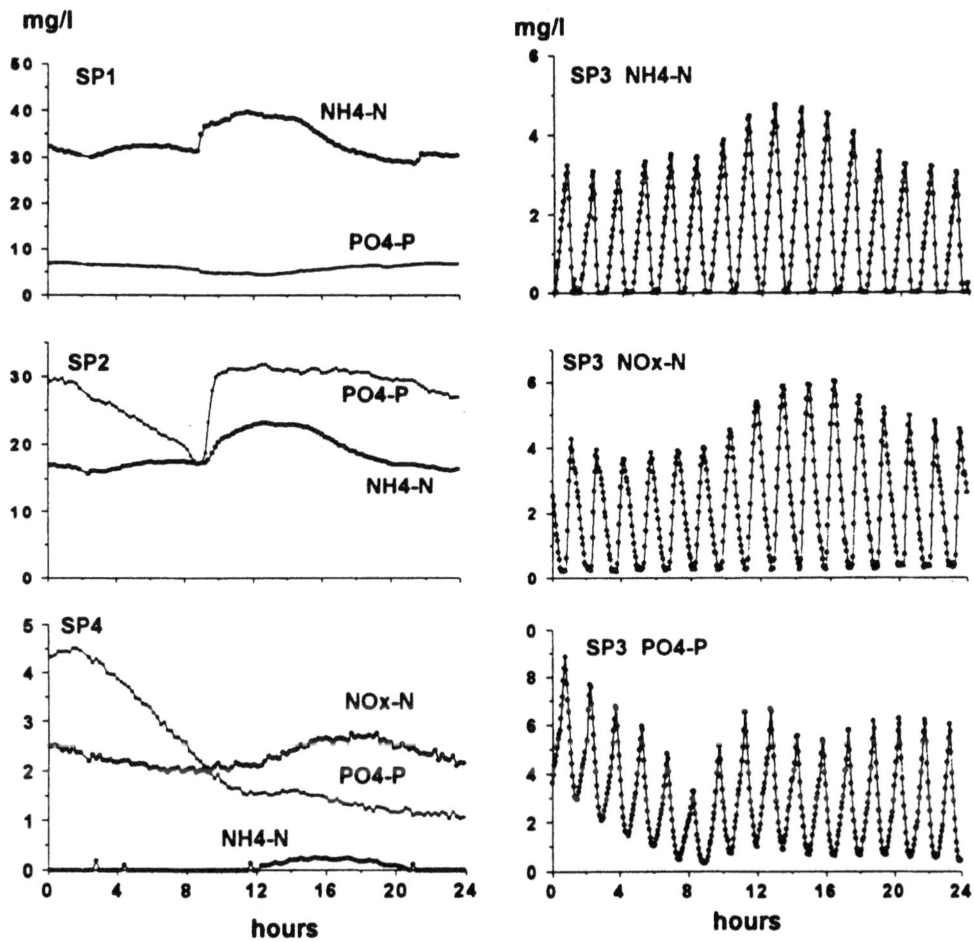

Figure 1.19: FIA measurements of the pilot plant process.

1.4 Bibliography

1. D.E. Thornberg, M.K. Nielsen and K.C. Andersen 1993. Nutrient Removal: On-line Measurements and Control Strategies. *Wat. Sci. Tech.*, 38, 549-560

2. J. Ruzicka and E.H. Hansen 1988. *Flow Injection Analysis*, 2nd edition, John Wiley & Sons, New York

3. S.H. Isaacs, H. Søeberg, L.H. Christensen and J. Villadsen 1992. A computational technique for simulating the dynamic response of a flow injection analysis system. *Chem. Eng. Sci.*, 47, 1591-1600

4. E. Arvin 1985. Biological Removal of Phosphorus from Wastewater. *CRC Critical Rev. Environ. Control*, 15, 25-64

5. J. Einfeldt 1992. The Implementation of Biological Phosphorus and Nitrogen Removal with the Bio-Denipho Process on a 265,000 PE Treatment Plant. *Wat. Sci. Tech.*, 25, 161-168

6. S.H. Isaacs, M. Henze, H. Søeberg and M. Kümmel 1994. External Carbon Source Addition as a Means to Control an Activated Sludge Nutrient Removal Process. *Water Research*, 28, 511-520

7. H. Zhao, S.H. Isaacs, H. Søeberg and M. Kümmel 1994. A Novel Control Strategy for Improved Nitrogen Removal in an Alternating Activated Sludge Process. Part I & II. *Water Research*, 28, 521-542

8. K.M. Pedersen 1988. Real-Tids Måling af Næringsalte i Biologisk Spildevandsrenseanlæg (*Real Time Monitoring of Nutrients in a Biological Waste Water Treatment Plant*). PhD thesis, Instituttet for Kemiteknik, Danmarks Tekniske Universitet, Lyngby (Denmark) (in Danish)

9. K.M. Pedersen, M. Kümmel and H. Søeberg 1990. Monitoring and control of biological removal of phosphorus and nitrogen by flow-injection analysers in a municipal pilot-scale waste-water treatment plant. *Analytical Chimica Acta*, 238, 191-199

10. Tecator Application Note AN 60/83 (8/6, 1983): *Determination of o-phosphate based on the stannous chloride method using flow injection analysis*

11. Tecator Application Note ASN 60-01/83 (27/1, 1983): *Determination of reactive phosphorus in water by flow injection analysis (stannous chloride method)*

12. Tecator Application Note AN 50-01,02,03,04/82 (21/12, 1982): *Determination of ammonia nitrogen by flow injection analysis and gas diffusion*

13. M. van Son, R.C. Schothorst and G. Den Boef 1983. Determination of total ammoniacal nitrogen in water by flow injection analysis and a gas diffusion membrane. *Anal. Chim. Acta*, 153, 271-275

14. W.E. van der Linden 1983. Membrane Separation in Flow Injection Analysis. Gas Diffusion. *Anal. Chim. Acta*, 151, 359-369

15. Tecator Application Note AN 62/83 (4/3, 1983): *Determination of the sum of nitrate and nitrite in water by flow injection analysis*

16. Tecator Application Note ASN 62-01/83 (4/3, 1983): *Determination of the sum of nitrate and nitrite in water by flow injection analysis*

17. Tecator Application Note ASN 62-02/83 (11/3, 1983): *Determination of the sum of nitrate and nitrite in water by flow injection analysis*

1.5 Appendix

The following are the recipes for the reagents employed in the analyzers. Prepared reagents last several months except where noted.

- Phosphate

  R1 5 g ammoniummolybdat (Merck 1182) plus 35 mL concentrated H_2SO_4 (Merck 731) in 1 liter deionized water.

  R2 37 mL concentrated HCl (Merck 317) plus 1 g stannouschloride (Merck 7815) plus 5 g hydroxylammoniumchloride (Merck 4616) in 1 liter deionized water. Lasts only about one week.

- Ammonia

  R3 4 g sodium hydroxide (Merck 6498) plus 1.6 g EDTA (Fluka 03680) in 1 liter deionized water.

R4 10 mL indicator stock solution plus 10 mL phosphate buffer stock solution in 1 liter deionized water. The solution is allowed to stabilize 24 hours, and then the absorbance at 590 nm is adjusted to 0.3 with either *NaOH* or *HCl*.

Indicator stock solution: 1 g ammonia indicator mixture (Tecator 5000-0295) plus 5 mL 0.1 M *NaOH* in 200 mL deionized water.

Phosphate buffer stock solution: 6.9 g sodium dihydrogen phosphate monohydrate (Merck 6346) in 500 mL deionized water.

- Nitrate

R5 10.7 g ammoniumchloride (Merck 1145) plus 10 g sodium tetraborate decahydrate (Merck 6308) plus 1 g EDTA plus 0.7 mL copper sulfate solution in 1 liter deionized water.

R6 10 g sulfanilamide (Merck 11799) plus 52 mL concentrated *HCl* in 1 liter deionized water.

R7 1.0 g N-(1-naphthyl)-ethylene-diammonium-dichloride (Sigma N-9125) in 1 liter deionized water (lasts approximately one to two weeks).

Copper sulfate stock solution: 10 g copper sulfate (Merck 2790) in 1 liter deionized water.

- Standard Calibration Solution

A stock solution containing 1000 mg/L each of PO_4-P, NH_4-N and NO_x-N and is prepared by dissolving 4.39 g potassiumdihydrogenphosphate (Merck 4873) plus 6.07 g sodiumnitrate (Merck 6537) plus 3.82 g ammoniumchloride (Merck 1145) in 1 liter deionized water. The standard solution is then prepared by diluting the stock solution 1:100 in deionized water to produce a solution containing 10 mg/L of each component.

- Preparation of cadmium column

Cadmium granules (Merck 16411) are treated with 0.5 M *HCl* for one minute. The granules are then rinsed with deionized water 2 times. The granules are then treated with copper sulfate stock solution (see R7 above) for approximately one minute. This is followed by thorough rinsing with deionized water. The granules are then ready to be packed in the plexiglass column.

Chapter 2

On-line Measurement of Viable Biomass

A.J.C. SPIERINGS
Technisch Buro I.K.S. b.v.
Postbus 143 4140 AC, Leerdam (THE NETHERLANDS)

ABSTRACT – *Bioprocesses are in general not very well controlled. This is particularly the case when people are interested in the on-line control of biomass. This chapter discusses an instrument called the biomass monitor (BM), which uses a capacitance measurement and can be used for on-line measurement of any cell type. Results from a brewing company show very little difference between measurements of viable cell concentration using methylene blue staining and those using the capacitance reading. Research carried out by a pharmaceutical company has shown the difference in detection between optical density (OD) measurement, respiration quotient (RQ), and capacitance. During a fermentation run, the time to the highest concentration of biomass can vary up to 10 hours. It has also been shown that the concentration of viable biomass dropped dramatically over those 10 hours.*

KEYWORDS – *Biomass Monitor, On-line Measurements, Capacitance Measurement, Viable Biomass*

J.F.M. Van Impe et al. (eds.), Advanced Instrumentation, Data Interpretation, and Control of Biotechnological Processes, 41-66.

2.1 Basic Theory of Biomass Measurement

2.1.1 Capacitance, conductance and electric fields

2.1.1.1 Electric fields and capacitance

An electric field is a force field which pushes electrical charges (e.g., ions in a solution) along. Thus if one places two metal electrodes into a beaker of aqueous *KCl* and uses them to apply an electric field to the solution then the effects shown in Figure 2.1 would be seen. The electric field runs from the positively charged electrode to the negatively charged one and it pushes the positive ions (K^+) towards the negative electrode and the negative ions (Cl^-) the opposite way.

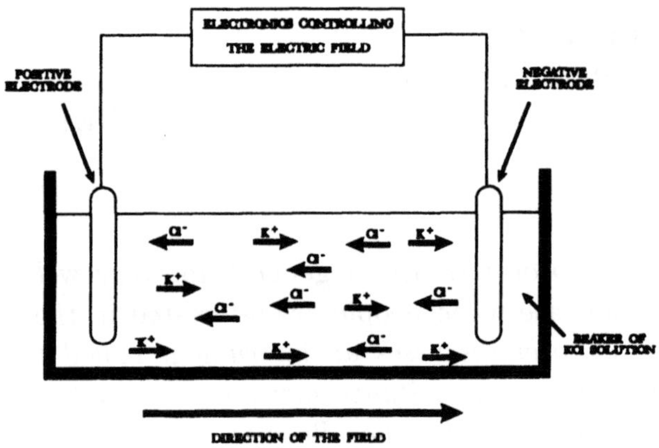

Figure 2.1: Effect of an applied electric field on the ions in a beaker of aqueous *KCl*. The field is created in the beaker by applying an electrical potential difference (voltage) between the two electrodes. The field runs from the positive electrode to the negative one and pushes the positive ions in the direction of the field and the negative ions the opposite way.

If some biological cells are placed into the *KCl* solution then the situation in Figure 2.2 results (for clarity the ions are shown just as positive and negative charges). The ions both inside and outside of the cells are pushed along by the field but when they bump into the cells' intact plasma membranes they can move no further. The result of this is that the membranes at the poles of the cells have become *coated* with layers of oppositely charged ions, i.e., the field has induced charge separations (polarizations) across them.

These polarizations are *stabilized* by the electrostatic attraction between the oppositely charged ions on either side of the membranes (remember that *like charges repel, opposite charges attract*).

Capacitance is a measure of the extent of the charge separations (polarizations) that are induced in a material (e.g., a cell suspension) by the application of an electric field. If the field induces a lot of charge separations in a material then the capacitance is high. Thus it is clear that the capacitance of the cell suspension in Figure 2.2 is high. If the concentration of cells in the suspension is increased (i.e., the cellular volume fraction is raised) then more intact membranes are polarized by the field and so the capacitance rises further. Thus *capacitance can be used to measure the amount of cells present in the suspension.*

Capacitance is measured in farad (F), but as 1 farad is an extremely large capacitance it is more convenient to work in picofarad (pF) where 1 pF equals 10^{-12} farad. A typical fermentation will give values in the order of a few pF to some 50 pF or more, depending on the concentration and size of the cells and on how the biomass monitor (BM) is set up.

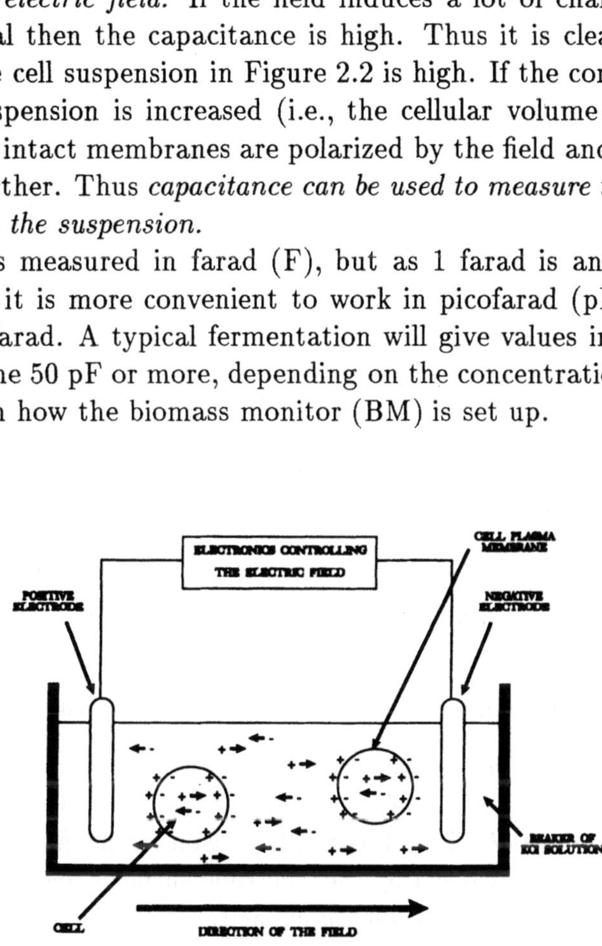

Figure 2.2: Effect of the addition of cells to the aqueous ionic solution shown in Figure 2.1. For simplicity the ions are shown as charges. The ions in the cell cytoplasm and the suspending medium move with the electric field until they bump into the cell's plasma membrane which stop them progressing further. The net result is the formation of charge separations (polarizations) across the plasma membrane at the poles of the cells.

2.1.1.2 Conductance

Apart from capacitance the biomass monitor also measures the conductance of the fermentation broth. *The conductance of a solution is a measure of the concentration of ions present, their charge and their mobilities.* Thus a solution containing a high concentration of highly charged and highly mobile ions will have a high conductance. Similarly a solution containing a low concentration of poorly mobile and singly charged ions will have a low conductance. Conductance is measured in siemens. However, as 1 siemens is a large value, conductance is normally expressed in millisiemens (1 mS = 10^{-3} S).

The conductance of an aqueous ionic solution is temperature dependent. As the temperature increases the mobility of the ions increases and so the conductance must rise. As a rule of thumb one can expect a 1 to 2 % increase in conductance for every 1 °C rise in the temperature of the solution being measured.

Although conductance is not used by the BM to measure biomass it is nevertheless a useful thing to measure as it reflects the production or utilization of charged species by the cells. In fermentations where a lot of ionic material is added during the course of a run the conductance will measure these additions also (e.g., pH control using acid/alkali). Depending on how the biomass monitor is set up typical fermentations will have conductances anywhere up to 10 mS, although some will have much higher values.

2.1.1.3 Measuring capacitances and conductances

The biomass monitor's probe has four metal electrode pins; the two outer ones produce the electric field while the two inner ones measure the charge separations that the field induces. The reason for using four electrodes is that it reduces artifacts caused by the electrode surfaces (electrode polarization).

In the previous discussions of capacitance and conductance it was mentioned that the exact values one measures depend on how the BM was set up. The reasons for this are twofold. The first reason is that both capacitance and conductance depend on the electrode geometry. The BM electrodes have been carefully designed to optimize these dimensions and there is little difference in these dimensions between probes of the same type. Thus provided one does not change electrode type (e.g., from 25 mm Fermentation probe to Mexican Hat probe) the effect of electrode geometry on capacitance and conductance is constant.

The second effect on the capacitance and conductance measured due to how the biomass monitor is set up depends on the range setting (high and low ranges) of the BM. The biomass monitor's electronics can only measure values up to 200 pF and 20 mS. If the BM was switched to its more sensitive setting (low range) and one (or both) of these values was exceeded then it would be useful to be able to scale the capacitance and conductance signals down by about a half so that they now would fall within the BM's limits. To do this one would switch to high range on the BM. A side effect of attenuating the signal by a half in this way is to halve the sensitivity to biomass, hence high range is a less sensitive setting than low range.

The result of this is that if a calibration curve of capacitance versus biomass concentration is done with a certain probe type and BM range setting, then the same probe type and range must be used for fermentation readings that will use that calibration curve.

Although the BM will measure conductances up to 20 mS it begins to lose accuracy at measuring capacitances (and hence biomass) at levels above about 10 mS in low range and about 18 mS in high range. Thus when setting up a BM for a new fermentation one should first try measurements on the fermentation broth in low range. If the broth's conductance is above 10 mS then one must switch to high range.

2.1.1.4 Summary

(a) The biomass monitor (BM) measures biomass by measuring the capacitance of the cell suspension. It displays capacitances in pF (picofarad).

(b) An electric field is a force field that pushes electric charges along, e.g., ions in aqueous solution.

(c) The capacitance of a cell suspension reflects the extent of the charge separations (polarizations) induced in it by the application of an electric field. If the field induces a lot of charge separations then the capacitance is high.

(d) The application of an electric field to a suspension of cells with intact plasma membranes results in moving up the ions inside and outside the cell to the plasma membrane and polarizing it. As capacitance gives a measure of the extent of such field induced polarizations the capacitance of the suspension is high.

(e) The higher the cell concentration (volume fraction), the more plasma membranes there are to polarize and so the capacitance will be higher. Thus capacitance increases as the cell concentration does.

(f) Conductance is also measured by the BM. The conductance of a solution is a measure of the concentration of ions present, their charge and their mobilities. Thus a solution containing a high concentration of highly charged and highly mobile ions will have a high conductance. The BM displays conductances in mS (millisiemens).

2.1.2 Frequency of an electric field

2.1.2.1 Effect of changing the direction of an electric field

In Figures 2.1 and 2.2 the electric field is only going in one direction; one can, of course, change the field direction. Figure 2.3 shows the effect of doing this on the cell suspension shown in Figure 2.2. As one can see the field polarizes the plasma membranes first one way round (Figure 2.3A), and then when it changes direction it causes a charge separation of the same magnitude but round the opposite way (Figure 2.3B). As the extent of the charge separations induced by the field is the same irrespective of which direction the field is going in, the capacitance of the suspension is also not changed by changing the *direction* of the field *per se.*

In addition to just changing the field direction one can also control the rate at which it does so. How many times one changes the field direction per second is the field's frequency. An electric field changing direction slowly has a low frequency, while one that is changing direction rapidly has a high frequency. Frequency is measured in hertz (Hz) or megahertz (MHz), where 1 MHz equals 10^6 Hz.

It is important to be able to set the frequency of the electric field because this enables one to optimize the biomass measuring regime. If a field was applied that was only in one direction or had a very low frequency then what the BM would measure would be dominated by artifacts due to effects at the electrode's surfaces (electrode polarization). By using frequencies between 0.2 and 10 MHz one can measure the biomass content of the cell suspension with minimal artifacts from the electrodes interfering with the readings.

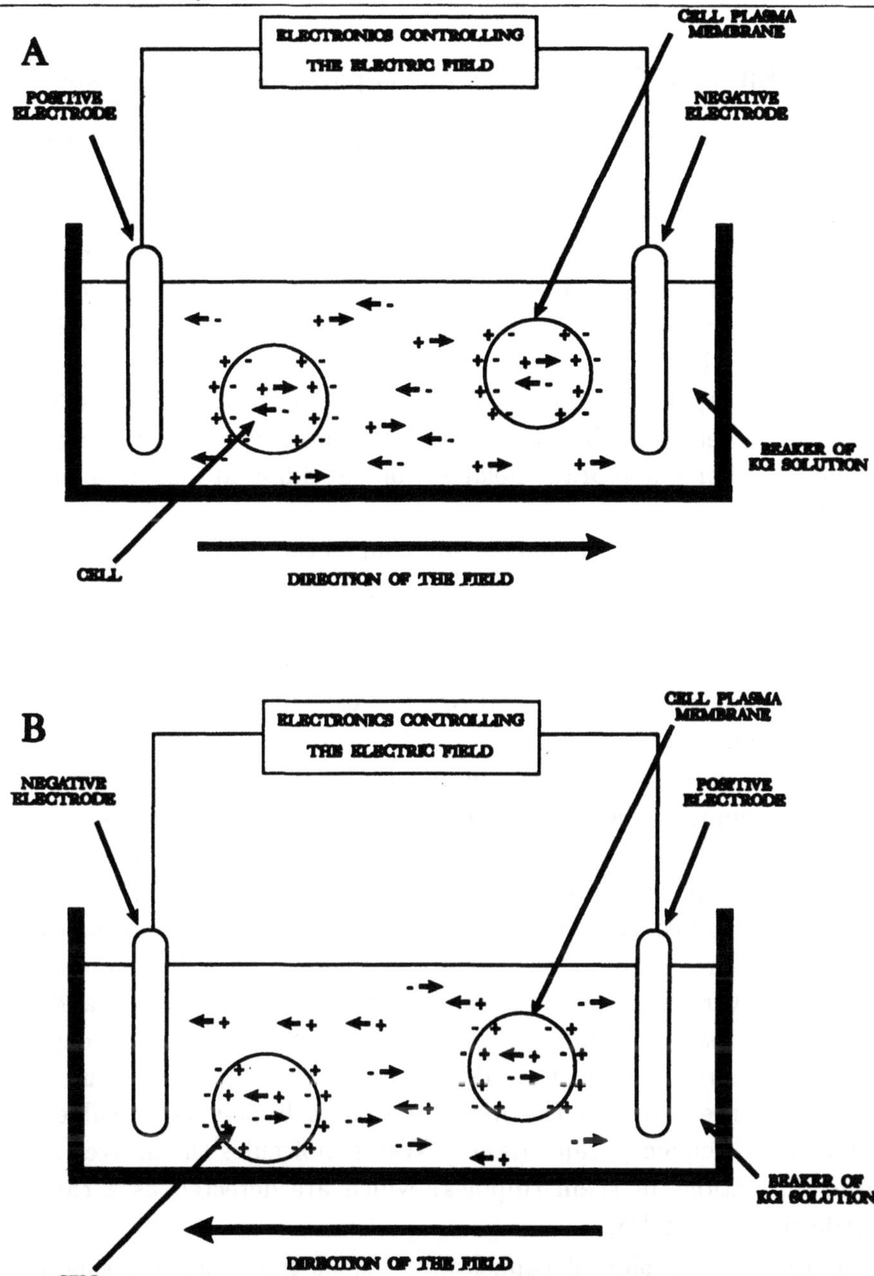

Figure 2.3: A and B demonstrate that reversing the direction of the electric field applied to a cell suspension merely reverses the direction of the membrane polarizations without affecting the magnitude of the polarization effect. Thus the capacitance of the cell suspension is not affected by the *direction* of the electric field *per se*.

2.1.2.2 Effect of frequency on the capacitance of a cell suspension

The *frequency* of the applied electric field has a marked effect on the capacitance of a cell suspension. As the frequency increases the capacitance falls because the extent of the charge separation drops. The reason for this is that the ions being pushed along by the field take a finite time to bump into the membranes and cause the polarizations. Thus as the field changes direction faster and faster (i.e., as the frequency increases), less and less ions have time to reach the membranes and so the extent of the charge separations drop and the capacitance falls. This point is illustrated in Figure 2.4 where an individual cell from the suspension in Figure 2.3 is shown along with the relative size of the membrane polarizations at different field frequencies and directions.

In Figure 2.4A the frequency is low and so the field is changing direction only relatively slowly. As can be seen a lot of ions have time to reach the membrane and polarize it before the field changes direction again and causes the membrane to be polarized the opposite way around. As there are a large amount of charge separations the capacitance of the suspension is correspondingly high.

In Figure 2.4B the frequency is much higher and less ions have time to reach the membrane before the field changes direction. As the amount of polarizations is less, so the capacitance of the suspension is reduced.

In Figure 2.4C the frequency is still much higher and the field is changing direction so fast that just as an ion begins to move in response to the field the field changes direction again and starts to push it the opposite way. Thus only a very few ions have time to reach the membranes and the polarizations across them are very small indeed. Therefore the capacitance of the suspension seen by the BM is almost entirely due to the medium. In fact it is the water in the medium that is largely responsible for this residual high-frequency capacitance. Water molecules themselves have tiny charge separations in them (dipoles) which are detected as a capacitance contribution by the BM.

Figure 2.5 is a plot of capacitance versus frequency corresponding to Figures 2.4A to C. As can be seen the capacitance goes from a high, low-frequency plateau (corresponding to where the cells are maximally polarized) to a low, high-frequency one (labeled C_∞, where the cells are hardly polarized by the field at all). C_∞ corresponds to the background capacitance of the water in the medium. *The drop in the capacitance due to this loss of induced plasma membrane polarization as the frequency is increased is called the*

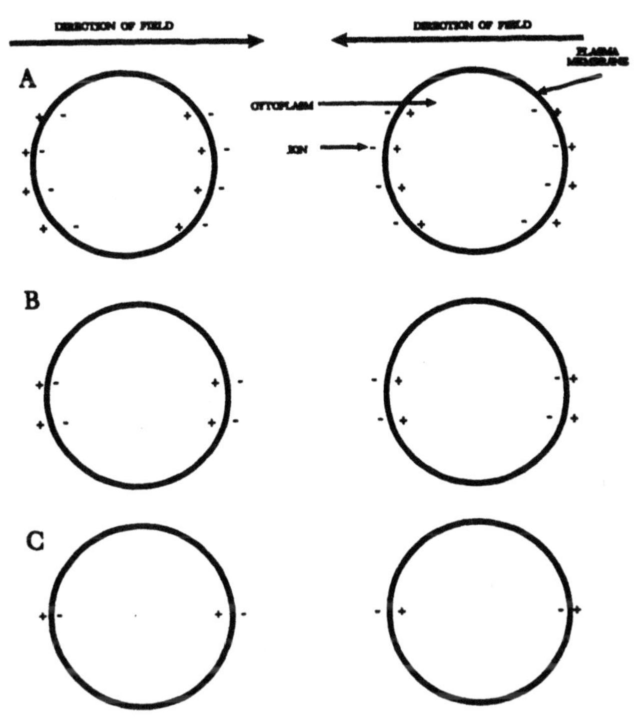

Figure 2.4: A, B, and C demonstrate the effect that increasing frequencies have on the magnitude of the charge separations induced across the plasma membrane of a cell from a suspension such as that shown in Figure 2.3. A is for low frequencies, B for higher frequencies and C for very high frequencies. At each frequency the cell is shown with the field going in either direction. For full details see the text.

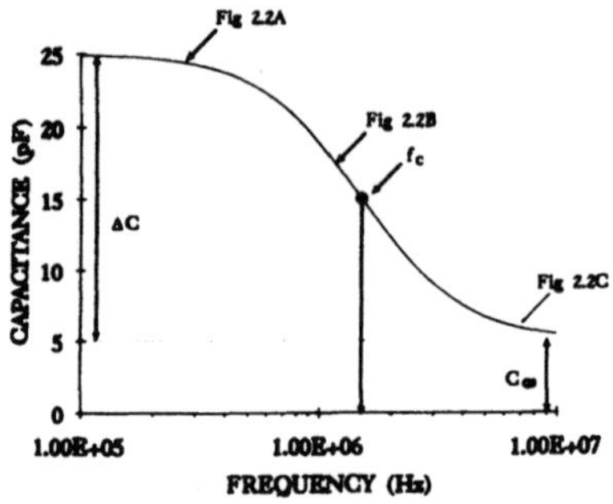

Figure 2.5: The capacitance of a cell suspension as a function of frequency (note the $\log_{10}$ frequency axis. 1.00E+5 on the x-axis labeling means 1.00×10^5, i.e., 10^5). At low frequencies the cell plasma membranes are maximally polarized (see Figure 2.4A) and so the capacitance is high. This gives a high, low-frequency plateau in capacitance. As the frequency is increased the capacitance of the suspension falls because the cell plasma membranes are not so heavily polarized (Figure 2.4B). At very high frequencies a low, high-frequency plateau is reached. Here the cell membranes are hardly polarized at all (Figure 2.4C) and the residual capacitance (C_∞) is mainly due to the water in the suspension. The height of the low-frequency plateau above C_∞ is called the capacitance increment (ΔC). The frequency at which ΔC has halved, i.e., at which the capacitance of the suspension equals $\Delta C/2 + C_\infty$ is called critical (characteristic) frequency (f_c). For this plot $\Delta C = 20$ pF, $C_\infty = 5$ pF and $f_c = 1.5$ MHz (the frequency at which capacitance equals $20/2 + 5 = 15$ pF).

beta-dispersion (β-dispersion) and its size is measured by the capacitance increment (ΔC, delta capacitance). The frequency at which the fall is half complete (i.e., when the capacitance equals $\Delta C/2 + C_\infty$) is called the critical (or characteristic) frequency (f_c).

Figure 2.6 is an actual plot of the capacitance of a yeast suspension as a function of frequency measured with a biomass monitor, clearly showing the inverted sigmoid shape shown in Figure 2.5.

2.1.2.3 The Cole-Cole alpha and the slope of the fall in capacitance with frequency

The fall in capacitance (during the β-dispersion) as a function of frequency in Figure 2.5 is quite steep. For some cell suspensions this slope can be quite shallow and so the capacitance only approaches its high- and low-frequency

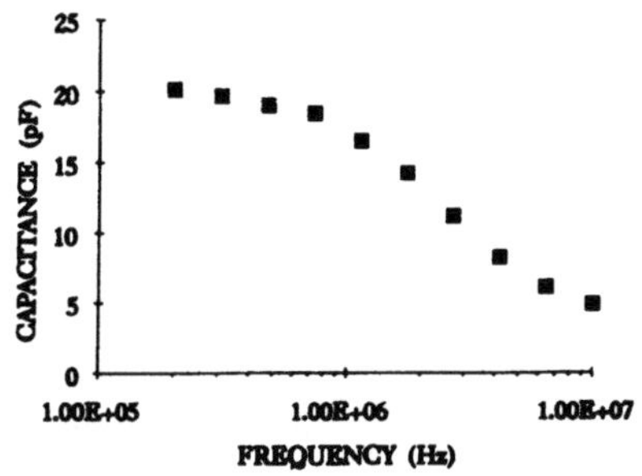

Figure 2.6: Capacitance of a suspension of yeast (9.7 10^8 cells/mL, 27 mg dry weight/mL) as a function of frequency as registered on a BM. The plot shows the *beta*-dispersion of the cells as described in the text and Figure 2.5.

plateaus quite slowly. The measure of this steepness used in this field of dielectric theory is called the *Cole-Cole alpha* (α). The alpha value is 0 for a steep fall (as in Figure 2.5) and would equal 1 for a dispersion that had an infinitely shallow fall in capacitance. The α value has no units and is a *measure* of the steepness of the fall but is *not* its slope. For biological cells the Cole-Cole alpha values of the β-dispersion are typically in the range up to 0.2, but for some yeast and bacteria can get as high as 0.3 to 0.4. The Cole-Cole alpha value is believed to increase as the distribution in cell electrical properties in the cell population widens. However the effect of cell heterogeneity does not fully account for the high alpha values seen in many cell suspensions, and remains something of a mystery.

The effect of a high Cole-Cole alpha on the β-dispersion as seen in the frequency range available to the BM is shown in Figure 2.7. As one can see for very high Cole-Cole alpha values the β-dispersion can fail to reach either of its plateaus within the frequency range available. This effect is seldom seen in real spectra but in those situations where it does occur it can complicate biomass measurements.

2.1.2.4 Effect of cell concentration on the capacitance versus frequency plots of the β-dispersion

As the main aim is to measure biomass with the biomass monitor it is important to see how the concentration of cells affects the shape of the capacitance versus frequency curve. This is illustrated in Figure 2.8. As can be seen, increasing the cell concentration has not changed the critical frequency (f_c) nor changed the value of the residual high frequency capacitance (C_∞). In

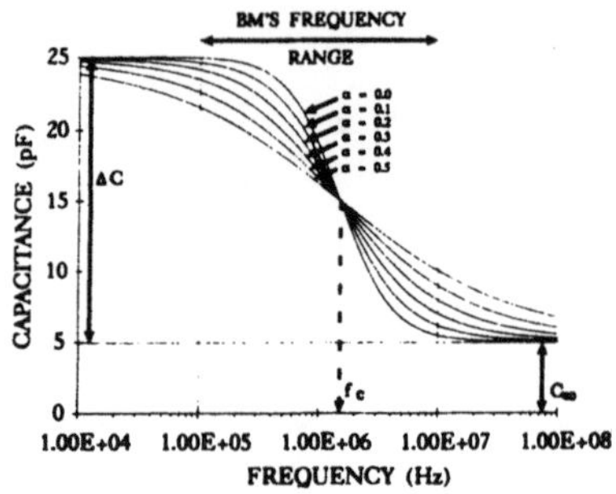

Figure 2.7: Effect of increasing the Cole-Cole alpha of the β-dispersion on the capacitance seen within the BM's frequency range. All the plots are for a β-dispersion with $\Delta C = 20$ pF, $f_c = 1.5$ MHz and $C_\infty = 5$ pF. The key points to note are that the f_c values for all the curves are the same and that at very high and very low frequencies all the curves reach the same high- and low-frequency plateau values. See the text for a full discussion.

fact, C_∞ does change very slightly with biomass content but this is usually too small to measure and so is not shown in Figure 2.8. What has changed with biomass content is the height of the low-frequency plateau above C_∞, i.e., the value of ΔC (Figure 2.8) is proportional to biomass. Thus, the problem of *measuring biomass reduces to finding ways of monitoring the ΔC of the suspension's β-dispersion.* Before going on to consider how this can be done, we need to consider how dead cells and non-cellular material affect ΔC and hence biomass measurements. This forms the subject of Section 2.1.3.

2.1.2.5 Summary

(a) How many times an electric field changes direction per second is measured by its frequency. Frequency is measured in Hz; the more rapidly the field changes direction, the higher the frequency.

(b) The biomass monitor displays its frequencies in MHz (1 MHz = 1 million Hz) and allows the user to alter the frequency of the electric field between about 0.2 MHz and about 10 MHz.

(c) The frequency of the electric field has a marked effect on the capacitance of a cell suspension. As the frequency is increased the capacitance of a cell suspension falls from a high low-frequency plateau to a

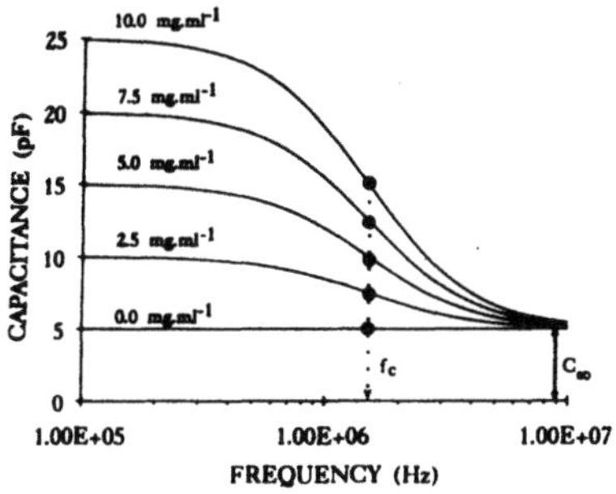

Figure 2.8: Capacitance as a function of frequency for suspensions of a hypothetical cell type at different cell concentrations (in mg dry weight/mL). The f_c value of the suspension does not change with biomass content and the C_∞ value does not change significantly. What does change as a function of the cell concentration is the height on the low-frequency plateau above C_∞, i.e., the ΔC values. Thus, to measure the biomass content of the cell suspension one needs to measure its ΔC value.

low high-frequency one (see Figure 2.5). This fall is due to the decrease in the extent of the field induced polarization of the plasma membrane as the frequency increases and is called the β-dispersion.

(d) How fast the capacitance falls during the β-dispersion is measured by the dispersion's Cole-Cole alpha value.

(e) The residual high-frequency capacitance (i.e., C_∞, see Figure 2.5) is due to the water in the suspending medium and effectively does not change with biomass content (see Figure 2.8). The height of the low frequency plateau above C_∞ is called the capacitance increment (ΔC) and this is proportional to biomass (see Figures 2.5 and 2.8).

(f) The problem of biomass measurement therefore reduces to one of measuring the ΔC of the β-dispersion of the cell suspension.

2.1.3 Effect of dead cells, non-biomass solids and immiscible liquids, bubbles and cell type

2.1.3.1 Effect of dead cells (necromass)

So far the biomass monitor has been said to measure the biomass content of cell suspensions but as yet no justification has been given. The biomass

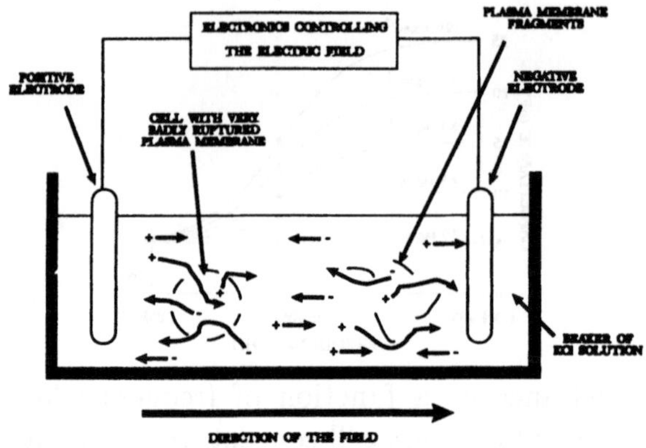

Figure 2.9: Movement of ions through and around the debris of dead (badly ruptured) cells. As can be seen the electric field fails to induce any charge separations across the membranes of the cells and so they are not seen by the BM.

monitor's *criterion for biomass is the volume fraction of cells with intact plasma membranes.* Figure 2.9 shows the effect if the cells in the suspension have *badly* ruptured plasma membranes (i.e., are dead). As can be seen the ions just move through the holes in the membranes or, if only membrane fragments are present, the ions move around the fragments under the influence of the field (the electrostatic attractions between the oppositely charged ions fail to prevent this). The result of this is that the dead (ruptured) cells do not get charge separations (polarizations) induced across their plasma membranes by the electric field (even at low frequencies) and so cannot give a high low-frequency plateau in capacitance (i.e., give a ΔC). A frequency scan of a suspension of dead (completely ruptured) cells would look like the 0.0 mg dry weight/mL plot on Figure 2.8. In fact cells with partially damaged membranes, as opposed to completely ruptured ones, will give a small residual ΔC contribution. Thus if one uses the ΔC value of the β-dispersion of a cell suspension to measure its cellular content then only living (not badly ruptured) cells are registered (i.e., biomass as opposed to necromass).

2.1.3.2 Effect of non-biomass solids, immiscible liquids and gas bubbles

Figure 2.10 shows the effect that non-biomass solids, immiscible liquids and gas bubbles have on the capacitance seen by the BM. In all cases because the non-biomass particles do not have bilayer membranes like cells, they

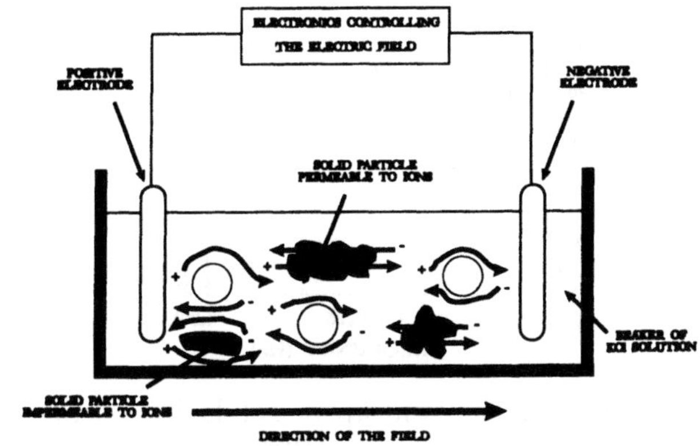

Figure 2.10: Movement of ions through and around non-biomass solids, immiscible liquids and gas bubbles in a cell growth medium. The amorphous shapes represent solids in the medium, e.g., protein flocks, chalk, ... The circles represent gas bubbles, oil droplets, ... As far as the BM is concerned micelles behave as very small oil droplets. In all cases the electric field fails to induce significant polarizations in the particles and so they do not contribute significantly to the capacitance seen by the BM.

are not significantly polarized by the electric field (even at low frequencies) and so do not contribute significantly to the capacitance (biomass) seen by the BM. If *a lot* of such non-polarizable species are present in a fermentor (e.g., if the gas hold up is *high*, or there is a *large* solids or immiscible liquid content) then their presence will noticeably reduce the capacitance slightly over the whole frequency range because one is replacing a finite volume fraction of polarizable material (cells and water) with non-polarizable material (bubbles, non biomass solids, ...). In this situation both ΔC and C_∞ for the suspension will be less than if the particles were not present (Figure 2.11).

It is important to note that the ΔC of the β-dispersion (the BM's measure of biomass) falls when non-biomass material reduces the biomass concentration in a cell suspension (see Figure 2.11). This ability to monitor biomass accurately in the presence of non-biomass material is in marked contrast to many other biomass measuring techniques (e.g., light scattering) where the non-biomass material would be registered as biomass.

Occasionally the non-biomass solids do get significantly polarized by the electric field and this can result in the solids actually adding to the capacitance measured by the BM rather than reducing it. This effect can be

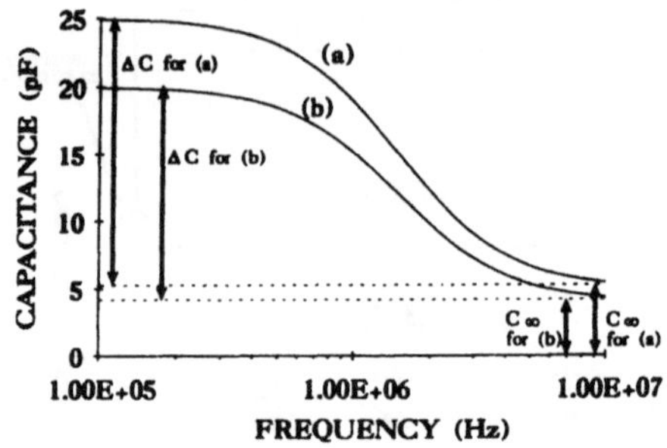

Figure 2.11: Effect of *high* gas hold-up or solids/immiscible liquid content, on the β-dispersion of a cell suspension. With no gas bubbles or non-biomass particles the usual ΔC is seen (a). With *a lot* of bubbles or particles in the medium the ΔC and C_∞ of the β-dispersion are both reduced (b). This effect occurs because one is replacing a finite volume fraction of polarizable material (water and cells) with non-polarizable material (air, solids, immiscible liquids). The reduced volume fraction of cells reduces the size of ΔC while the reduced volume fraction of water reduces C_∞.

detected by doing a polarization control experiment. In cases where non-biomass material does contribute a significant capacitance to the cell suspension it usually only causes a problem if the capacitance from the biomass is low (e.g., at the start of a fermentation) and the non-biomass material's concentration varies with time.

There are a *few* situations where the effects of non-biomass materials can become a problem. If the BM's probe is placed into a vessel where a vortex is likely to be present at the probe's tip then this will exacerbate any interference in the biomass readings from air bubbles, oil droplets, ... In media with high oil contents, if the probe is placed in the fermentor in a top-down position, then sometimes one finds that as the oil floats to the surface it becomes trapped under the end of the BM's probe resulting in anomalous readings. This effect can also occur with some anti-foams. If the medium contains large solid particles (be they biomass or non-biomass) one sometimes gets incorrect readings from the BM due to these becoming lodged on the probe's gold pins.

A final situation where non-biomass solids can cause problems is where the probe tip is placed very close to the walls of the measuring vessel (e.g., a fermentor or glass beaker). In this case one ends up measuring the liquid of interest plus a small contribution from the container's wall which is seen

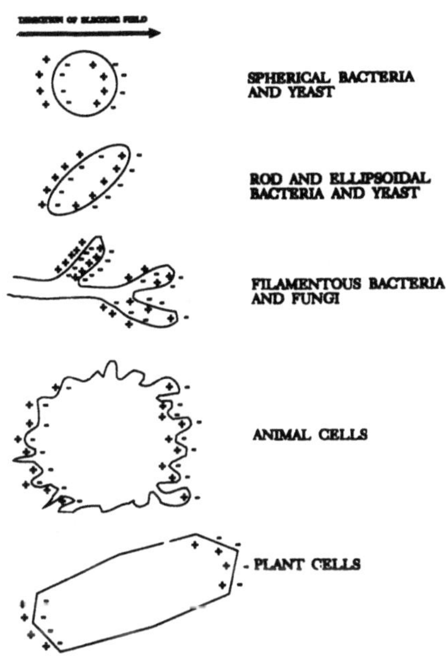

Figure 2.12: An electric field can polarize the plasma membranes of cells of any shape and so the BM can (in principle) be used to measure the biomass content of suspensions of any cell type. The cell wall surrounding many cells has no effect on the β-dispersion and hence on biomass measurement.

by the BM as non-biomass material. This effect is usually only seen when measurements are being made in small beakers. If one ensures that the probe pins are not within say a couple of cm from the vessel walls then the effect is usually not important. If measuring near the vessel wall is unavoidable then one should keep the proximity effects constant by keeping the probe in the same relative position to the wall for all the readings. If this is done then the wall effect results in a constant *error* (note that the wall effect changes the BM's cell constant).

2.1.3.3 Effect of cell type and shape on biomass measurements

As the biomass monitor detects intact plasma membranes by using an electric field to induce charge separations across them it can detect any type of cell irrespective of its shape. This point is illustrated for a variety of cell types and shapes in Figure 2.12. In all of these cases as the membranes are polarized by the field one can get β-dispersion curves like those on Figure 2.8. By measuring the ΔC of these curves one can measure the biomass content of the suspensions exactly as described previously.

Some cell types can occur in chains or clumps. It has been found experimentally that the walls of bacteria and yeast do not impede the motion of ions up to the cell's plasma membranes when an electric field is applied. This has two important consequences. First, the β-dispersion is not affected by the presents of the walls. The second is that the β-dispersion of a clump of walled cells is not effected by the clumping because of the walls acting as channels for the ions to move through. The situation for clumps of non-walled cells is more complex but what will probably be found is that the ΔC of the suspension is slightly less than expected.

2.1.3.4 Summary

(a) Cells with badly ruptured plasma membranes (i.e., dead cells) do not get polarized by an applied electric field (even at very low frequencies) and so do not cause a high low-frequency plateau in capacitance (i.e., they do not contribute a ΔC, see Figure 2.9). Thus the measurement of the ΔC of a cell suspension enables one to measure biomass as opposed to necromass.

(b) Non-biomass solids, immiscible liquids and gas bubbles do not have intact plasma membranes and so do not contribute to the capacitance

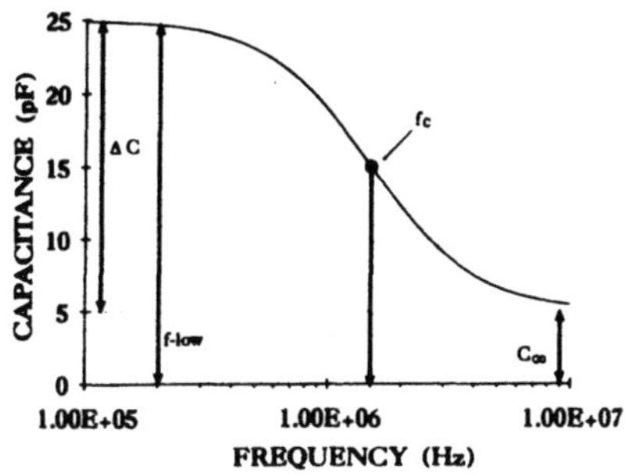

Figure 2.13: Capacitance of a cell suspension in the frequency range of the β-dispersion. The ΔC value is 20 pF and the C_∞ is 5 pF. The capacitance at the frequency marked *f-low* on the figure is approximately equal to $\Delta C + C_\infty$ (in this case 25 pF).

(ΔC) measured by the biomass monitor during biomass measurements (see Figure 2.10).

(c) The biomass monitor measures biomass by polarizing the plasma membranes of cells with an electric field. As this polarization effect occurs irrespective of cell type and shape one can in principle measure the biomass of any cell suspension.

2.1.4 Practical methods of monitoring cellular biomass

2.1.4.1 Introduction

In Section 2.1.2.4 it was shown that measuring the cellular content of a suspension can be done by measuring the ΔC of its β-dispersion. In Section 2.1.3 it was further shown that this method detects living cells (biomass) as opposed to dead ones (necromass). In addition, it was demonstrated that ΔC is relatively insensitive to non-biomass solids and bubbles. The next stage is to explain the two common methods by which one measures the ΔC of a cell suspension and hence its biomass content.

2.1.4.2 Measuring the ΔC of the β-dispersion using a single spot frequency

Figure 2.13 shows a frequency scan of a cell suspension. It is the magnitude of ΔC that is proportional to the biomass content of the suspension (in this case ΔC = 20 pF). C_∞ does not change significantly with biomass content

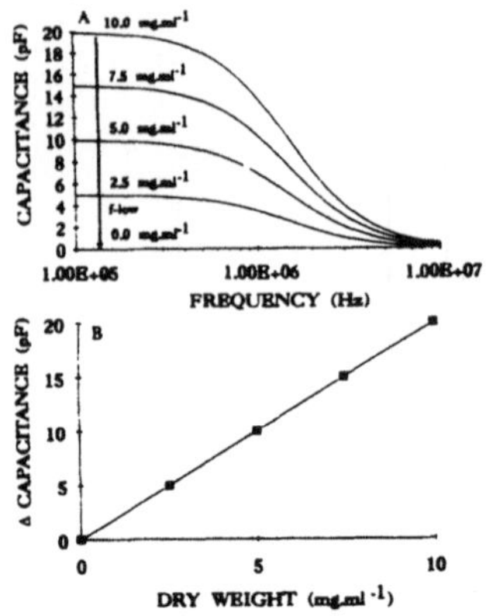

Figure 2.14: (A) Capacitance of cell suspensions at different biomass levels as a function of frequency. The biomass levels are dry weights (mg/mL) but any biomass measurement method may equally well be used. The capacitances at high frequencies for the β-dispersions shown are all equal to zero, i.e., the C_∞ term has been set to zero. This was achieved by backing the capacitance of the suspending medium to zero when no cells were present. This means that the capacitances for each suspension measured at *f-low* are now approximately equal to ΔC. Thus monitoring the capacitance at *f-low* as a function of biomass content of the suspension gives a calibration curve like (B).

and in this figure it is equal to 5pF. Thus, what is needed is a convenient means of estimating the ΔC of the suspension.

If one picks a *single frequency on the low-frequency plateau of the β-dispersion* (marked as *f-low* on Figure 2.13) then the capacitance at that frequency approximately equals $\Delta C + C_\infty$ (i.e., 25 pF). Of that 25 pF only 20 pF corresponds to ΔC, and hence to the biomass content of the suspension. What is needed is a means of backing off the C_∞ term.

As described earlier (Section 2.1.2.2) C_∞ is a capacitance mainly due to the medium alone and not to the cells. As the capacitance of the suspending medium does not vary much during a fermentation one can measure the capacitance of the medium at *f-low* prior to inoculating the medium and back this value to zero pF. With C_∞ backed off this way any capacitance increase measured at the spot frequency (*f-low*) will reflect an increase in ΔC and hence an increase in the biomass content of the suspension.

This idea is illustrated in Figure 2.14. In (A) the capacitance of the medium in which the cells were suspended was backed to zero pF (at *f-low*) prior to the addition of any cells. This had the effect of setting C_∞ to zero pF. As the biomass content increased the ΔC of the β-dispersion also increased. The capacitance at *f-low* is approximately equal to ΔC and so a plot of the capacitance at *f-low* versus biomass content gives a calibration curve such as Figure 2.14B.

In a real fermentation one would take a sample of the cells from the fermentor at the maximum biomass content. Next one would do a frequency scan of the cells to get a plot like Figure 2.13 (a polarization control experiment should also be done). From this plot one would pick a suitable spot frequency value (*f-low*) on the low-frequency plateau of the β-dispersion.

The next stage would be to produce a calibration curve like Figure 2.14B. To do this the frequency of the biomass monitor is set to *f-low* and *left there.* A fresh sample of medium is taken and its capacitance backed to zero pF. Aliquots of cells are then added to the medium and the capacitance at *f-low* recorded (approximately equal to ΔC). A calibration curve like Figure 2.14B can then be produced.

The biomass monitor probe can then be inserted into the fermentor and the fermentor sterilized. The BM is set to *f-low* (and *left there*) and immediately prior to inoculation (at the fermentors working temperature and with the gas regime to be used on) the capacitance of the medium is backed to zero pF. Any capacitance changes now recorded by the BM will reflect changes in the biomass concentration. The calibration curve of capacitance at *f-low* (approximately equal to ΔC) versus biomass can then be used to convert the measured capacitance values for the fermentation back to biomass concentration units.

2.1.4.3 Measuring the ΔC of the β-dispersion using two spot frequencies

A better way to measure ΔC is to measure capacitance simultaneously at two frequencies (Figure 2.15). One of the frequencies sits on the high low-frequency plateau of the β-dispersion. This frequency is called *f-low* in the figure and *corresponds to the* f-low *used in single frequency measurements of* ΔC. The second frequency is situated on the low high-frequency plateau of the β-dispersion and is labeled as *f-high* in the figure. The capacitance measured at *f-low* approximately equals $C_\infty + \Delta C$ while the capacitance measured at *f-high* approximately equals C_∞. Obviously, if one measures the capacitance of the cell suspension at *f-low* and *f-high* simultaneously and then subtracts the capacitance at *f-high* (C_∞) from the capacitance at *f-low*

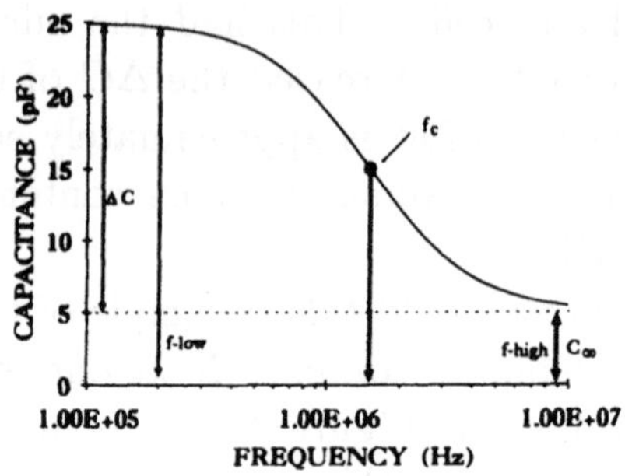

Figure 2.15: Capacitance of a cell suspension in the frequency range of the β-dispersion. To measure the biomass content of the suspension one needs the value of ΔC (in this case 20 pF). The capacitance at the spot measuring frequency on the high-frequency plateau of the dispersion (i.e., the frequency *f-high*) approximately equals C_∞ (5 pF). The capacitance at the spot measuring frequency on the low-frequency plateau of the dispersion (i.e., the frequency *f-low*) approximately equals $C_\infty + \Delta C$ (25 pF). Thus, to measure ΔC one measures the capacitance at *f-high* (C_∞) and subtracts it from the capacitance measured at *f-low* ($C_\infty + \Delta C$). This forms the basis of the dual frequency method of biomass measurement.

($C_\infty + \Delta C$) then the resulting capacitance is ΔC (which is proportional to the biomass concentration). The biomass monitor allows the user the options of measuring biomass at a single frequency (*single frequency mode*) or using two frequencies (*dual frequency mode*).

The procedures for setting up a biomass monitor for dual frequency measurements in a real fermentation are very similar to the ones used for single frequency measurements. First, one would take a sample of the cells from the fermentor at the maximum biomass content. Next one would do a frequency scan of the cells to get a plot like Figure 2.15 (it is recommended that a polarization control is also done). From this plot one would pick a suitable spot frequency value (*f-low*) on the low-frequency plateau of the β-dispersion. The *f-high* frequency is factory set at about 10 MHz.

The next stage would be to produce a calibration curve like Figure 2.14B. However, the capacitance used is the difference in the capacitances measured at *f-high* and *f-low* (approximately equal to ΔC). To do this the frequency of the biomass monitor is set equal to *f-low* and *left there.* Then the biomass monitor is set to dual frequency mode. The capacitances now displayed are the difference capacitances required. A fresh sample of medium is taken and aliquots of cells are then added to the medium and the capacitances recorded (approximately equal to ΔC). A calibration curve like Figure 2.14B can then be produced.

The biomass monitor probe can then be inserted into the fermentor and the fermentor sterilized. Immediately prior to inoculation the BM is set to *f-low* (and *left there*) and set to dual frequency mode. Any capacitance changes now recorded by the BM will reflect changes in the biomass. The calibration curve of capacitance versus biomass concentration can then be used to convert the measured capacitance values for the fermentation back to the biomass concentration units required.

2.1.4.4 Comparison of the single and dual frequency methods

The only real difference between the single and dual frequency measurements of ΔC is how one removes the C_∞ term from the capacitance measured at *f-low*. In the single frequency measurements one measures C_∞ at the start of the experiment and backs it to zero manually. The underlying assumption is that C_∞ does not change during a fermentation. In the dual frequency measurements C_∞ is measured and subtracted from the signal each time a reading is taken. This means that no assumption about a constant C_∞ term is made. As C_∞ can vary with gas hold up to some extent (see Section 2.1.3.2) the dual frequency method offers some improvements in signal stability over single frequency measurements.

The *f-low* frequency for both the single and dual frequency modes are identical. In both cases all that is required is that the frequency sits well on to the low-frequency plateau of the β-dispersion at a frequency above any low-frequency tip-ups in capacitance not due to the β-dispersion. The selection of a suitable *f-low* frequency is the most important aspect of setting up a biomass monitor for a new application.

2.1.4.5 Summary

(a) To measure the biomass content of a cell suspension one has to measure the ΔC of its β-dispersion.

(b) There are two methods that can be used to measure ΔC on the biomass monitor. Both methods use a spot frequency on the high, low-frequency plateau of the β-dispersion. This frequency is called *f-low* (see Figures 2.13 and 2.15). The only real difference between the two methods lies in how the C_∞ term is removed from the capacitance measured at *f-low* so that ΔC is measured rather than $\Delta C + C_\infty$.

(c) In the single frequency method of measuring ΔC the C_∞ term is physically offset to zero pF at the start of the fermentation prior to inoculation (Figure 2.14).

(d) In dual frequency measurements C_∞ is measured at a spot high frequency (*f-high*) on the low high-frequency plateau of the β-dispersion (Figure 2.15). Each capacitance measurement consists of measuring the capacitance simultaneously at *f-low* and *f-high*, and then subtracting the capacitance at *f-high* from that at *f-low* to give ΔC.

2.2 Experimental Results

2.2.1 Biomass monitor reading versus consistency measurement

2.2.1.1 Introduction

In the brewing industry it is important to start each brew with the same amount of viable yeast cells. This is not only important for planning (each brew will then take the same time), but also important for the quality and flavor of the beer.

Normally a brewer will add the yeast by weight or volume, after checking the viability of the yeast by methylene blue staining in the laboratory. This check in the lab is time consuming, not completely accurate and only provides an indication of the viability of one sample of the yeast at a certain time. By adding a biomass monitor to the system, a brewer will not only know how much yeast he puts into the brew, but will also know how much *viable* yeast he puts into it.

The amount of cells in the yeast slurry is usually measured by brewers by centrifugation and stated in % spun solids. That is why the following test compares the biomass monitor measurement with % spun solids.

2.2.1.2 Results

Figure 2.16 shows the correlation between the readings in spun solids (figure from the laboratory) and the biomass monitor readings. It is clear that both readings are almost the same over a wide range. It must also be remembered

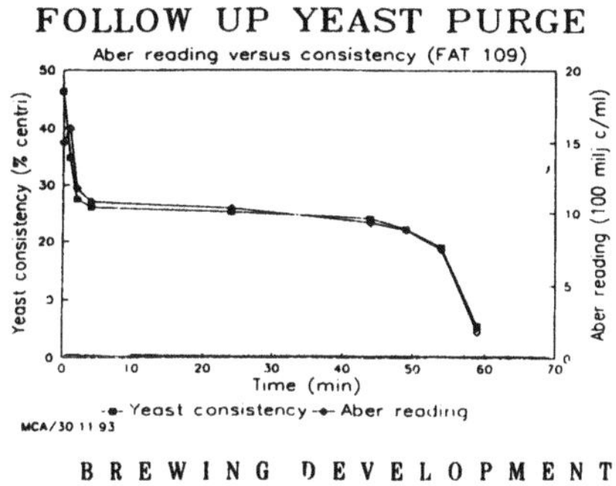

Figure 2.16: Biomass monitor reading versus consistency measurement in % after centrifugation.

that methylene blue staining has an inaccuracy of about 15 %, while that of the biomass monitor is only 3 %.

Nowadays this principle of controlling the yeast pitching by capacitance is used by a lot of breweries. As well as yeast pitching, breweries can use the biomass monitor for yeast dosing into bottles and cans and also for yeast recovery after a brew.

2.2.2 Biomass monitor reading versus *OD* and *RQ* measurements

2.2.2.1 Introduction

Knowing the concentration of viable biomass is not only important to the brewing world. Biotechnological and pharmaceutical companies, and organizations involved in biological research are also very interested in real time biomass measurement. By using a biomass monitor the influence on the viability of the cells by changing different parameters can be seen. The biomass monitor gives an instantaneous indication of the effect of a change in a parameter on the viability of the cells.

In addition, it is often important to know the optimal moment to harvest cells. The biomass monitor will indicate this on-line.

In the following the correlation between several on-line biomass measuring methods and off-line analysis is investigated. Three methods have been compared: (i) capacitance, (ii) respiration quotient (*RQ*), and (iii) optical density (*OD*).

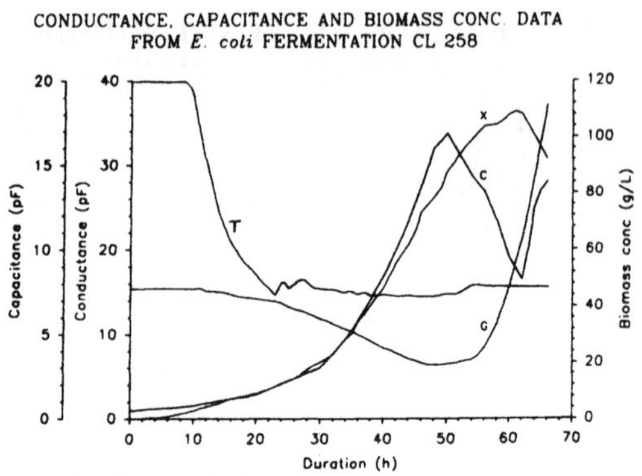

Figure 2.17: Conductance, capacitance, and biomass concentration data from *E. coli* fermentation CL258. T: respiration quotient; C: capacitance; X: *OD* measurement; and G: conductance.

2.2.2.2 Results

The results from the different techniques were compared with the off-line dry weight and methylene blue staining. For this test, an *E. coli* fermentation was used. The temperature was 30 °C, the pH was 5.8 and the experiment was performed to find out when cell lysis occurs. Figure 2.17 shows the results from the different methods (data obtained from a pharmaceutical company). Looking at the highest concentration of cells the *OD* measurement gives the highest value at 61 hours, the *RQ* shows an increase at 54 hours and the capacitance reading gives the highest value at 50 hours. The off-line measurement indicated that the viable biomass actually decreased after 50 hours. Therefore the highest concentration of viable biomass occurred at 50 hours. This means that although the total biomass concentration was largest at 61 hours (measured by *OD* measurement) the concentration of viable biomass had already been declining for 10 hours.

2.2.3 Discussion

Both sets of results show that the capacitance reading of the biomass monitor can be used for on-line measurement of viable biomass. It can be used for all kinds of cell types at a concentration from about 10^6 cells per mL (depending on cell size). The conductivity of the media should not be higher than 25 mS/cm. Provided this requirement is met, the biomass monitor system will give answers (and new questions) to a lot of questions on fermentation. In summary, it provides a better understanding of what is happening with viable cells during a fermentation.

Chapter 3

Membrane Inlet Mass Spectrometry for the Characterization and Monitoring of Biotechnological Processes

F.R. LAURITSEN
Odense University Institute of Biochemistry
DK-5230 Odense M (DENMARK)
email: frants@dou.dk

ABSTRACT – *Volatile metabolites produced by microorganisms can be measured directly and on-line in microbial media by membrane inlet mass spectrometry. A polymer membrane is used as the sole separation between the media and a mass spectrometer. Gases and volatile organic compounds dissolve in the membrane, diffuse through it and vaporize into the mass spectrometer, where they are ionized and analyzed according to their m/z ratio. In this way microbial media can be analyzed directly, without any pre-treatment, a facility which has made on-line process control possible. Here the main principles of membrane inlet mass spectrometry as well as several practical examples of its use are presented.*

KEYWORDS – *Mass spectrometry, Membrane Inlet, On-line Monitoring, Process Control, Fermentation*

J.F.M. Van Impe et al. (eds.), Advanced Instrumentation, Data Interpretation, and Control of Biotechnological Processes, 67-103.

3.1 Introduction

The complex composition of microbial media makes the analysis of such samples a very difficult task. A number of sensors exists, which can be used directly in the medium. However, these sensors are mostly for physical variables. Most sensors for the measurement of organic substrates and products require filtered samples as a minimum. Consequently measurements are either carried out off-line or the sensor must be combined with a filtration module. Several analytical techniques have been used for on-line process control in combination with filtration modules. Typical examples are enzymatic or chemical detection, high performance liquid chromatography (HPLC), infrared spectroscopy (IR) and mass spectrometry. The technique, membrane inlet mass spectrometry (MIMS), described in this chapter is a sensor that can be used without any pretreatment of the samples. The membrane itself acts as a strong filter allowing only volatile compounds to pass from the reaction mixture into the mass spectrometer. The principle of using a polymer membrane as the separation between a reaction mixture and the vacuum of a mass spectrometer was invented by Hoch and Kok in 1963 in connection with photo-synthetic studies. Shortly thereafter, the method was adopted by physiologists, who used miniaturized membrane probes for the in-vivo measurement of blood gases. In 1975 Reuß inserted a membrane probe into a fermentor for the on-line monitoring of dissolved gases and small volatile organic compounds, an experiment which has led to the study of many microbial processes. With the introduction of chemical ionization and tandem mass spectrometry in 1985 by Cooks the capabilities of MIMS were extended to include a strong analytical method for the identification of volatile metabolites directly from microbial media. Extremely low detection limits ($<$ 1 microgram/liter) for some environmentally important compounds have made screening of environmental samples one of todays largest applications. Recently, methods were developed which made it possible to measure volatile compounds directly from organic solutions. All these applications have been reviewed [4][5][9][16][18][19][23]. In the past, almost all measurements of organic compounds in aqueous solution have been carried out with silicone membranes. Silicone membranes are highly hydrophobic in nature and hydrophobic organic compounds dissolve easily in the membrane, whereas the hydrophilic water does not. This opens the possibility of enrichment of sample as compared to water by the transport through the membrane. Extremely low detection limits can be obtained for hydrophobic compounds, whereas polar compounds are more difficult to measure. Recent work by Lauritsen et al. [13] has demonstrated that microporous and/or hydrophilic membranes can be used for the measurement of polar, volatile compounds. With such membranes the limitation in the technique seems no longer to be found in the transport through the membrane, but in the evaporation of sample from the membrane into the vacuum of the mass spectrometer. Currently, techniques using different kinds of stimulated desorption is under development at Purdue University, Indiana and Odense University, Denmark. Experiments have actually been carried out [10], where glucose was monitored directly from aqueous solution by the use of a hydrophilic, porous membrane, but further work still needs to be done to produce stable and quantitative signals. However, these recent developments bring much hope that in the near future it will be possible to measure non-volatile compounds such as penicillins and glucose directly from microbial media.

3.2 Mass Spectrometry

In mass spectrometry ions are separated according to their mass/charge ratio (m/z), a property which has made mass spectrometry one of the most versatile analytical techniques. However, before a sample can be mass analyzed, it must be introduced into a high vacuum (10^{-6} torr), vaporized and ionized. The ionized molecules are focused onto the mass analyzer, where they are separated according to their m/z ratio and finally the current of the mass selected ions is amplified and recorded. In this way a characteristic signal from individual compounds in a sample can be obtained. Figure 3.1 shows the components of a mass spectrometer. An introduction to practical mass spectrometry can be found in reference [22].

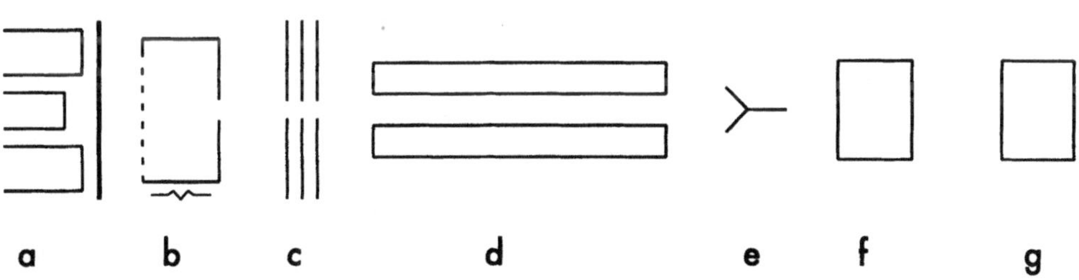

Figure 3.1: Components of a mass spectrometer: (a) inlet; (b) ion source; (c) focusing lenses; (d) mass analyzer; (e) detector; (f) amplifier; (g) computer.

A wide variety of techniques have been developed for the introduction of sample into a mass spectrometer. Among the common techniques are capillary leaks for gas analysis and membrane inlets or spray techniques such as thermospray, electrospray and ionspray for the analysis of aqueous samples. In the spray techniques a filtered sample is introduced into the mass spectrometer through a capillary tube, which is either at high temperature (≈ 250 °C) or at high voltage (≈ 4 kV) and the sample is vaporized and ionized simultaneously. Solid samples can often be introduced directly into the vacuum and then vaporized either by heating or bombardment with light, ions or atoms. For characterization or monitoring of biotechnological processes only methods for direct analysis of aqueous samples are of interest. Of these methods, membrane inlets are suited for the introduction of volatile compounds, whereas the spray techniques are suited for non-volatile compounds.

3.2.1 Mass spectrometers

Mass analysis can be carried out either by deflection in a magnetic field, time of flight or movement in an oscillating electric field. The main principles of the most common analyzers are illustrated in Figure 3.2.

In magnetic field instruments (Figure 3.2a) ions are accelerated to a well defined energy, before they enter a homogeneous magnetic field. Here they will follow a circular path with radius

$$R = 144 \cdot \frac{\sqrt{V}}{B} \cdot \sqrt{\frac{m}{z}} \tag{3.1}$$

where R [cm] is the radius of curvature, V [volt] is the accelerating potential, B [gauss] the magnetic field strength and the mass to charge ratio (m/z) is expressed as atomic mass units [a.m.u.] per number of charges. Mass spectra can be recorded either by varying the accelerating voltage or the magnetic field strength, where the latter is the most commonly used method. Often magnetic analyzers are used in combination with an energy analyzer (the so-called two sector instruments) in order to obtain very high mass resolution. Two sector instruments are suited for accurate mass analysis and have a relatively high upper mass limit (≈ 20,000 a.m.u.). A few examples of the use of magnetic mass analyzers in combination with membrane inlets have been published, but in general they are large and unhandy instruments as compared to the quadrupole mass analyzers used.

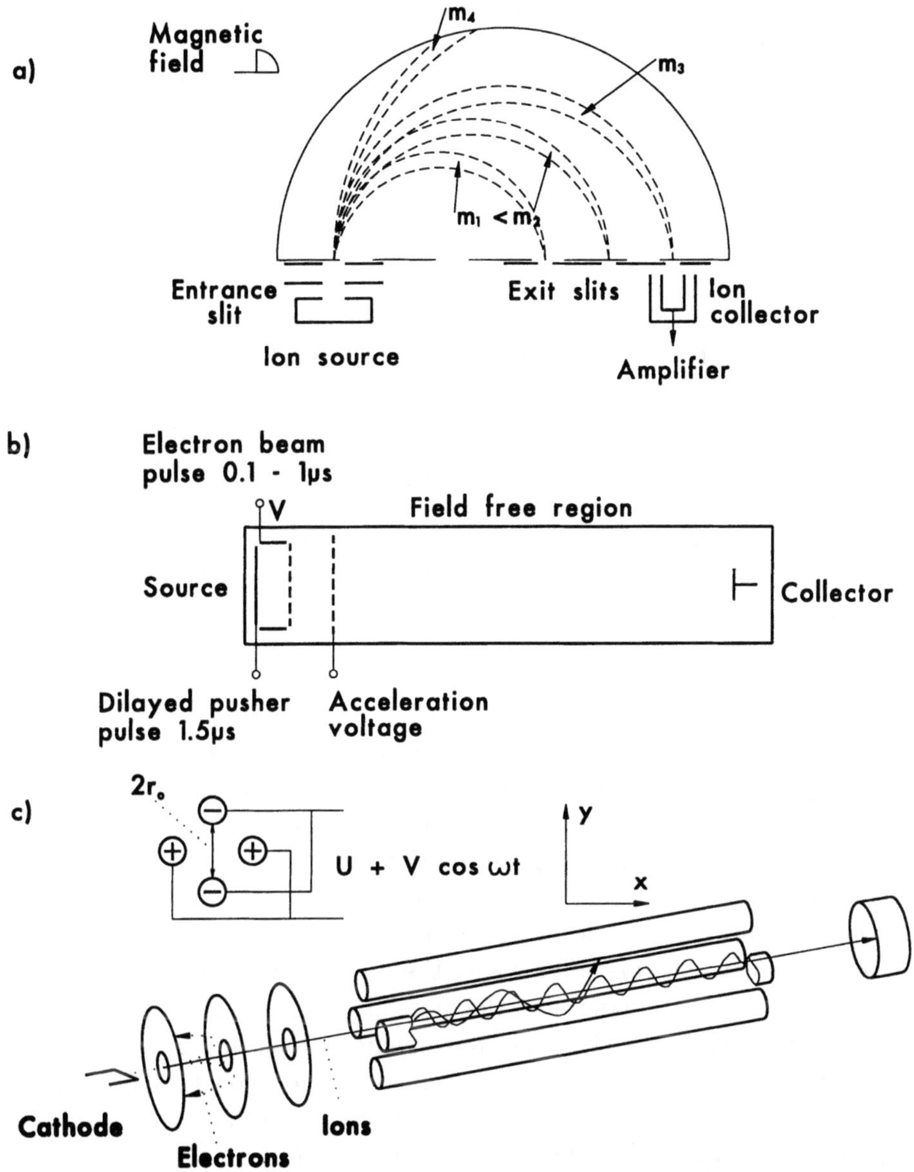

Figure 3.2: Mass spectrometers: (a) magnetic field analyzer; (b) time of flight instrument; (c) quadrupole mass filter.

Time of flight instruments (Figure 3.2b) are based on the time an ion with a well defined energy takes to pass through a vacuum tube of a certain length. The flight time, t [μsec] is related to the m/z ratio as follows

$$t = 72 \cdot \frac{L}{\sqrt{V}} \cdot \sqrt{\frac{m}{z}} \tag{3.2}$$

where L [m] is the length of the tube. In principle, time of flight instruments have an unlimited upper mass range and with the newly developed reflector instruments mass resolution is also quite high. Time of flight instruments have found widespread use for the analysis of peptides and proteins, but as yet they have not been used in conjunction with membrane inlets.

Quadrupole mass spectrometers (Figure 3.2c) have a limited mass range (2 to 4,000 a.m.u.) and resolution ($\approx$ 2,000), but they dominate the low mass range market because of their relatively low cost and easy operation. Quadrupole mass spectrometers are sufficient for the applications described in this chapter. The most important aspects of quadrupole mass spectrometry are described below.

Quadrupole Mass Spectrometers

A quadrupole mass spectrometer consists of 4 hyperbolic (round) rods arranged in parallel, so that they make a two-dimensional quadrupole (Figure 3.2c). An electrical potential composed of a constant and an oscillating voltage [$\phi = U + V\cos(\omega t)$] is applied to the rods and creates the electrical quadrupole field. Ions enter the quadrupole along the axis of the rods and pass through the quadrupole mass spectrometer undisturbed in this direction. By contrast, ions are affected by the oscillating electrical fields in the quadrupole plane ($x-y$ plane) and the mass filtering action of the quadrupole mass spectrometer is based on the movement of ions in this plane. The x axis is conventionally defined as the direction with a constant positive bias [$\phi_x = +(U + V\cos(\omega t))$], whereas the electrical potential in the y direction has a constant negative bias [$\phi_y = -(U + V\cos(\omega t))$].

The filtering action of a quadrupole mass spectrometer is based on the fact that light ions are easier affected by an oscillating electrical field than heavier ions. Along the x axis positive ions are repelled by the constant positive potential $+V$ and they would have stable paths if no alternating electrical field $V\cos(\omega t)$ was present. Since light ions are easier affected by the RF-field than heavier ions, a smaller RF-amplitude is needed to expel the light ions from the quadrupole. The x direction can therefore be considered as a high-pass mass filter. Along the y direction the constant negative bias $-V$ on the rods attracts positive ions and, were it not for the RF-field all positive ions would be expelled from the quadrupole. However, operational parameters of the quadrupole can be found, where the RF-field can refocus ions back to the center of the quadrupole. Once again light ions are more easily affected by the RF-field than the heavier ions, and a smaller RF-amplitude is needed to guide the light ions back to the mass filter. In this way the y-direction acts as a low-pass mass filter. The combined action of the high- and low-pass mass filter is a narrow window allowing only ions with a certain m/z ratio to pass through the quadrupole. A full description of the theory of quadrupole mass spectrometers can be found in reference [3].

From the above discussion it is evident that the mass of the stable ion depends on the amplitude of the applied fields, and it can be shown that there exists a linear relationship between the amplitude of the applied fields (U and V) and the m/z ratio. The transmitted mass under optimal conditions of resolution ($V/U = 5.958$) is

$$\frac{m}{z} = \frac{V}{7.2 \cdot f^2 \cdot r_0^2} \tag{3.3}$$

where V [volt] is the zero to peak amplitude of the RF-field, f [MHz] is the RF-frequency and r_0 [cm] the internal radius of the quadrupole. A mass spectrum can be recorded in two ways. First, the voltages U and V are increased linearly with the U/V ratio kept constant. This gives mass spectra with constant resolution (defined as $m/\Delta m$) over the complete spectrum. Second, the spectrum is scanned with the DC/RF ratio slightly increasing with mass. This results in an increased mass resolution with mass, giving spectra with a constant peak width (Δm constant). Most commercial instruments use the constant Δm mode, and this has the disadvantage that the transmission efficiency through the quadrupole of the ions decreases with increasing mass, since a higher mass resolution (defined as $m/\Delta m$) is needed to obtain the same Δm at high mass as at low mass.

3.2.2 Ionization of volatile compounds

Before a mass analysis can be carried out sample molecules have to be ionized. In connection with MIMS, where the samples are volatile, two ionization techniques are suited. Electron impact ionization (EI), where the gaseous molecules are bombarded with electrons, and chemical ionization

(CI), where molecules are ionized by charge transfer from a reaction gas to the sample molecules. Both techniques have their advantages and disadvantages. With EI a characteristic and often rather complicated spectrum results. The spectrum gives useful structural information of the compounds in the sample that may led to identification, but often the lack of molecular ions $[M^+]$ or the interference between peaks from different compounds in the sample makes it impossible to identify the constituents with EI alone. CI on the other hand gives high intensity molecular ions and very simple spectra. The two techniques are therefore complementary and the combined use of EI and CI often leads to identification. A good text-book for interpretation of EI mass spectra is reference [20], whereas reference [8] can be recommended for CI mass spectra. When MIMS is used as a technique for the continuous monitoring of reaction systems the composition of the sample is well known and long term stability rather than analytical capabilities is of importance. In general EI is a *clean* ionization technique and can be used without a continuous calibration, whereas CI needs calibration.

3.2.2.1 Electron impact ionization

When gas molecules are hit by energetic electrons (70 eV), ions can be created by different mechanisms. The most common mechanisms are creation of singly and doubly charged positive ions, ions created by fragmentation of the charged ions and ions created by a rearrangement of the atoms in the molecule:

$$ABC + e^- \rightarrow ABC^+ + 2e^- \tag{3.4}$$

$$ABC + e^- \rightarrow ABC^{++} + 3e^- \tag{3.5}$$

$$ABC + e^- \rightarrow A^+ + BC + 2e^- \tag{3.6}$$

$$ABC + e^- \rightarrow AC^+ + B + 2e^- \tag{3.7}$$

The mass spectrum is composed of a mixture of the possible ions. Spectra from small organic compounds and aromatics normally contain ions in the molecular mass region at relatively high intensity and the spectra are reasonably simple. In contrast larger organic compounds (mass > 100 a.m.u.) often show severe fragmentation with only low intensity ions in the molecular mass region. As examples, Figures 3.3a and 3.3b show the EI mass spectrum of p-methoxybenzaldehyde and ethylhexanoate respectively.

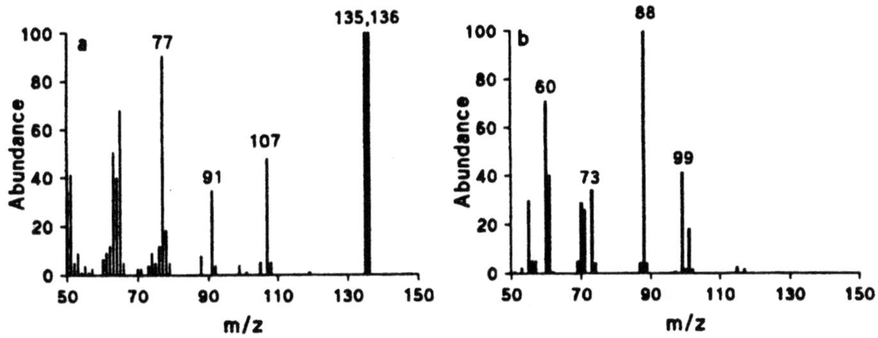

Figure 3.3: Examples of electron impact mass spectra: (a) p-methoxybenzaldehyde; (b) ethylhexanoate.

3.2.2.2 Chemical ionization

When a gas at low pressure ($< 10^{-3}$ torr) is bombarded with electrons a set of ions is created. This results in the characteristic electron impact mass spectrum.

At higher pressures (0.1 to 1 torr) the ions created by electron bombardment will collide many times with neutral gas molecules before they leave the ion source. Through these collisions a set of new ions is generated, and these ions will not react further with the reaction gas. Chemical ionization is based on the transfer of charge from this set of stable ions to gaseous sample molecules. Such ions together with possible fragments thereof, and occasional reaction products between these ions and the reaction gas, constitute a chemical ionization mass spectrum. The resulting spectrum deviates from the EI spectrum in two crucial ways: (i) with the right reaction gas the ionization is soft, which means that only very few fragments arise; (ii) with the right reaction gas the ionization process can be selective, meaning that only certain classes of compounds are ionized. The charge transfer from the set of reaction ions to the sample molecules can take place after several mechanisms. The most common are the use of Brønsted acid reaction systems and charge transfer systems. In this chapter only Brønsted acid reaction systems will be used.

The transfer of protons from a Brønsted acid to sample molecules is the generally used form for chemical ionization. Almost any volatile organic compound can be used as reaction gas, but methane, isobutane and ammonia are the standard gases used in analytical chemistry. Here methane will be used as an example to demonstrate the ionization process.

When the methane reaction gas is bombarded with electrons a set of primary ions are created:

$$CH_4 + e^- \rightarrow CH_4^+,\ CH_3^+,\ CH_2^+,\ \ldots \tag{3.8}$$

At high pressure (0.1 to 1 torr) the ions will typically collide with neutral gas molecules 20 times before they leave the ion source. The primary ions are often extremely reactive and will react with the neutral gas molecules and hereby create a secondary set of ions, which are stable in the reaction gas:

$$\begin{aligned} CH_4^+ + CH_4 &\rightarrow CH_5^+ + CH_3 \\ CH_3^+ + CH_4 &\rightarrow C_2H_5^+ + H_2 \\ CH_2^+ + CH_4 &\rightarrow C_2H_4^+ + H_2 \\ CH_2^+ + CH_4 &\rightarrow C_2H_3^+ + H_2 + H \end{aligned} \tag{3.9}$$

$$C_2H_3^+ + CH_4 \rightarrow C_3H_5^+ + H_2 \tag{3.10}$$

With a gas pressure of ≈ 1 torr the relative intensities of the individual ions are CH_5^+: 48 %, $C_2H_5^+$: 41 %, and $C_3H_5^+$: 6 %. In addition to these ions several low intensity ions can be observed in the spectrum arising by other ion molecule reactions than those mentioned above and by ionization of impurities in the reaction gas.

The secondary set of ions can react with sample molecules in different ways:

$$\text{Transfer of a proton:}\quad BH^+ + M \rightarrow MH^+ + B \tag{3.11}$$

$$\text{Hydride ion abstraction:}\quad BH^+ + M \rightarrow (M - H)^+ + BH_2 \tag{3.12}$$

$$\text{Charge transfer:}\quad BH^+ + M \rightarrow M^+ + BH \tag{3.13}$$

$$\text{Cluster formation:}\quad BH^+ + M \rightarrow (M + BH)^+ \tag{3.14}$$

where BH^+ is the reaction ion (Brønsted acid) and M the sample molecule. Transfer of protons is the most abundant reaction, and will take place with high efficiency whenever the reaction is exothermic, that is whenever the proton affinity (PA) of the sample is higher than the proton affinity of the reaction ions. PA values for a large number of compounds can be found in reference [17]. If the reaction is endothermic [PA(M) $<$ PA(B)], no significant transfer of protons will take place. This opens the possibility of selective ionization, for example ammonia (PA 205 kcal/mole) is used for selective ionization of nitrogen containing compounds. The proton affinity of some commonly used reaction gases are H_2 (101 kcal/mole), CH_4 (CH_5^+ 131 kcal/mole and $C_2H_5^+$ 164 kcal/mole), H_2O (173 kcal/mole), $i-C_4H_{10}$ (197 kcal/mole), and NH_3 (205 kcal/mole). Fragmentation is observed whenever the excess energy [PA(M) - PA(B)] in the charge transfer process is sufficiently high.

Figure 3.4 shows a comparison of the EI and the methane CI spectrum of benzylealcohol. In the spectrum the protonated molecular ion $[M+H]^+$ is not the most abundant ion. Instead $m/z = 91$ dominates the spectrum. This ion is created by protonation of benzylealcohol followed by loss of the neutral and stable water molecule $[M+H-H_2O]^+$. The relative intense ion at $m/z = 77$ is created in a similar fashion, except that the stable methanol molecule is lost. Compounds containing a functional group, Y, often lose the neutral and stable HY compound. Many compounds (aliphatic molecules in particular) react with the BH^+ ion through hydride abstraction to give the $[M-H]^+$ ion. This is seen in the relative abundant $m/z = 107$ ion. In some cases cluster ions are observed, where the addition product of the reaction ion and the sample molecule is observed.

3.2.3 Tandem mass spectrometry

In cases where similar or otherwise related compounds are present in a mixture the various analytical techniques often give rise to interfering signals. This makes the analysis of mixtures rather complicated and the use of techniques with two steps of separation prior to detection can be necessary. A well known example is the combination of chromatography and mass spectrometry, where molecules first are separated according to their retention time in the chromatograph and then mass analyzed in the mass spectrometer. In this way characteristic signals from each individual compound in the mixture can be obtained. Two separation steps can also be obtained with mass spectrometry alone in the so called tandem mass spectrometers. Here two mass analyzers are connected in series with a collision cell between the analyzers. Mass selected ions (precursor ions) passing through the first mass spectrometer are collided with noble gas molecules in the collision cell with

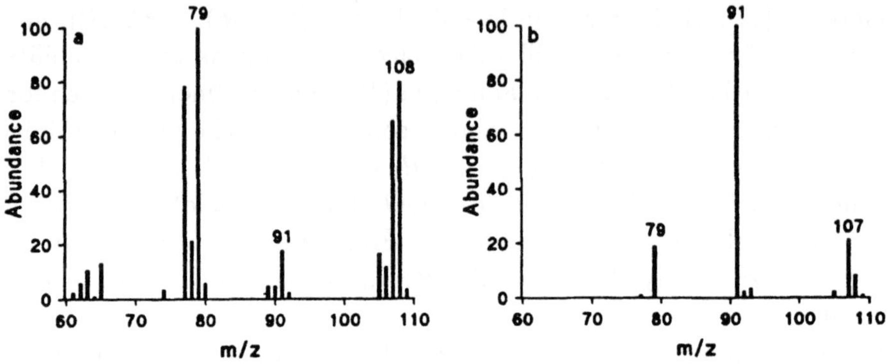

Figure 3.4: Comparison of (a) electron impact ionization and (b) methane chemical ionization mass spectra of benzylealcohol.

fragmentation as result. The fragment ions (product ions) are then analyzed with the second mass spectrometer. Tandem mass spectrometry (MS/MS) can be carried out with any kind of ionization technique, but often it is an advantage to use a soft ionization (e.g., CI) giving high intensity molecular ions, which are then selected in the first mass spectrometer. In general the selectivity obtained with combined chromatography/mass spectrometry is higher than with MS/MS. However, the analysis time is much shorter with MS/MS and continuous monitoring is possible. As a good text-book on tandem mass spectrometry reference [21] can be recommended.

Many types of tandem mass spectrometers exist, but here only the most common type, i.e., the triple quadrupole mass spectrometer, will be mentioned. The outline of a triple quadrupole mass spectrometer is shown in Figure 3.5. The first and third quadrupole are used as normal quadrupole mass spectrometers, whereas the second quadrupole is used without the constant voltage added to the rods [RF only quadrupole, $\phi = V\cos(\omega t)$]. Under RF only conditions a quadrupole acts as a strong focusing device containing all ions with a m/z ratio above a certain limit. In this way ions scattered in the low energetic collisions (0 to 100 eV) are focused back to the quadrupole axis and can be collected and analyzed in the third quadrupole. At the entrance and exit from the collision cell a set of focusing lenses ensures an efficient transfer of ions from one quadrupole to another. A prerequisite for fragmentation in the collision cell is collision energy, i.e., kinetic energy of

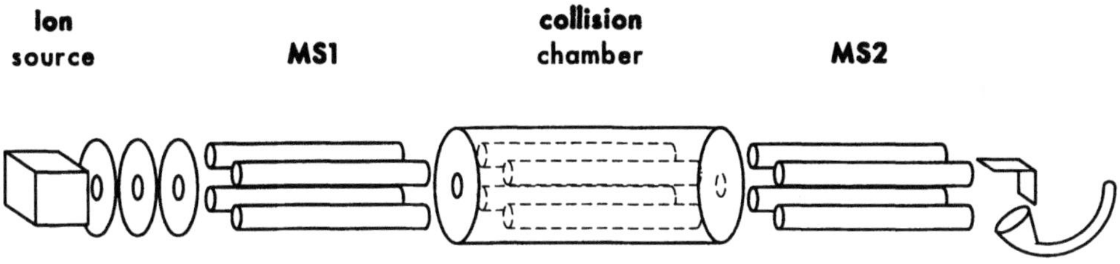

Figure 3.5: Triple quadrupole mass spectrometer.

the ions. The kinetic energy of the parent ions is simply the potential difference between the ion source and the average potential (the bias) of the quadrupole in the collision cell and it is easily adjusted by changing the bias of the quadrupole. Not all kinetic energy is available for internal excitation and fragmentation since both energy and momentum must be conserved. The energy available for internal excitation, E_{CM}, can be calculated by a transformation of the laboratory reference system to a center-of-mass system and is given by:

$$E_{CM} = E_{LAB}\frac{m_g}{m_g + m_p} \tag{3.15}$$

where E_{LAB} is the kinetic energy of the parent ion in the collision cell, m_g the mass of the collision gas and m_p the mass of the parent ion. Equation 3.15 shows that the available excitation energy increases with the mass of the collision gas and is proportional to the kinetic energy of the parent ion. With argon, the typical collision gas, the energy available for excitations is just 20 % of the kinetic energy for a $m/z = 160$ parent ion. The higher the internal energy the more fragmentations can be expected and at sufficiently high energies more than 1 chemical bond can be broken.

The pressure of the collision gas has a large influence on the observed product ion spectra. This is illustrated in Figures 3.6a and 3.6b, which show the product ion spectrum of $m/z = 135$ from m-methoxybenzaldehyde and the variation in intensity of the various product ions as a function of collision gas pressure respectively. The product ions originate either directly from collisions between the parent ion and gas molecules or from further degradation of the initially produced fragments. At low pressures ($< 10^{-4}$ torr) collisions between the parent ion and gas molecules are rare and only a few low intensity fragments are observed (Figure 3.6b $m/z = 107, 92$ and 77). As the pressure is increased collisions become more abundant, and in the pressure range around 10^{-4} torr, precursor ions will on average collide once with the collision gas.

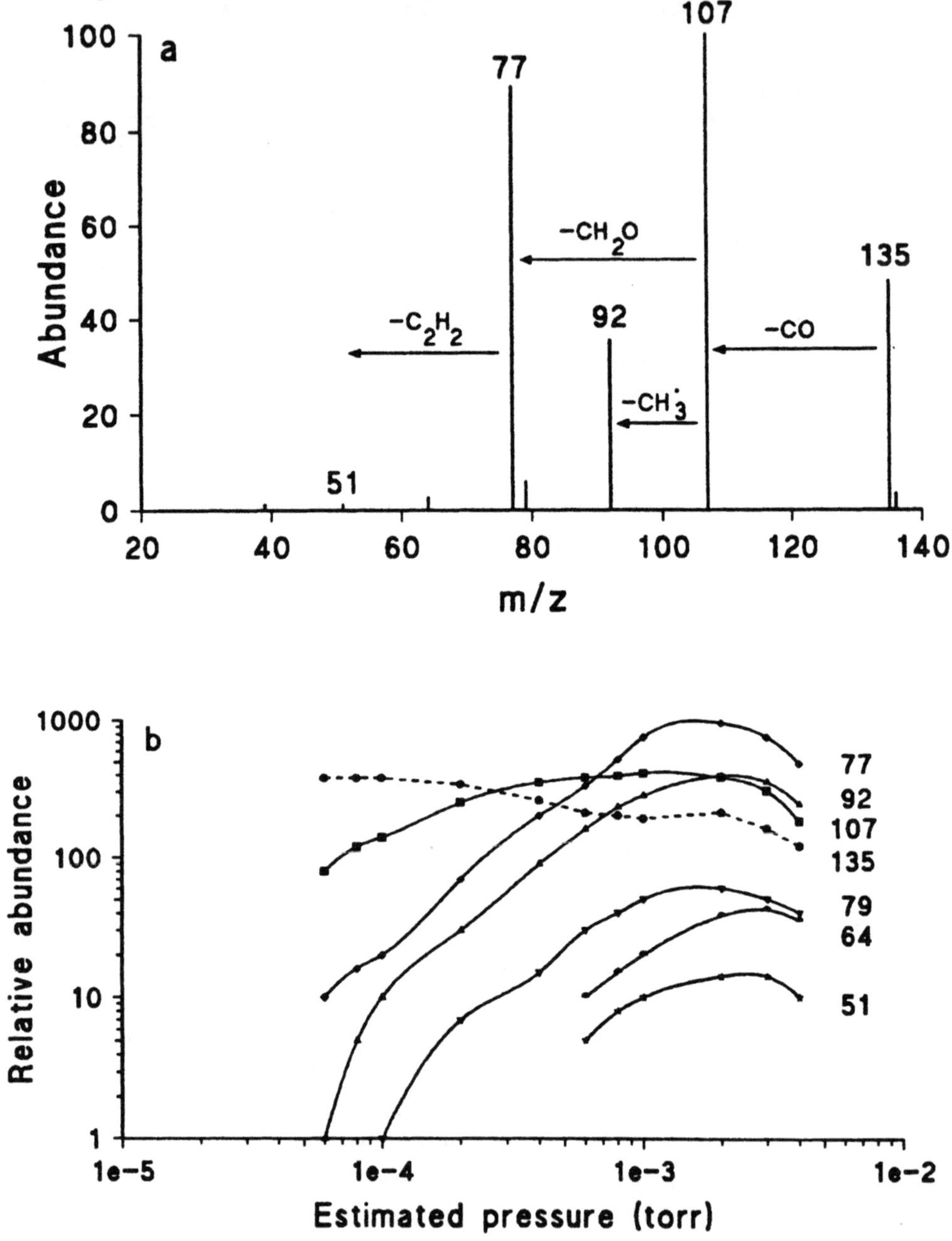

Figure 3.6: (a) Product ion spectrum of $m/z = 135$ from m-methoxy-benzaldehyde; (b) relative intensities of the product ions from $m/z = 135$ as a function of collision gas pressure.

This means that many product ions are created and some of these will again collide with the collision gas resulting in the production of secondary fragments. In Figure 3.6b the intensity of the $m/z = 107$ primary fragment becomes smaller than the intensity of the secondary fragments $m/z = 77$ and 92 in the upper 10^{-4} torr range. At even higher pressures ($> 10^{-3}$ torr) the intensity of all ions drop towards zero as a result of strong scattering in the collision cell.

3.3 Membrane Inlet Mass Spectrometry

The demands on the membranes used in membrane inlet mass spectrometry (MIMS) are quite high. The membrane is the only barrier between a liquid at atmospheric pressure and a high vacuum of 10^{-9} atmosphere. At the same time it must ensure a fast and selective transport of sample molecules into the vacuum. Traditionally MIMS is carried out with non-porous membranes like polyethylene, Teflon, natural rubber and silicone rubber. These materials show excellent performances for the measurement of gases and volatile organic compounds of hydrophobic character [5]. Hydrophilic membranes, like polyethylene terephtalate, have been used for the detection of water in hydrophobic organic solvents [1]. Recently, microporous membranes and solvent chemical ionization were introduced for the measurement of volatile organic compounds dissolved in organic solvents [14], and for the measurement of polar organic compounds in aqueous solutions [13]. In this chapter only the use of non-porous membranes will be discussed and the word *membrane* will be used as a synonym for such membranes.

The transport of compounds through a polymer membrane is a complicated three step process. First, the compound must dissolve in the polymer, then it passes through the membrane by simple diffusion, and finally it must evaporate from the membrane surface into the vacuum. This process (evaporation of a liquid through a polymer membrane) is called *pervaporation.* In such a pervaporation process the relative flow of different molecules often differs and the possibility of enrichment exists. With the right choice of membrane, it is possible to increase the ratio of solute to solvent by the transport through the membrane. Aqueous solutions are always measured with hydrophobic membranes, which discriminate against water, whereas hydrophilic membranes can be used for the measurement of water in hydrophobic organic solvents.

3.3.1 Theory

As mentioned above pervaporation of molecules through a polymer membrane is a diffusion process and as such it can be described by Fick's two laws. The flow density at depth x in the membrane at time t is given by the equation

$$J(x,t) = -D(c)\frac{\partial C(x,t)}{\partial x} \tag{3.16}$$

and the concentration at depth x and time t is given by

$$\frac{\partial C(x,t)}{\partial t} + \frac{\partial}{\partial x}(-D(c)\frac{\partial C(x,t)}{\partial x}) = 0 \tag{3.17}$$

where $J(x,t)$ is the flow density, $C(x,t)$ is the concentration, and $D(c)$ is a concentration dependent diffusion constant.

In connection with gas measurements the diffusion constant can be considered as concentration independent, since the interactions of gas molecules with each other and with the polymer molecules are insignificant. Organic molecules on the other hand behave differently and both mutual interactions and interactions between sample molecules and polymer molecules can be expected. As a result the observed MIMS signal can not in general be expected to be linear with concentration. In some cases the presence of one compound changes the flow rates of other compounds either directly through interactions between the two compounds or indirectly through modifications to the polymer. At low concentrations the interactions can be neglected and linear signals are observed. In the following it is assumed that the concentration is sufficiently low for the diffusion constant and that the solubility constant is independent of concentration.

At steady state Fick's first law (3.16) can easily be integrated and the steady state flow density is found as

$$J = D\frac{C_l - C_v}{l} \tag{3.18}$$

where C_l is the concentration at the liquid side of the membrane, C_v the concentration at the vacuum side of the membrane, and l the thickness of the membrane. Using Henry's law $[C(x,t) = S \cdot p(x,t)]$, the connection between flow and partial pressure is found as

$$J = D \cdot S \cdot \frac{p_l - p_v}{l} \tag{3.19}$$

with

J	:	flow density	[mole/(sec cm^2)]
D	:	diffusion constant	[cm^2/sec]
S	:	solubility constant	[mole/(torr cm^3)]
p_l, p_v	:	partial pressure on the liquid and the vacuum side of the membrane	[torr]
l	:	membrane thickness	[cm]

The product between the diffusion constant and the solubility constant is called the permeability constant, $\sigma = S \cdot D$. In connection with MIMS, $p_v \ll p_l$, and the steady state flow of a compound through the membrane becomes directly proportional to its partial pressure on the liquid side of the membrane.

The flow through the membrane before steady state is reached can be calculated from Fick's second law (3.17) using appropriate initial and boundary conditions. For a sudden change in concentration at the membrane surface we obtain the following conditions:

$$\begin{aligned} t > 0 &: \quad p(0,t) = p_t \\ t > 0 &: \quad p(l,t) = 0 \\ t = 0 &: \quad p(x,0) = p_0[1 - (x/l)] \end{aligned} \tag{3.20}$$

where p_0 and p_t represent the partial pressures at the liquid side of the membrane before and after a sudden change in concentration respectively. The partial pressure inside the membrane at time t, is calculated from Fick's second law using the method of separation of variables. However, the partial pressures inside the membrane are not of interest, it is the flow through the membrane which is important. This flow can be calculated from Fick's first law (3.16) using the solution found to Fick's second law and solving the equation at the vacuum side of the membrane. The flow density through the membrane as a function of time is

$$\begin{aligned} J(x,t) &= -\frac{\sigma\, p_t}{l} \cdot (1 + 2\sum_{n=1}^{\infty}(-1)^n \exp(-\frac{n^2}{G}t)) \\ G &= \frac{l^2}{\pi^2\, D} \end{aligned} \tag{3.21}$$

assuming that the initial pressure, p_0, was equal to 0. Figure 3.7 shows a comparison between the calculated and measured transients for N_2 through a 12 μm Teflon film and for hexanol at low concentration through a 25 μm silicone membrane respectively.

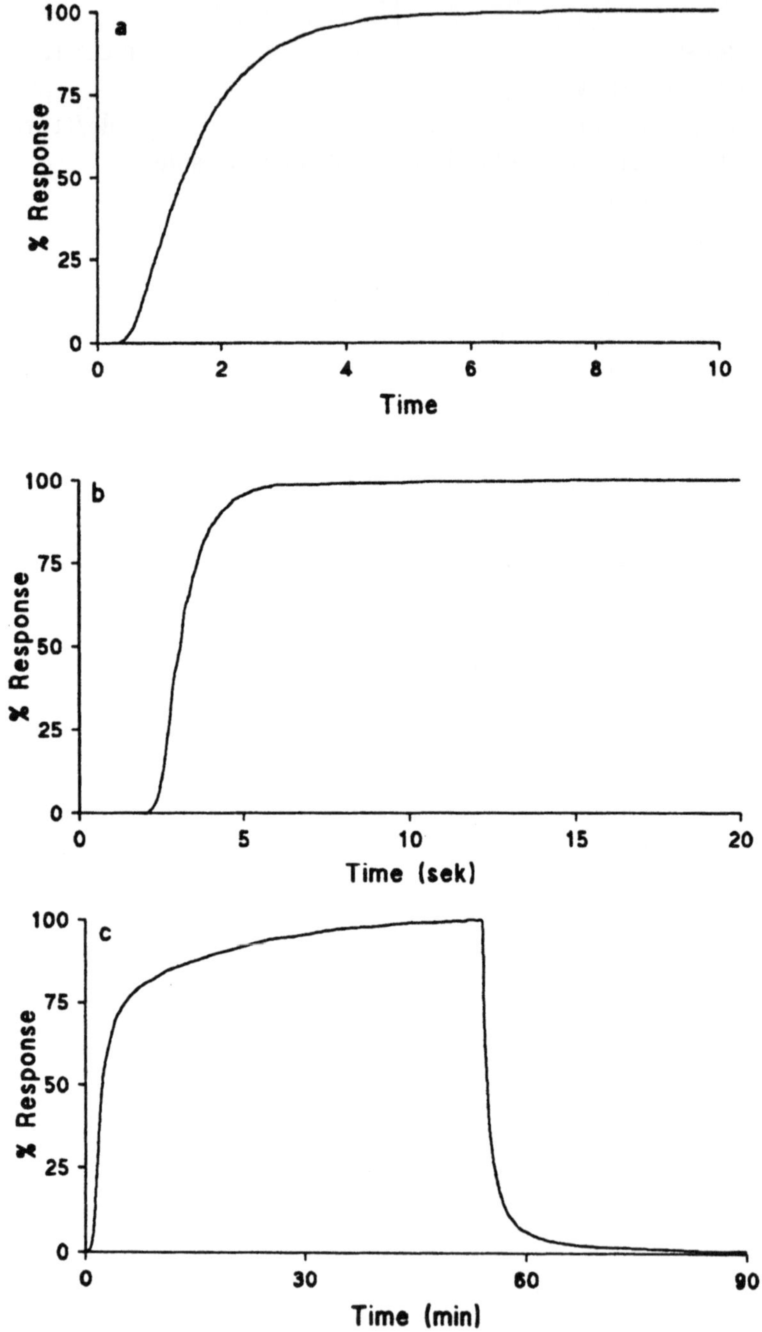

Figure 3.7: Comparison between (a) calculated membrane transient, (b) transient for N_2 through a 12 μm thick Teflon membrane, and (c) transient for hexanol through a 25 μm thick silicone film.

A very good agreement is found between the calculated profile and the measured profile for nitrogen. However, a disagreement is found between the calculated profile and the profile measured for hexanol. The hexanol signal does not come to a steady state. A slow continuous increase in signal is observed over an extended period of time. This slow continuous increase can be observed when organic compounds with a high boiling point are measured at low concentration, and it is caused by interactions between the compound and vacuum surfaces [11].

The response time of a measuring device is normally defined as the 10 to 90 % rise time, and for the membrane it can be calculated from Equation (3.21) as

$$t_{resp} = t_{90} - t_{10} = 2.342 \frac{l^2}{\pi^2 D} \tag{3.22}$$

with

t_{resp} : membrane response time [sec]

It is important to note that the response time increases with the square of the thickness and decreases with an increasing diffusion constant. Furthermore, Equations (3.19) and (3.22) show that, whereas the steady state flow depends on the product of the diffusion and solubility constant, the response time is independent of the solubility.

The diffusion of molecules through a polymer membrane is a temperature dependent process, which can be described by Clausius-Clapeyron's equation for activated processes

$$D = D_0 \exp(-\frac{E_d}{RT}) \tag{3.23}$$

where D_0 is a gas and membrane constant and E_d is the activation energy for diffusion.

The solubility in the membrane follows an Arrhenius equation

$$S = S_0 \exp(-\frac{\Delta H}{RT}) \tag{3.24}$$

where S_0 is a gas and membrane constant and ΔH is the heat of solution.

The membrane permeability is the product between the solubility and the diffusion constants

$$\sigma = S_0 \; D_0 \exp(-\frac{\Delta H + E_d}{RT}) = \sigma_0 \exp(-\frac{E_\sigma}{RT}) \tag{3.25}$$

where E_σ is the activation energy for permeation.

An increase in temperature will cause larger fluctuations in the polymer chains. This results in more free space between the polymer chains and in an increase in the diffusion constant with temperature ($E_d > 0$). For most compounds heat will be liberated by solvation and the heat of solution is negative ($\Delta H < 0$). In organic membrane materials the energy required to move polymer chains is larger than the heat of solution. The activation energy for permeation is therefore positive and the flow through the membrane increases with temperature. In some silicone materials the opposite situation has been observed. Weak interactions between the polymer chains and the highly branched structure of silicone cause a small energy of diffusion. This can result in negative energies of permeation, and the flow through the membrane decreases with increasing temperature. However, the common silicone membranes used in connection with MIMS exhibit an increase in flow with temperature although the effect is relatively small.

3.3.2 Practical use of silicone membranes for the measurement of volatile organic compounds

The amorphous structure of silicone materials makes silicone membranes well suited for the measurement of volatile organic compounds with MIMS. Silicone materials exist in many different forms, each form having its own transport properties. The generally used material is polydimethylsiloxane, which exists in different shapes; as hollow fibers, larger tubings and as sheets. The transport properties are similar for all three types and the use of each type will be discussed in Section 3.3.3 on membrane inlet design. Here only a few guidelines concerning detection limits and response times will be given.

Polydimethylsiloxane is a highly hydrophobic polymer and the detection limit for an organic compound depends primarily on its hydrophobicity and its dipole momentum. Hydrophobic compounds dissolve easily in the polymer and extremely low detection limits (nanogram/liter, ppt) can be obtained for compounds such as small chlorinated aliphatic hydrocarbons and methyl- and chlorobenzenes. The introduction of polar groups like aldehyde, ketone, ester, ether, or alcohol increases the detection limit by 1 to 3 orders of magnitude. Highly polar groups like acids and amines can only be measured in the non-dissociated form at limits in the upper ppb to the lower ppm range. Table 3.1 shows typical detection limits for a variety of organic compounds, but it should be kept in mind that these limits are only guidelines. The detection limit depends on the particular membrane inlet design used. Large variations in detection limits within a certain group of

Compound	Detection limit [ppb]
Chloroform	< 0.1
Carbontetrachloride	< 0.1
Dichloroethylenes	< 0.1
Trichloroethylene	< 0.1
Tetrachloroethylene	< 0.1
Trichloroethane	< 0.1
Benzene	< 0.1
Toluene	< 0.1
Xylenes	< 0.1
Naphthalene	< 1
Chlorobenzene	< 1
Dichlorobenzenes	< 1
Heptane	1 to 10
Tetrahydrofurane	1 to 10
Acetone	1 to 10
Butanal	1 to 10
Ethylhexanoate	1 to 10
Nitrobenzene	1 to 10
o-Cresol	1 to 10
p-methoxybenzaldehyde	1 to 10
Naphthol	1 to 10
1-Butanol	10 to 1,000
1-Octanol	10 to 1,000
Acetic acid	10^3 to 10^4
Hexanoic acid	10^3 to 10^4
1,4-Butanediol	10^4 to 10^5
Phenylacetic acid	10^4 to 10^5
Phenoxyacetic acid	10^5

Table 3.1: Typical detection limits for organic compounds as measured with a silicone membrane.

compounds can be expected. For example the detection limit of octanol with its long hydrocarbon chain is about 40 times lower than that for ethanol. It is difficult to put an upper limit to the size of compounds which can be measured with MIMS. However, in general, it is possible to detect hydrophobic compounds with a boiling point of less than 250 °C and molecules as large as 300 a.m.u. have been measured.

The membrane response time for organic compounds is typical in the range of 1 to 5 minutes, depending on molecular size and hydrophobicity. Hydrophobic compounds tend to interact with the polymer, softening it, and in this way the activation energy for diffusion is lowered. As mentioned earlier, less volatile compounds (boiling point above 150 °C) interact with surfaces in vacuum. The response time observed for these compounds therefore reflects a combination of the membrane response time and a response time due to surface interactions. Unless special precautions are taken in the membrane inlet design, transients as slow as several hours can be observed.

3.3.3 Membrane inlet design

Membrane inlets exist in many different designs, each design suited for a particular application or measuring problem. However, all designs have to deal with the same basic transport problems, i.e., diffusion of sample through an unstirred liquid layer in front of the membrane, transport through the membrane and transport of the vaporized sample from the inside surface of the membrane to the ionization region of the mass spectrometer. In the design of a membrane inlet one must compromise between an ideal solution to the transport problems and practical considerations regarding the particular application. Four different types of membrane inlets are used today: (i) membrane probes mounted directly in a reactor (Figure 3.8a); (ii) stirred sample cells (Figure 3.8b); (iii) flow-through cells (Figure 3.8c); (iv) helium-purge inlets (Figure 3.8d).

Membrane inlets inserted directly into reactors (Figure 3.8a) have been used for gas measurements in many applications such as monitoring of fermentation processes, and in-vivo measurements of blood gases. The typical membrane probe used for the monitoring of fermentation processes consists of a perforated steel tube covered with a silicone tube. The steel tube is closed at one end and connected to the mass spectrometer through the other end. Normally stirring in the fermentor is sufficient to ensure a minimal unstirred liquid layer in front of the membrane, and the only limitation is found in condensation of sample in the vacuum tube. Condensation effects make this type of membrane inlet inadequate for compounds with a boiling point of more than 100 °C and even with highly volatile compounds like ethanol slow transients can be observed. A great advantage of these inlets is that they can be sterilized. In physiology miniaturized ($<$ 1 mm diameter) membrane probes are used, which can be inserted directly into the blood stream of live animals. These measurements are complicated by a very slow sample flow in front of the membrane and special precautions have to be taken. One possibility is to use organic membrane materials, which have low permeability and/or supports with only a few holes. In this way the gas consumption of the membrane probe can be reduced.

Sample cells (Figure 3.8b) are the inlets of choice for most kinetic studies. They consist of a reaction vessel with a volume of 1 to 10 mL and have a membrane covered port through one of the vessel walls. The vessel is mounted as close as possible to the ion source of the mass spectrometer and most volatile compounds can be measured without disturbing condensation effects. Mechanical stirring is necessary in order to minimize the unstirred liquid layer in front of the membrane and thermostating of the vessel is also needed. Sample cells have been designed in many variations depending on the particular type of reaction to be studied. They are also useful for the characterization of microbial media. In this application growth media or a filtrate thereof are simply transferred to the measuring cell and mass spectra and/or tandem mass spectra can be recorded after a few minutes delay time.

Flow-through membrane probes (Figure 3.8c) are the most versatile membrane inlets and they are used extensively for the monitoring and characterization of microbial processes as well as the analysis of environmental samples. A pump is used to pass a continuous stream of sample through a small cell with the membrane. The membrane cell is mounted in the immediate vicinity of the ionization region in order to minimize problems with condensation effects for volatile compounds. After passage through the measuring cell the sample can either be returned to the reactor or discharged. The flow through the membrane cell ensures sufficient stirring and thermostating is obtained either directly by flushing the cell with thermostated water or indirectly by thermostating the sample liquid itself prior to passage through the membrane cell. A great advantage of flow cells is that samples can be modified prior to detection. For example an adjustment of pH can bring acids and amines from a dissociated form to a non-dissociated form and thereby allowing them to pass the membrane.

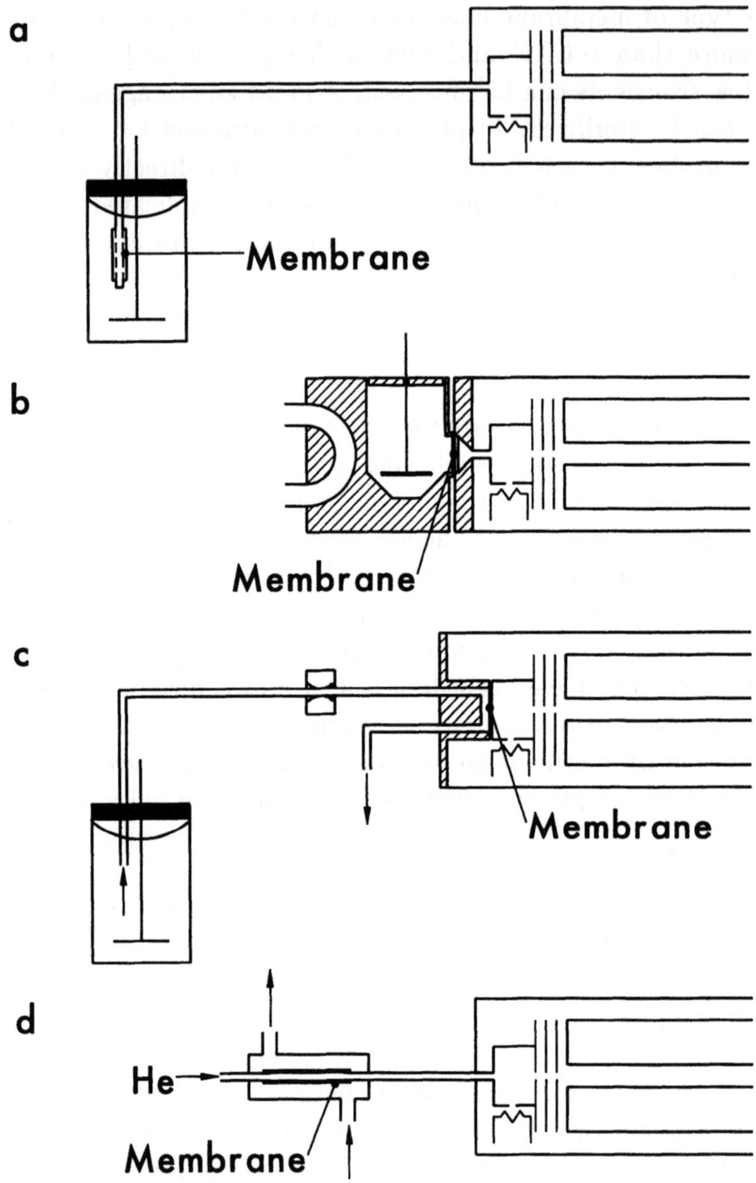

Figure 3.8: Different types of membrane inlets: (a) membrane probe mounted directly in a reactor; (b) stirred sample cell; (c) flow-through cell; (d) helium-purge inlet.

The latest type of membrane inlet (Figure 3.8d) uses a silicone tube through which helium is purged continuously, while aqueous sample flows continuously over the exterior membrane surface. The helium purge gas removes sample molecules from the inside of the membrane and carries the sample molecules to the ion source. This type of inlet has shown excellent performance for the measurement of highly volatile, hydrophobic compounds which are detected in the nanogram/liter concentration range.

3.4 Applications of Membrane Inlet Mass Spectrometry

3.4.1 Microbial degradation of chlorinated aliphatic compounds

Small chlorinated aliphatic compounds have become a serious threat to ground water reservoirs. Such compounds, which are not usually degraded in the environment, are believed to be harmful to human health even at very low concentrations (1 μg/L). Chlorinated aliphatic hydrocarbons can be degraded biologically by methanotrophic bacteria. These bacteria have a non-specific enzyme, methane monoxygenase, which is capable of oxygenating a number of the potentially polluting chlorinated hydrocarbons [2]. Studies of the microbial degradation of chlorinated hydrocarbons are normally carried out by gas chromatography, often in combination with *purge and trap* techniques in order to measure very low concentrations (ng/L). This approach makes fast data acquisition complicated and fundamental kinetic studies become difficult.

We have used [6] MIMS to study the kinetics of the aerobic degradation of chlorinated ethenes in pure cultures of *Methylococcus capsulatus* (Bath). With MIMS it is possible to measure the concentrations of chlorinated ethenes on-line at extremely low concentrations (Figure 3.9) with response times at the order of 1 minute.

Furthermore, the technique gives linear signals over many orders of magnitude (Figure 3.10). It is therefore possible to study directly the kinetics of the degradation at a concentration range comparable to those observed in water reservoirs. Cultures of *Methylococcus capsulatus* (Bath) were grown in shake flasks at 45 °C for two days in the presence of atmospheric air and methane. A sample of the growing culture was transferred to a small closed reactor and a chloroethene substrate was added.

A piston pump was used to extract continuously a sample from the reactor, passing it through a MIMS flow cell and then recycling it back to the reactor. In this way concentrations of dichloroethene substrates and their oxygenated products could be monitored simultaneously during aerobic degradation.

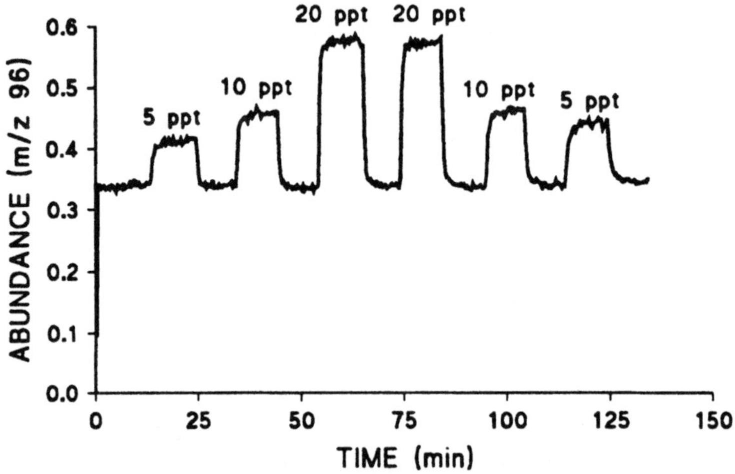

Figure 3.9: Successive injections of t-dichloroethylene at extremely low concentrations (nanogram/liter).

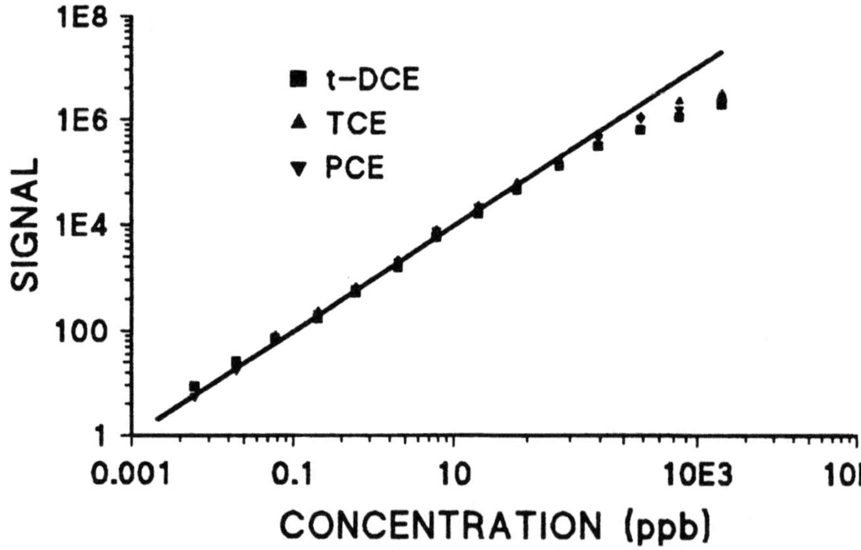

Figure 3.10: Linearity of the MIMS technique.

Figure 3.11 shows an example such a monitoring of t-dichloroethene substrate and t-dichloroethene oxide product following an injection of 5 ppb t-dichloroethylene. From experiments similar to that in Figure 3.11 it was possible to obtain initial velocities for the degradation at many different concentrations. The important Michaelis rate constant, V_{MAX}/k_M, could be calculated. Figure 3.12 shows an example of such a measurement for trichloroethene. The rate constant was calculated as 2 L/(min · gram d.w.).

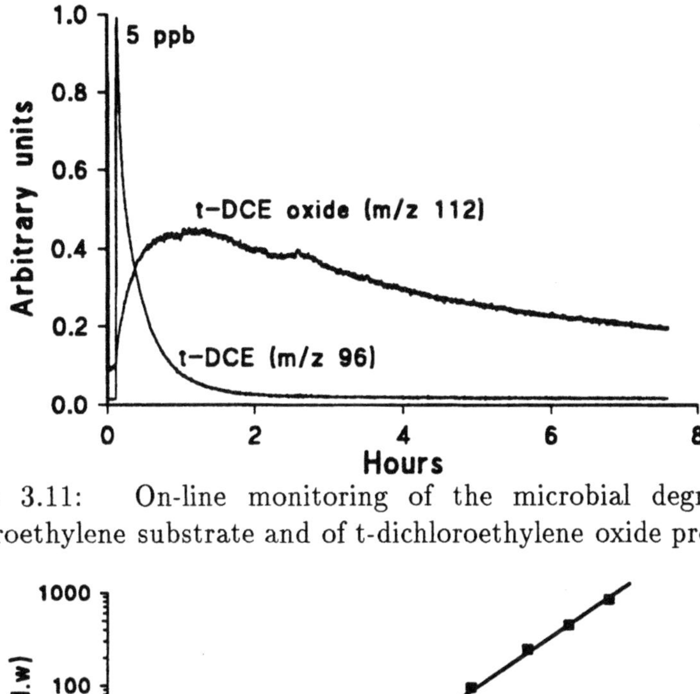

Figure 3.11: On-line monitoring of the microbial degradation of t-dichloroethylene substrate and of t-dichloroethylene oxide product.

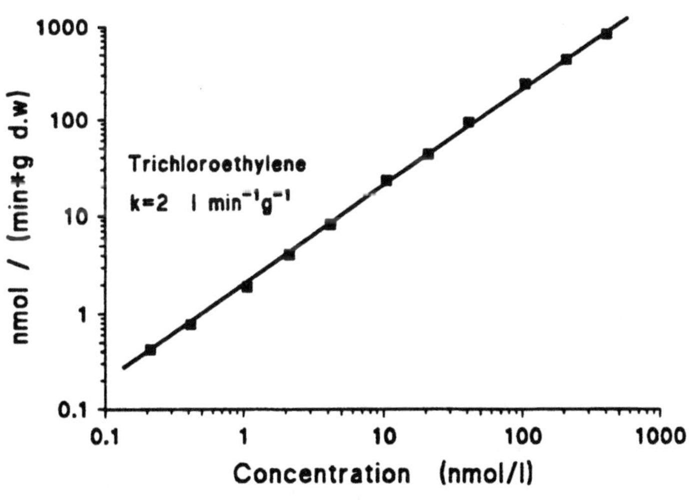

Figure 3.12: Initial velocities for the degradation of trichloroethylene as a function of substrate concentration.

3.4.2 Fermentation of *Penicillium chrysogenum*

The different types of penicillin differ only in the side chain at the β-lactam ring. Addition of certain precursor molecules during the fermentation determines the type of penicillin produced. For example, addition of phenoxyacetic acid results in penicillin-V, whereas addition of phenylacetic acid results in penicillin-G. An on-line monitoring of the precursor molecules is therefore an obvious possibility for controlling the production.

On-line measurement of phenoxyacetic acid during a penicillin-V fermentation is a very difficult task even for membrane inlet mass spectrometry. Phenoxyacetic acid contains a carboxylic acid, which can only pass the membrane on the non-dissociated form. It is therefore necessary with an on-line acidification of the sample. With a boiling point of 285 °C, the compound can hardly be considered volatile. A further complication is to withdraw the sample from the fermentor in sterile form. In order to solve these problems a sampling system has been designed [7], based upon a microfiltration module mounted in the fermentor and an infusion pump (Figure 3.13). The infusion pump shifts such by turning liquid from the fermentor or from a standard solution of phenoxyacetic acid through a MIMS detection system. On its path through the system the sample is acidified with a *HCl* solution, thermostated to 70 °C in a heat exchanger, and finally, passed through a MIMS flow cell.

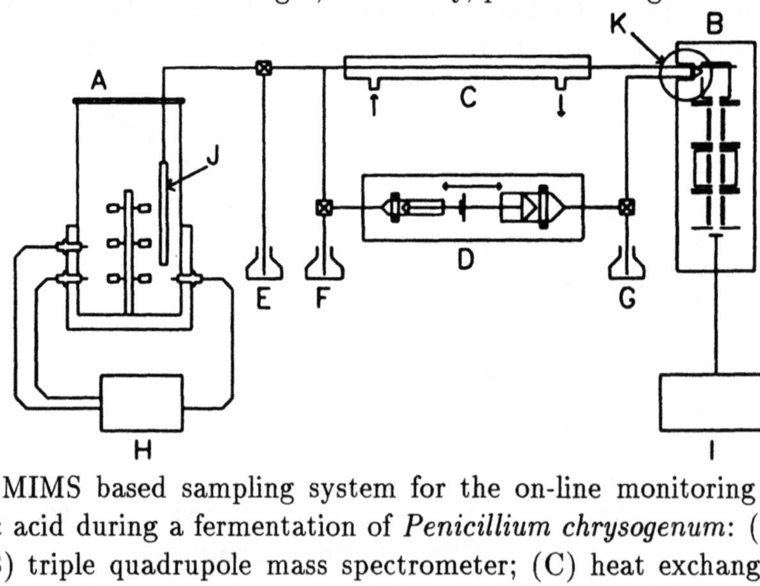

Figure 3.13: MIMS based sampling system for the on-line monitoring of phenoxyacetic acid during a fermentation of *Penicillium chrysogenum*: (A) fermentor; (B) triple quadrupole mass spectrometer; (C) heat exchanger; (D) syringe pump; (E) standard solution; (F) acid; (G) waste; (H) sensors; (I) computer; (J) filtration module; (K) membrane inlet.

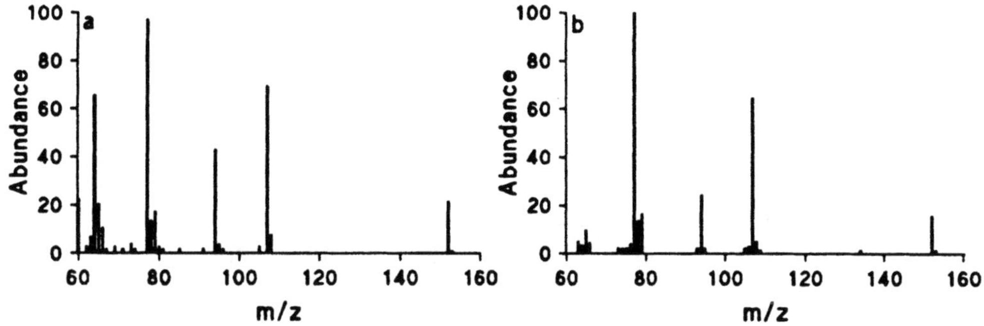

Figure 3.14: EI mass spectra of (a) *Penicillium chrysogenum* fermentation broth during production phase and (b) standard solution of phenoxyacetic acid.

Figure 3.14a shows an EI mass spectrum of the growth medium after 75 hours of fermentation, and, for comparison, Figure 3.14b shows the standard spectrum of phenoxyacetic acid. Apart from the peaks at $m/z = 60$ and 64, the two spectra are identical and the concentration of phenoxyacetic acid can be measured at the intense $m/z = 107$ peak without interference. By the help of tandem mass spectrometry it was possible to identify the compound giving rise to $m/z = 64$ as sulphur dioxide, which was created by the acidification of the sodium sulfite containing medium. $m/z = 60$ was also created by the acidification of the medium, but an unequivocal identification was not possible. Tandem mass spectrometry demonstrated that the compound contains sulphur and the spectrum suggested carbonylsulphide as a possible candidate.

The result of an on-line monitoring of phenoxyacetic acid during a 200 hours long fermentation is shown in Figure 3.15. After 30 hours the medium is shifted from a growth phase medium to a production medium containing phenoxyacetic acid. The phenoxyacetic acid was consumed and incorporated into penicillin as reflected in the spectrophotometric measurements of penicillin-V. After 120 hours and 180 hours phenoxyacetic acid is again added to the medium and this is nicely catched by the MIMS detection system.

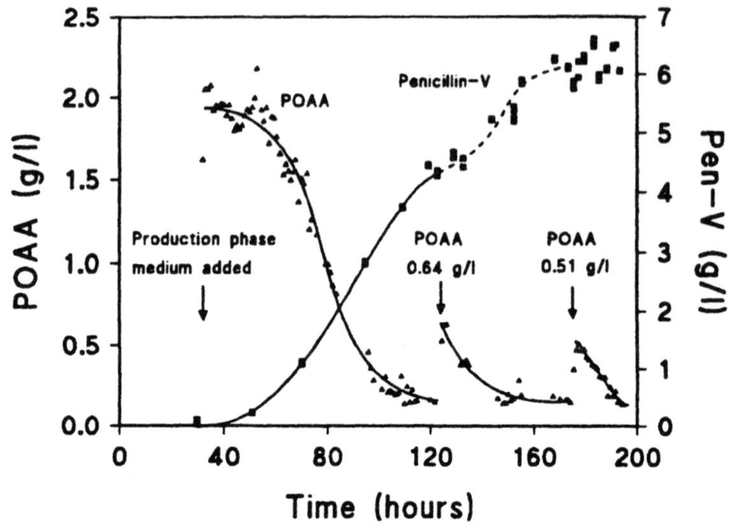

Figure 3.15: On-line monitoring of phenoxyacetic acid during a penicillin-V fermentation.

3.4.3 Identification of metabolites in microbial media

The analysis of microbial media for volatile compounds is a rather difficult task, which is normally carried out by gas chromatography and combined gas chromatography/mass spectrometry. These chromatographic methods offer a very high sensitivity and selectivity, but in many cases they demand time consuming and occasionally complicated sample pretreatments prior to analysis. Although the sensitivity is higher and the selectivity lower with membrane inlet tandem mass spectrometry than with chromatographic methods, it is an alternative possibility, since direct analysis from untreated media can be carried out. In this section we will show examples of the use of membrane inlet tandem mass spectrometry for the identification of major metabolites directly from growth media or filtrates thereof.

3.4.3.1 The parasitic protozoe *Trichomonas vaginalis*

The human parasite *Trichomonas vaginalis* causes inflammation of the vagina. There is therefore an interest in the metabolism of this organism in the hope that the production of unusual metabolites could provide new avenues for chemotherapeutic attack. We have used membrane inlet tandem mass spectrometry for the identification of volatile metabolites produced by the organism when grown in shake flasks [12].

After a few days of growth 2 to 5 mL of the growth media was centrifuged and carefully decanted prior to mass spectrometric analysis. Although direct measurements in the growth media are possible, this was avoided due to the risk of infection. The mass spectrometric analysis was carried out with a small measuring cell.

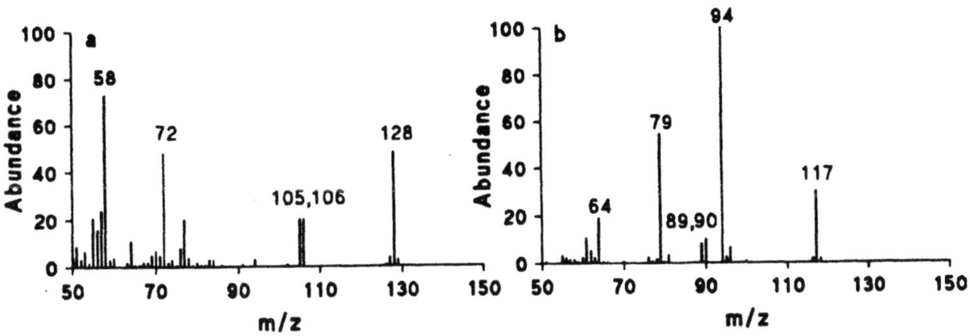

Figure 3.16: EI mass spectra of filtrates of growth media from *Trichomonas vaginalis*: (a) uninoculated media; (b) after 4 days growth.

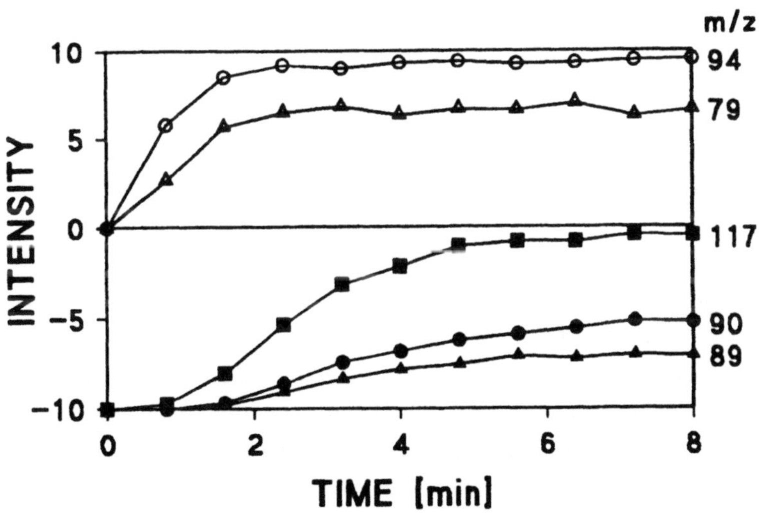

Figure 3.17: Intensity of different fragments in the EI-MS spectrum as a function of time after injection of growth medium into the measuring cell.

Figure 3.16 shows the EI-MS spectrum of the growth medium prior to inoculation and after 4 days of growth. The uninoculated growth media contain some volatile compounds measurable with MIMS (Figure 3.16a), but these compounds were consumed during growth and a new characteristic spectrum (Figure 3.16b) is observed after 4 days of growth. Chemical ionization of the 4 days sample showed two intense peaks at $m/z = 118$ and at $m/z = 95$, suggesting that the EI-MS spectrum was composed of at least two compounds. By recording successive EI-MS spectra as fast as possible after injection into the measuring cell, it was possible to divide the major peaks in the EI-MS spectrum into fragments with fast response time and fragments with slow response time (Figure 3.17). This difference is explained by the fact that compounds with different molecular properties have different transport rates through the membrane. Ions with similar response times may originate from the same compound, whereas ions with different response times must originate from different compounds.

On the basis of the measurements indole has been identified, a typical degradation product of the amino acid tryptophan. Dimethyl disulfide which can be created by autoxidization of methanethiol has also been found. The latter is known to be a product in the metabolism of trichomonads. The concentrations in the growth media were estimated as 4 μM and 20 μM for dimethyl disulfide and indole respectively.

3.4.3.2 The yeast *Brettanomyces anomalus*

The yeast *Brettanomyces anomalus* is highly resistant to food preservatives and sometimes contaminates fermented beverages. It produces strong spicy, smoke-like, woody or phenolic odours leading to spoilage of wine, beer and cider. We have used [12] membrane inlet tandem mass spectrometry to study volatile metabolites produced by biotransformation in growth media of *Brettanomyces anomalus. Brettanomyces anomalus* was grown for 21 days in the presence of either coumaric acid, ferulic acid, syringic acid or vanillic acid as precursors for metabolic products. After 21 days the growth media was filtrated and transferred to a MIMS measuring cell, where EI and CI mass spectra and tandem mass spectra were recorded.

CI mass spectra of all samples with precursor added showed an intense peak at $m/z = 89$. Intense CI peaks were further observed at $m/z = 153$ and $m/z = 123$ in the samples grown with ferulic acid and coumaric acid respectively. A blind sample without any precursor added showed no peaks in the CI mass spectrum and none of the added precursors gave any mass spectrometric signals, since they could not pass the membrane. Figure 3.18a-c shows the product ion mass spectra of m/z = 153, 123 and 89 respectively. The compounds giving rise to m/z = 153 (CI) and m/z = 123 (CI) were identified as 4-ethyl-2-methoxyphenol and 4-ethyl-phenol respectively. These are simple degradation compounds of the ferulic acid and cuomaric acid precursors added to the growth media. The m/z = 89 ion (CI) found in all samples with precursor added was identified as ethyl acetate, a compound not previously identified as a product from these precursors. The concentration of ethyl acetate was estimated as 350 μM in the samples grown with coumaric acid, ferulic acid and syringic acid, whereas the concentration was approximately 10 times less in the sample grown with vanillic acid.

Acknowledgments

I thank professor D. Lloyd, University of Wales College of Cardiff, and professor H. Degn, Dr. T. Kotiaho, K.F. Hansen, and S. Gylling, Physical Biochemistry Group, Odense University, for many fruitful discussions and their supply of material and data.

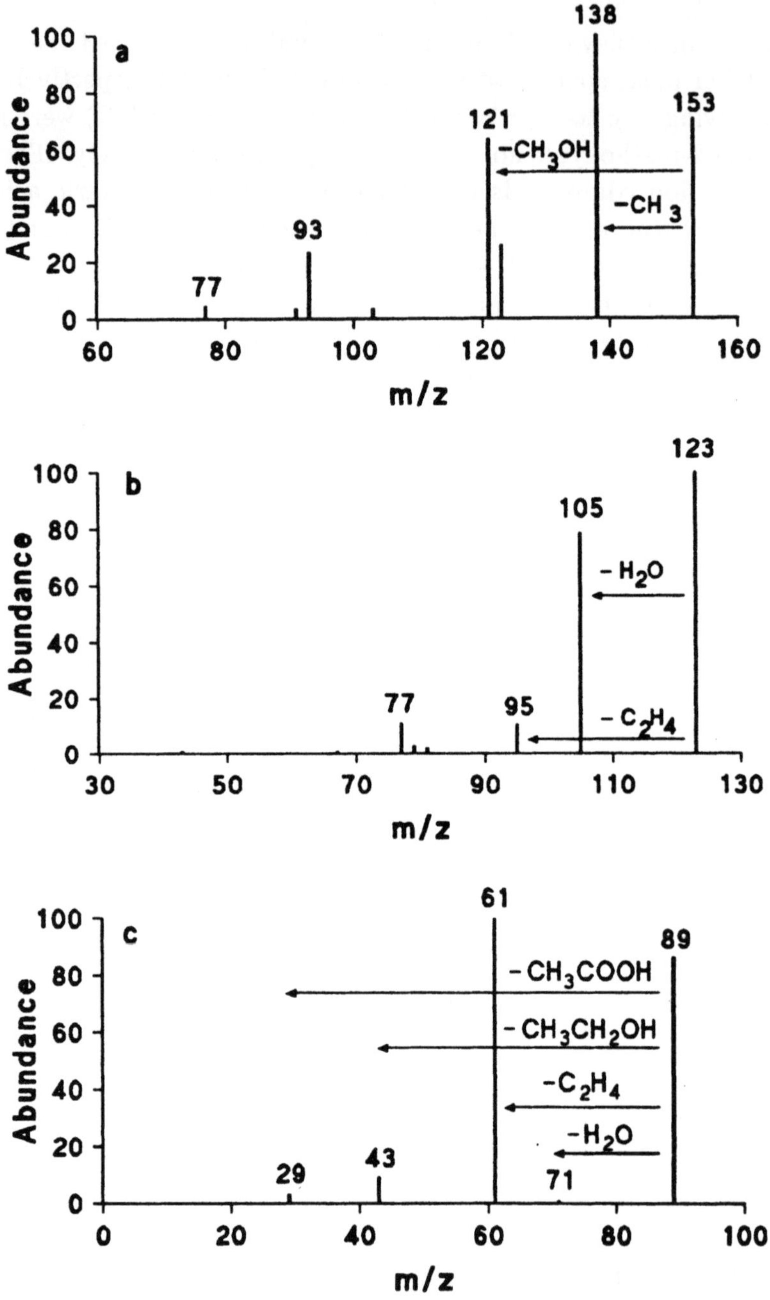

Figure 3.18: Product ion mass spectra of (a) $m/z = 153$, (b) $m/z = 123$, and (c) $m/z = 89$ from the samples grown in the presence of ferulic acid, coumaric acid, and syringic acid respectively.

3.5 Bibliography

1. S. Bohátka and H. Degn. Continuous measurement of water in organic solution by membrane inlet mass spectrometry. *Rapid Commun. Mass Spectrom.*, 5, 433-436

2. J. Colby, D.I. Stirling and R. Dalton 1977. The soluble methan monoxygenase of *Methylococcus capsulatus* (Bath). Its ability to oxygenate n-alkanes, n-alkenes, ethers and alicyclic, aromatic and hetero cyclic compounds. *Biochem. J.*, 161, 333-344

3. P.H. Dawson 1976. *Quadrupole mass spectrometry and its application.* Elsevier Science Publishers, Amsterdam

4. H. Degn, R.P. Cox and D. Lloyd 1985. Continuous measurements of dissolved gases in biochemical systems with the quadrupole mass spectrometer. *Methods. Biochem. Anal.*, 31, 165-194

5. H. Degn 1992. Membrane inlet mass spectrometry in pure and applied microbiology. *J. Microbiol. Methods*, 15, 185-197

6. S. Gylling, I. Virkajarvi, T. Kotiaho and F.R. Lauritsen 1994. Microbial degradation of aliphatic chlorinated hydrocarbons studied by membrane inlet mass spectrometry. In: *Proceedings of the 42nd American Society of Mass Spectrometry Conference*, Chicago

7. K.F. Hansen, F.R. Lauritsen and H. Degn 1994. An on-line sampling system for fermentation monitoring using membrane inlet mass spectrometry. Application to phenoxyacetic acid monitoring in penicillin fermentation. *Biotechn. Bioeng.*, 44(3), 347-353

8. A.G. Harrison 1992. *Chemical ionization mass spectrometry.* CRC Press Inc., Boca Raton

9. T. Kotiaho, F.R. Lauritsen, T.K. Choudhury, R.G. Cooks and G.T. Tsao 1991. Membrane introduction mass spectrometry. *Anal. Chem.*, 63, 875A-883A

10. T.K. Kotiaho 1992. *Methodology and applications of membrane introduction mass spectrometry.* PhD thesis, Purdue University, Indiana

11. F.R. Lauritsen 1990. A new membrane inlet for on-line monitoring of dissolved volatile organic compounds with mass spectrometry. *Int. J. Mass Spectrom. Ion Process.*, 95, 259-268

12. F.R. Lauritsen, L.T. Nielsen, H. Degn, D. Lloyd and S. Bohátka 1991. Identification of volatile metabolites in microbial cultures by membrane inlet tandem mass spectrometry. *Biol. Mass Spectrom.*, 20, 253-258

13. F.R. Lauritsen, T.K. Choudhury, L.E. Dejarme and R.G. Cooks 1992. Microporous membrane introduction mass spectrometry with solvent chemical ionization and glow discharge for the direct detection of volatile organic compounds in aqueous solution. *Anal. Chim. Acta*, 266, 1-12

14. F.R. Lauritsen, T. Kotiaho, T.K. Choudhury and R.G. Cooks 1992. Direct detection and identification of organic compounds dissolved in organic solvent by reversed-phase membrane introduction mass spectrometry. *Anal. Chem.*, 64, 1205-1211

15. F.R. Lauritsen, T. Kotiaho and D. Lloyd 1993. Rapid and direct monitoring of volatile fermentation products in the fungus *Bjerkandera adusta* by membrane inlet mass spectrometry. *Biol. Mass Spectrom.*, 22, 585-589

16. F.R. Lauritsen and D. Lloyd 1994. Direct detection of volatile metabolites produced by microorganisms: Membrane inlet mass spectrometry. In: C. Fenselau (Ed.), *Mass spectrometry for the characterization of microorganisms*, 91-106, ACS Symposium series, 541, American Chemical Society, Washington DC

17. S.G. Lias, J.F. Liebman and D.L. Rhoda 1984. *J. Phys. Chem. Ref. Data*, 13, 695-808

18. D. Lloyd and R.I. Scott 1983. Direct measurement of dissolved gases in microbiological systems using membrane inlet mass spectrometry. *J. Microbiol. Methods*, 1, 313-328

19. D. Lloyd, S. Bohátka and J. Szilágyi 1985. Quadrupole mass spectrometry in the monitoring and control of fermentations. *Biosensors*, 1, 179-212

20. F.W. McLafferty 1980. *Interpretation of mass spectra*, University Science Books, Mill Valley, California

21. F.W. McLafferty (Editor) 1983. *Tandem mass spectrometry.* Wiley-Interscience, New York

22. F.A. White and G.M. Wood 1986. *Mass spectrometry: Applications in science and engineering.* Wiley-Interscience, New York

23. S. Woldring 1970. Biomedical application of mass spectrometry for monitoring partial pressures. *J. Assoc. Advan. Med. Instr.*, 4, 43-56

Chapter 4

Flow Cytometry

P. BREEUWER
Wageningen Agricultural University
Department of Food Science
Food Chemistry-Microbiology Section
Bomenweg 2 6703 HD, Wageningen (THE NETHERLANDS)
email: pieter.breeuwer@lmc.lmt.wau.nl

ABSTRACT – *Flow cytometry is a rapid and sensitive method to measure physiological parameters of cell populations. Among other things, flow cytometry can measure the fluorescence of (previously stained) cells. This chapter focuses on the utilization of fluorescence methods to measure cell viability and intracellular pH. Recent developments take advantage of the existence of a membrane potential or reducing activity in live cells. Likewise, novel fluorescent probes such as BCECF and carboxy-SNARF-1 have been developed to measure intracellular pH. It is most likely that further progress in the development of instruments and fluorescent probes should contribute to a prominent place of FCM in biological research.*

KEYWORDS – *Flow Cytometry, Fluorescent Probes, Intracellular pH, Viability, Membrane Potential*

J.F.M. Van Impe et al. (eds.), Advanced Instrumentation, Data Interpretation, and Control of Biotechnological Processes, 105-120.

4.1 Introduction

Flow cytometry (FCM) can be considered as *advanced instrumentation*, although essentially it is nothing more than an automated microscope. The idea to make life of the cytologists and microbiologists easier has since long attracted scientists, but it turned out to be very complex to replace the human eye by another photodetector. The first flow cytometers were capable of counting and differentiating relatively big cells (which was a lot at that time), but were lacking the sensitivity for more sophisticated applications. However, over the last decade, flow cytometers are rapidly getting more performant. Nowadays, simultaneously two or more characteristics of individual cells can be analyzed, which make these instruments very useful for counting and discriminating differences within (mixed) cell populations, determining physiological indicators as intracellular pH and calcium [12][20], following processes like phagocytosis and endocytosis [12], detection of yeasts in yoghurt [9] and even analyzing relatively small cells like bacteria [3]. Consequently, flow cytometry is now recognized as an important technique in various fields such as medical microbiology, food microbiology and biotechnology.

First the flow cytometer is introduced, followed by a more detailed presentation of assessment of cell viability and measurement of intracellular pH with flow cytometry.

4.2 The Flow Cytometer

Flow cytometry is the measurement of physical and/or chemical characteristics of (biological) particles while passing in a fluid stream through a measuring apparatus [20]. The most common parameters measured with a flow cytometer are fluorescence and light scattering. A simple representation of a single parameter flow cytometer for measuring the fluorescence of particles is given in Figure 4.1.

Basicly, the instrument consists of three following parts which will be discussed shortly:

1. The *fluidic system.* The general idea is that cells should pass one by one in a laminar flow past the measuring point. Generally, the velocity of the sample stream is regulated by use of a sheath fluid (Figure 4.2).

2. The *optical system.* The optical system basicly consists of an excitation source, optical filters, and a photomultiplier (PMT). Common

excitation sources are a mercury arc lamp or an argon ion laser. The excitation and emission wave-lengths are selected with various optical filters, depending on the application. The PMT finally converts the optical signal into an electrical signal.

3. The *electronic system.* The electrical signals which are received from the PMT are normally represented in a histogram (Figure 4.3) where the y-axis represents the number of fluorescent particles and the x-axis is divided into *channels* (mostly 256) relative to the fluorescence intensity of the signals. At present, almost all flow cytometers have a built-in logarithmic amplifier. This devise produces an output signal proportional to the logarithm of the input signal. In this way a wider range of signals can be measured in one histogram.

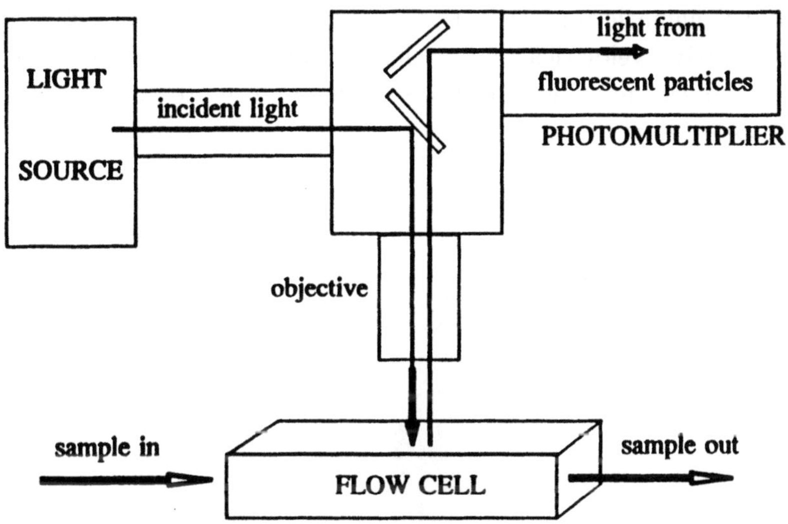

Figure 4.1: Schematic representation of a flow cytometer. The sample flow is illuminated by the incident light beam and the absorption of light by fluorescent particles results in the emission of fluorescent light which is detected by a photomultiplier. Figure provided by Chemunex S.A., Maisons-Alfort (France).

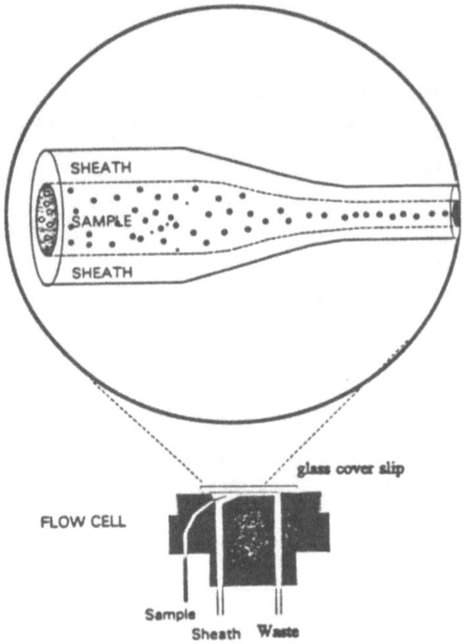

Figure 4.2: Schematic representation of the fluidic system of a flow cytometer. The relative velocities of the sheath fluid pump and the waste fluid pump determines the velocity of the sample. Figure provided by Chemunex S.A., Maisons-Alfort (France).

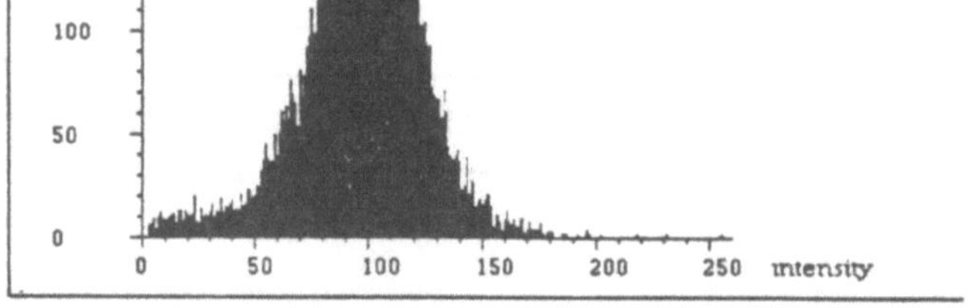

Figure 4.3: Example of a histogram. The x-axis (divided in channel 0 to 256) represents the relative fluorescence intensity of the analyzed particles. The y-axis represents the number of fluorescent particles. Figure provided by Chemunex S.A., Maisons-Alfort (France).

The use of fluorescence is one of the reasons why FCM is a popular tool for measuring cell characteristics. The principal reason for this are the sensitivity and specificity of fluorescent methods. It is possible to detect specifically very small numbers of several different fluorescent molecules at the same time. Obviously, a key requirement for fluorescence measurements is a sufficient (fluorescence) staining of the target cells. Discussion of all the possibilities and methods to stain cells is far beyond the scope of this chapter, but the following list mentions some of the most important groups of probes.

1. *Immunoreagents.* Fluorescent probes as fluorescein isothyocyanate (FITC), phycoerythrin, texas red, lucifer yellow are used most often to make conjugates with specific antibodies. For example, FITC antibodies are (often in combination with a DNA/RNA dye) applied to detect several bacteria with FCM [6][22].

2. *DNA and RNA probes.* Some dyes among these are the hoechst dyes and 4'6-diamidino-2-phenylindole (DAPI). These dyes, which fluoresce blue, have a strong preference for the A-T base pairs, ethidium bromide and propidium iodide (PI). PI and EB form complexes with DNA and RNA by intercalating between base pairs. They both are fluorescent in the red region. Mitramycin and Chromomycin are dyes specifically for G-C rich regions of DNA.

3. *Physiological indicators.* This is a rather wide group which includes dyes as fura-2, indo-1, quin-2, ..., for intracellular calcium measurements; carboxy-seminaphthorhodafluor (carboxy SNARF-1), dicyanohydroquinone (DCH) and 2',7'-bis-(2-carboxyethyl)-5-(and-6)-carboxyfluorescein (BCECF) for intracellular pH measurements; cyanine dyes (e.g., dihexyloxacarbocyanine [DiOC6(3)]), styryl dyes, and oxonols for membrane potential measurements, and many other dyes.

4. *Enzyme substrates.* Generally this group consists of specific substrates which become fluorescent upon the action of the enzyme. Some examples are the non-fluorescent fluorescein diacetate and fluorescein β-D-galactopyranoside which become highly fluorescent after hydrolysis by non-specific esterases and β-D-galactosidase, respectively [10].

This is only a short introduction about FCM. For extensive information about flow cytometry and staining procedures reference is made to the books of Shapiro and Melamed *et al.* [12][20].

4.3 Applications

4.3.1 Assessment of cell viability and vitality

Viable or non-viable, that is the question which has fascinated many researchers over the years. But what is the definition of viability? A major difficulty is that all methods have different criteria to decide if cells are viable (or not). Usually the ability of cells to reproduce is considered as the reference method [8], but this method has several disadvantages. Cells which do not reproduce are not necessarily non-viable and reproduction of cells takes normally several days. Obviously, there is considerable interest in more rapid methods. Some rapid techniques use the criterion of membrane integrity for viability, e.g., the ability of cells to exclude dyes as methylene blue and propidium iodide or to accumulate dyes as fluorescein [20]. A combination of two dyes like fluorescein diacate to stain intact cells and propidium iodide to stain dead cells can provide more information. This principle is the basis of a newly commercialized kit of Molecular Probes (Eugene, USA) for the determination of the number of viable cells in bacterial populations, such that viable and dead cells stain green and red, respectively. However, this criterium of membrane integrity has also some drawbacks. Cells with intact membranes but which have lost the ability to reproduce are nevertheless considered viable. Moreover, active efflux of probes may also frustrate the fluorescent viability staining [2][15].

In the following some methods will be discussed which use other characteristics of viable cells, such as the existence of a proton motive force and reducing capacity (e.g., electron donors coupled to the electron transport chain).

4.3.1.1 Proton motive force

Most (aerobic) bacteria possess membrane bound proton pumps and/or an electron transport chain which extrude protons from the cytoplasm to generate a transmembrane electrochemical gradient of protons, the proton motive force [1]. It is known already for some time that bacterial cells can accumulate Rhodamine 123 [8][20]. Rhodamine 123 is a lipophilic water soluble cation which is thought to be membrane permeable. The accumulation is most likely related to maintenance of the membrane potential Ψ (Figure 4.4), (i.e., addition of an uncoupler (CCCP), which dissipates the proton motive force, induced efflux of the dye). Kaprelyants and Kell used this property of cells to assess cell *viability* with FCM [8]. They showed that starved cells did not accumulate Rhodamine 123, whereas *viable* mid-log cells did (Figure 4.5). The cells were measured by low angle light scattering and a fluorescence parameter, and the data were represented in three-dimensional figures.

In 1966 Rotman and Papermaster [17] reported the use of non-fluorescent precursor molecules which were taken up by mammalian cells and cleaved by intracellular enzymes to give a fluorescent product. This staining by fluorescein diacetate or related compounds is principally based on the assumption that only cells which have an intact plasma membrane are able to accumulate the fluorescent probe. Recently it was shown however, that in *S. cerevisiae* efflux of (intracellular) carboxyfluorescein (cF) is an energy-dependent, carrier mediated extrusion which is most likely driven by the membrane potential and/or the proton gradient [2], suggesting that the fluorescence intensity of the cells is not necessarily correlated to the viability. To determine the viability a two-step procedure was developed whereby the yeast cells were first loaded with cF by incubation with cFDA, and subsequent determination of the efflux of cF (in washed cells). This efflux of cF could be analyzed by FCM (Figure 4.6).

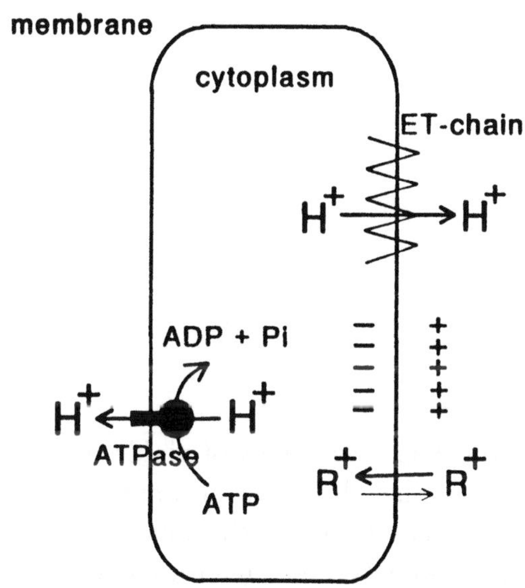

Figure 4.4: Schematic representation of a bacterial cell. The accumulation of Rhodamine 123 is driven by the membrane potential ($\Delta\Psi$). Legend: ET-Chain, electron transport chain; R, Rhodamine 123; Pi, intracellular Phosphate; - , negative charge.

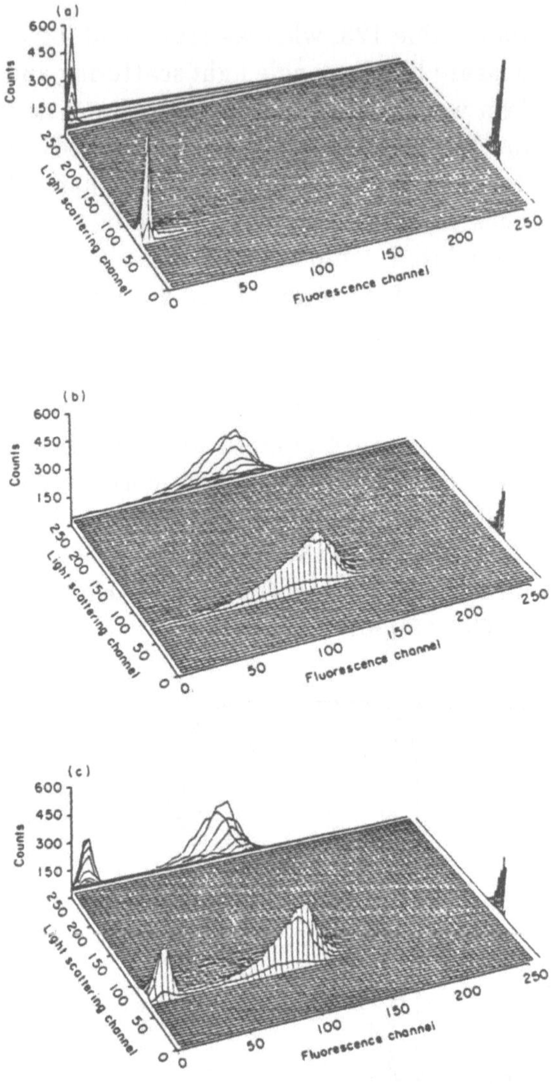

Figure 4.5: Flow cytometric assessment of viable and non-viable cells and their mixtures, based on the ability to accumulate Rhodamine 123. (a) *Micrococcus luteus* cells were grown to stationary phase in batch culture in lactate minimal medium and starved by storage at 4 °C for 1 month and then stained with Rhodamine 123. (b) Cells were grown, taken from mid-logarithmic phase of the culture, and immediately stained as in (a). (c) Cells from (a) and (b) were mixed in the ratio 1:2 and treated as in (a) and (b). Reproduced from [8] with permission of Blackwell Scientific Publications LTD, Oxford (UK).

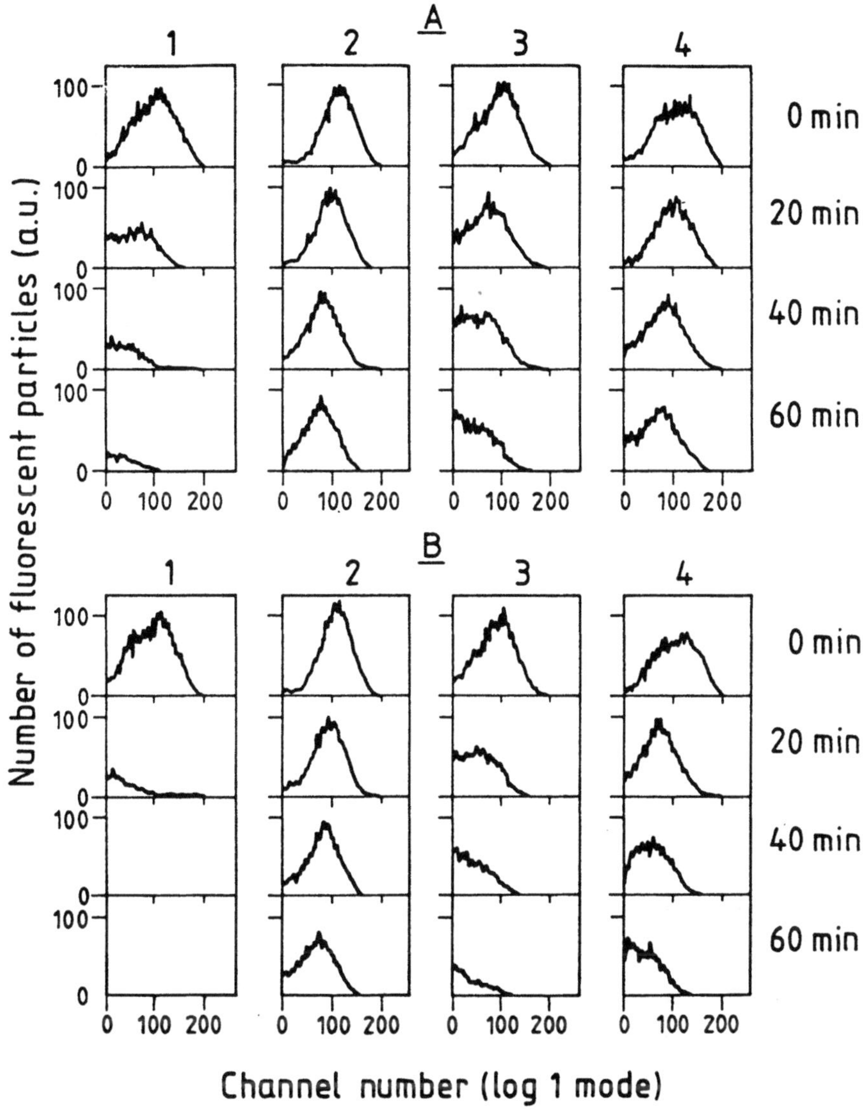

Figure 4.6: Histograms of a *S. cerevisiae* population. Cells were loaded with 0.043 mM cFDA at 40 °C. Subsequently, loss of fluorescence at 40 °C after 0, 20, 40, and 60 minutes was measured with a flow cytometer in non-energized cells (A) and in cells energized with glucose (10 mM) (B). Cells were pre-treated as follows: no treatment (1), heat treatment at 60 °C for 90 seconds (2), pre-incubation with benzoic acid (10 mM) (3), and pre-incubation with DNP (1 mM) (4). Reproduced from [2] with permission of American Society for Microbiology.

In viable (not treated) cells the histogram shifted to the left as a result of the loss of intracellular fluorescence. Heat treated cells lost their ability to export cF, most likely because they were unable to maintain a proton-motive force. Otherwise treated cells (with benzoic acid or dinitrophenol) were still able to extrude cF after addition of glucose. Analysis was performed with a single parameter flow cytometer. The extent to which a yeast population looses its fluorescence was correlated in flow cytometry experiments with yeast *viability*. Viable and non-viable cells (according to the plate count method) were mixed at different ratios. As demonstrated in Figure 4.7, an excellent correlation was found between the viable counts as determined by the plate count method (counting after 5 days incubation at 30 °C on Sabouraud agar) and the fraction of yeast cells which had lost approximately 100 % of cF after incubation for 30 min at 40 °C, such that they could not longer be detected with the flow cytometer.

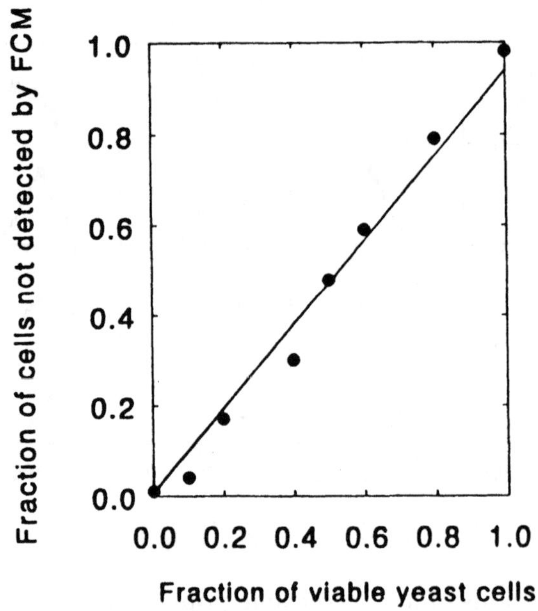

Figure 4.7: Comparison of the fraction viable cells and the fraction of *S. cerevisiae* L115 cells which lost all fluorescence. The viable cell counts were determined by the plate count method. The cells were loaded with cF by incubation with 0.043 mM cFDA in McIlvaine buffer pH 4.0, washed and resuspended in McIlvaine buffer pH 7.3. The efflux assay was performed at 40 °C. The fluorescent cells were counted at time 0 and after 30 minutes. Reproduced from [2] with permission of American Society for Microbiology.

4.3.1.2 Reducing capacity

A simple method for the detection of electron transport chain activity in cells is the application of tetrazolium dyes as nitroblue tetrazolium (NBT) and 2-(p-iodophenyl)-3-p-nitrophenyl)-5-phenyltetrazolium chloride (INT) [16][18], which function as alternative electron acceptors (Figure 4.8). These compounds are then reduced to insoluble colored formazan crystals. The relatively new tetrazolium dye 5-cyano-2,3 ditolyl tetrazolium chloride (CTC), has the quality that the reduced formazan product is fluorescent, which potentially allows the use of flow cytometry. Respiring cells (i.e., viable cells), reduce CTC to the formazan product (CTF), which precipitates and thereby accumulates intracellularly [13][19]. CTC has an excitation optimum around 450 nm and is fluorescent in the red region around 620 nm. Severin and Seidler [19] applied this technique in Ehrlich ascites tumor cells in combination with FCM, using a double parameter flow cytometer. The total cell population was determined with DNA staining (DAPI staining), and the percentage of dehydrogenase-positive cells (i.e., cells with redox activity) was determined from the gated DNA-positive particles. Cells which were negative, i.e., exhibited no red fluorescence (formazan), were considered non-viable.

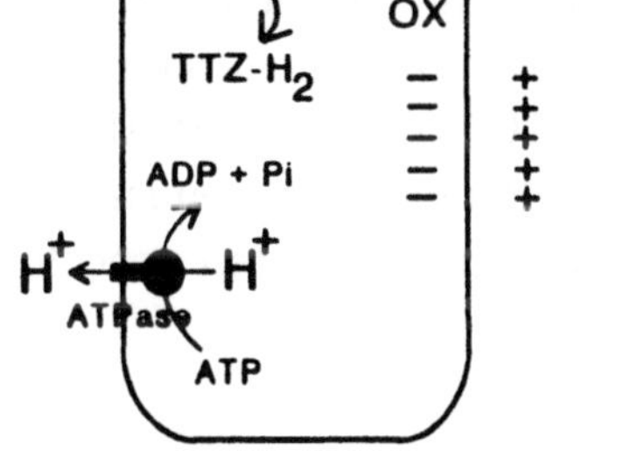

Figure 4.8: Schematic representation of a bacterial cell. The tetrazolium dye acts as an alternative electron acceptor. The reduced formazan product is insoluble in water and precipitates inside the cell. Legend: ET-Chain, electron transport chain; TTZ, Tetrazolium dye; $TTZ\text{-}H_2$, reduced tetrazolium dye (formazan product); DH, (NADH) dehydrogenase complex; OX, oxidase complex; Pi, intracellular Phosphate.

4.3.2 Measurement of intracellular pH

It is generally accepted that the cytoplasmic pH is a very important factor in cell metabolism [1]. Most cells try to maintain their intracellular pH in a (small) range of pH 7 to 8. Determination of this parameter provides information about energy status and vitality of the cell. However, what method should be used to measure? Methods such as nuclear magnetic resonance (NMR) and distribution of (radio-labeled) weak acids have a limited time resolution. In this respect, fluorescent probes offer the advantage of a short time response and, in combination with FCM, individual analysis of cells. Obviously, the use of fluorescent probes has also some disadvantages, i.e., (passive or active) efflux of probe to the external environment and calibration of the intracellular fluorescence signal can be troublesome [1][7][15].

Fluorescein and fluorescein derivatives are highly pH dependent fluorescent dyes [11], have relatively high extinction coefficients and have therefore been used extensively. Thomas *et al.* [21] reported already in 1979 the use of carboxyfluorescein for the measurement of the cytoplasmic pH in Ehrlich Ascite cells. Currently, BCECF is the most popular pH indicator. This fluorescein derivative with a pK_a of nearly 7, has four to five negative charges at physiological pH, which makes its retention in cells better than fluorescein or carboxyfluorescein [5][14]. BCECF is normally introduced in cells via its acetoxymethyl ester (BCECF-AM). However, it has been reported recently that carboxyfluorescein and BCECF are actively extruded from *S. cerevisiae* and *Lactococcus lactis*, respectively [2][15]. This may limit the application of these probes for intracellular pH measurements. A novel pH probe, carboxy-SNARF-1 was used for measurement of the internal pH in Chinese Hamster Ovary (CHO) cells with FCM [23]. The advantage of this pH indicator is the larger pH dependent emission spectral shift compared to BCECF. So far, this probe has not been used for measurement of pH in yeast or bacteria.

The ratio technique and calibration of the fluorescence signal

How can we translate the obtained fluorescence signal to a pH value? Nearly all used pH sensitive probes have absorption spectra with a pH sensitive and pH insensitive wave-length. The ratio between these two wave-lengths is an indicator for the pH and has the advantage that it is essentially independent of probe concentration and other phenomena such as dye bleaching.

The pH is related to the ratio by the Henderson-Hasselbalch equation [4]:

$$pK_a = \text{pH} - \log \frac{R - R_{min}}{R_{max} - R} - \log \frac{F_a}{F_b}$$

where R is the ratio at given pH, R_{min} and R_{max} represent the ratios at the acid and base extremes respectively, and F_a and F_b represent the fluorescence at the pH-insensitive wave-length at the acid and base extremes respectively.

Furthermore, for a proper correlation between the ratio and the intracellular pH, the intracellular pH has to be calibrated. Generally this is achieved by equilibrating the intra- and extracellular pH with nigericin (a H^+/K^+ antiporter) in a K^+ medium. A disadvantage of this technique is that in some cell types this equilibration proceeds very slowly or does not occur at all (e.g., yeast cells) [2][7].

4.4 Conclusion

Flow cytometry has proved to be an effective tool for measuring physiological parameters of individual cells. Flow cytometers are getting more and more sensible, can measure several parameters at the same time and have nowadays sophisticated software for data treatment. Moreover, FCM is a very rapid and sensitive technique in that it can still measure at very low cell densities. Nevertheless, an adequate fluorescence staining remains a major requirement for the use of FCM. Especially for the fluorescence staining of relatively small cells such as bacteria there is a significant need for (novel) probes which are highly fluorescent (high extinction coefficient and high quantum yield) and sensible to specific physiological changes in cells. Furthermore, as illustrated in the previous paragraphs a good understanding of the interaction of the fluorescent probe and the physiology of the microorganism is critical.

In conclusion, notwithstanding the efforts which have to be made, it may be expected that with the continuous development of instruments and probes, FCM should play an eminent role in the understanding of complex biological problems.

Acknowledgments

The author thanks Tjakko Abee for critical reading of the manuscript.

4.5 Bibliography

1. I.R. Booth 1985. Regulation of cytoplasmic pH in bacteria. *Microbiol. Rev.*, 49, 359-378

2. P. Breeuwer, J.L. Drocourt, F.M. Rombouts and T. Abee 1994. Rapid assessment of yeast viability by carboxyfluorescein and flow cytometry. *Appl. Environ. Microbiol.*, 60, 1467-1472

3. P. Fouchet, C. Jayat, Y. Héchard, M.H. Ratinaud and G. Frelat 1993. Recent advances of flow cytometry in fundamental and applied microbiology. *Biol. Cell.*, 78, 95-109

4. M.L. Graber, D.C. DiLillo, B.L. Friedman and E. Pastoriza-Munoz 1986. Characteristics of fluoroprobes for measuring intracellular pH. *Analytic. Biochem.*, 156, 202-212

5. R.P. Haugland 1992. In: K.D. Larison (Ed.), *Handbook of fluorescent probes and research chemicals.*, Molecular Probes Inc. Eugene, USA

6. K.A. Hoff 1988. Rapid and simple method for double staining of bacteria with 4', 6-Diamidino-2-Phenylindole and Fluorescein Isothiocyanate-labeled antibodies. *Appl. Environ. Microbiol.*, 54, 2949-2952

7. M.R. James-Kracke 1992. Quick and accurate method to convert BCECF fluorescence to pHi: calibration in three different types of cell preparations. *J. Cell. Physiol.*, 151, 596-603

8. A.S. Kaprelyants and D.B. Kell 1992. Rapid assessment of bacterial viability and vitality by rhodamine 123 and flow cytometry. *J. Appl. Bacteriol.*, 72, 410-422

9. C. Laplace-Builhé, K. Hahne, W. Hunger, Y. Tirilly and J.L. Drocourt 1993. Application of flow cytometry to rapid microbial analysis in food and drink industries. *Biol. Cell.*, 78, 123-128

10. M. Manafi, W. Kneifel and S. Bascomb 1991. Fluorogenic and Chromogenic substrates used in bacterial diagnostics. *Microbiol. Rev.*, 55, 335-348

11. M.M. Martin and L. Lindqvist 1975. The pH dependence of fluorescein fluorescence. *J. Luminescence*, 10, 381-390

12. M.R. Melamed, T. Lindmo and M.L. Mendelsohn (Ed.) 1990. *Flow cytometry and sorting*, 209-225, Wiley-Liss Inc., New York, USA

13. Ph. Métézeau, R. Miglierina and M.H. Ratinaud 1994. *La cytometrie en flux: guide practique de la préparation à l'analyse des cellules.* Presses de l'Université de Limoges, Limoges, France

14. D. Molenaar, T. Abee and W. Konings 1991. Continuous measurement of the cytoplasmic pH in *Lactococcus lactis* with a fluorescent pH indicator. *Biochim. Biophys. Acta*, 1115, 75-83

15. D. Molenaar, H. Bolhuis, T. Abee, B. Poolman and W.N. Konings 1992. The efflux of a fluorescent probe is catalyzed by an ATP-driven extrusion system in *Lactococcus lactis. J. Bacteriol.*, 174, 3118-3124

16. G.G. Rodriguez, D. Phipps, K. Ishiguro and H.F. Ridgway 1992. Use of a fluorescent redox probe for direct visualization of actively respiring Bacteria. *Appl. Environ. Microbiol.*, 58, 1801-1808

17. B. Rotman and B.W. Papermaster 1966. Membrane properties of living mammalian cells as studied by enzymatic hydrolysis of fluorogenic esters. *Proc. Natl. Acad. Sci. USA*, 55, 134-141

18. G. Schaule, H.C. Flemming and H.F. Ridgway 1993. Use of 5-Cyano-2,3-Ditolyl Tetrazolium Chloride for quantifying planctonic and sessile respiring bacteria in drinking water. *Appl. Environ. Microbiol.*, 59, 3850-3857

19. E. Severin and E. Seidler 1992. Calibration of a flow cytometric assay of glucose-6- phosphate dehydrogenase activity. *Cytometry*, 13, 822-826

20. H.M. Shapiro 1988. *Practical flow cytometry.* Alan R. Liss, Inc. New York

21. J.A. Thomas, R.N. Buchsbaum, A. Zimniak and E. Racker 1979. Intracellular pH measurements in Ehrlich ascites tumor cells utilizing spectroscopic probes generated in situ. *Biochemistry*, 18, 2210-2218

22. R.L. Tyndall, R.E. Hand, R.C. Mann, C. Evans and R. Jernigan 1985. Application of flow cytometry to detection and characterization of *Legionella* spp. *Appl. Environ. Microbiol.*, 49, 852-857

23. E.D. Wieder, H. Hang and M.H. Fox 1993. Measurement of intracellular pH using flow cytometry with carboxy-SNARF-1. *Cytometry*, 14, 916-921

Chapter 5

Microcalorimetric Characterization of Bacterial Inocula

H. VANDENHOVE
Katholieke Universiteit Leuven
Faculty of Agricultural and Applied Biological Sciences
Laboratory of Soil Fertility and Soil Biology
Kardinaal Mercierlaan 92 B-3001 Heverlee (BELGIUM)

ABSTRACT – *The central theme of this chapter on microcalorimetric characterization of bacterial inocula is to show that the survival ability of bacterial inocula can be enhanced by using microcalorimetry to sample inocula at specific physiological growth stages. When assessing the influence of soil temperatures on the survival of Pseudomonas fluorescens inocula, we were confronted with a non-trivial variability of survival rates. Searching for a method that would reduce this variability, we used microcalorimetry as a control variable to determine sampling times of the inocula. Good correlations between heat production and other growth parameters were found; moreover, the growth stage of the inocula (determined microcalorimetrically) had an important impact on the growth lag in fresh medium and on the survival of Pseudomonas fluorescens in soil. The selection of inocula with equivalent microcalorimetric power outputs resulted in a sizeable reduction of the experimental variability. The measurement of different growth parameters together with the heat production rate made clear that the shape*

J.F.M. Van Impe et al. (eds.), Advanced Instrumentation, Data Interpretation, and Control of Biotechnological Processes, 121-158.

of the power-time curve is influenced by the type of metabolic activity and can be related to the different physiological states of the bacteria. From the above follows that standardization and optimization of bacterial inocula by using the microcalorimetric power output as a control variable for reproducible sampling of bacterial cultures may be important for inoculation experiments and practice. The variation that we are confronted with in inoculation trials is to some extent due to physiological differences between inocula and may be reduced through timely sampling.

KEYWORDS – *Microcalorimetry, Survival, Inoculum Standardization*

5.1 Introduction

Bacteria have been inoculated into soil for various purposes, such as the promotion of plant growth by PGPR (Plant-Growth Promoting Rhizobacteria) [18][21][41][42][53], the suppression of soil-borne plant pathogens [8][24][26] [34][86], or the degradation of pesticides [63].

Although successes in these agricultural trials cannot be denied, poor reproducibility and considerable variability of the results remain a bottleneck [18] hampering a large-scale use of inoculation technology. A key factor may be the varying degree of establishment and survival of the introduced bacterial populations [2][41][42].

Soil moisture and temperature regimes, pH and texture as well as O_2 and nutrient availability have been suggested as the major abiotic factors that govern the survival of bacteria introduced into soil [22][24][52][91]. Predation by protozoa [1][2][22][91], microbial antagonism and competition [2][24] and the physiological state of the introduced bacteria [79][82][85] have been suggested as the main biological factors.

When investigating the influence of preincubation and incubation temperatures on the survival of *Pseudomonas fluorescens* [89] we were confronted with a non-trivial variability of the survival rates of different batches of inocula; the results suggested that the physiological state of the inocula may play a major role. Searching for a method that would enable us to reproducibly sample inocula, we measured the power output of the cultures microcalorimetrically and used it as a control variable to determine the sampling times. The rationale for this approach was that the heat production rate, or power output, is a measure for the activity of the bacteria and that, moreover, microcalorimetric measurements are continuous, non-destructive and reproducible. The results presented in this chapter will show that microcalorimetry can, indeed, be of use to reproducibly sample standardized inocula.

After a state-of-the-art overview of microcalorimetry where, apart from a description of the general principles involved, some thermodynamic aspects and a discussion of important applications of microcalorimetry are covered, we will continue with the survival of *Pseudomonas fluorescens* inocula and the influence thereupon of the physiological state of the inocula. In former research [89] where the influence of soil preincubation and soil incubation temperature was assessed, an influence of temperature on survival could be noted. The results, however, showed that the survival variability between different inoculum batches could not be neglected. The reduction of this variability by inoculum standardization in terms of physiological criteria will be the central theme of this research. First, we analyzed how microcalorimetry can be of use in determining some of these criteria. Then we studied the influence of the growth stage of *Pseudomonas fluorescens* inocula (determined microcalorimetrically) on their subsequent survival in soil.

The fundamentals of the microcalorimetrical phenomena observed for *Pseudomonas fluorescens* are explored more comprehensively. Data are provided on a series of growth parameters (such as the ATP content of the cells and the rates of biomass production and of malate and O_2 consumption) that were monitored simultaneously with the microcalorimetric measurements. The results will be related to the previous findings where possible to gain more insight into how the growth stage of the inocula influences their survival in soil.

5.2 A View on Microcalorimetry

Virtually every physical, chemical or biological process involves heat production or heat absorption, and the quantities of heat that are produced or absorbed are proportional to the extent of the process. Calorimetric measurements are, therefore, of potential analytical interest.

Bacterial cultures produce heat quantities equal to the enthalpy changes resulting from their metabolism (excluding chemical side reactions); the enthalpy change is the sum of the energetic contributions of all the reactions that occur during metabolic processes, that is the catabolic and the anabolic reactions as well as the reactions involved in transport processes, movement and maintenance.

Calorimetric methods have an inherently low specificity. For complex biochemical or microbiological processes, this can sometimes be advantageous: a gross value for a process or sum of processes can be obtained, through which unknown and undefined phenomena may be detected. Another advantage of calorimetric methods is that they allow a continuous registration of life processes without any disturbance (such as, for example, the addition of reagents). Furthermore, the technique is independent of turbidity or color of the bacterial suspensions.

5.2.1 Main types of microcalorimeters

The term *microcalorimetry* can be defined only vaguely: there is no sharp difference between ordinary calorimetry (*macrocalorimetry*) and *microcalorimetry*. For both macro- and microcalorimetry, several calorimetric principles are in use, and the design of calorimeters varies considerably. Monk and Wadsö [61][62] and Forrest [33] give detailed descriptions of microcalorimeters; we will limit ourselves in this section to some basic principles.

There are two basic categories of calorimeters: heat conduction calorimeters and adiabatic calorimeters. The ideal adiabatic calorimeter excludes any heat exchange with its surroundings and measures temperature changes: the heat produced during the experiment (Q, expressed in joules) is proportional to the observed temperature change (ΔT) and to the heat capacity of the calorimetric system (ε, in J/°C):

$$Q = \varepsilon \cdot \Delta T \tag{5.1}$$

On the other hand, heat conduction calorimeters measure heat fluxes (dQ/dt, in watts). The heat from the system under study is quantitatively transferred from the vessel in which the reaction takes place through thermopiles to a surrounding heat sink (Figure 5.1). The thermopiles consist of a number of thermocouples that create an electrical potential difference (V, in volts) over the thermopile wall proportional to the temperature difference over the wall (the sensitivity to temperature differences between the sample vessel and the surrounding heat sink is about 10^{-6} °C for microcalorimeters). The calorimeter records this potential difference, which is directly proportional to the heat flux:

$$V = c \cdot \frac{dQ}{dt} \tag{5.2}$$

Thus, the heat conduction calorimeter measures a heat flux, the measuring unit of which is watts; hence, the recorded output of this type of calorimeter is generally called a *power-time curve*. Integration of the power output over time yields the (cumulative) heat production (also called *heat evolution* or *heat output*).

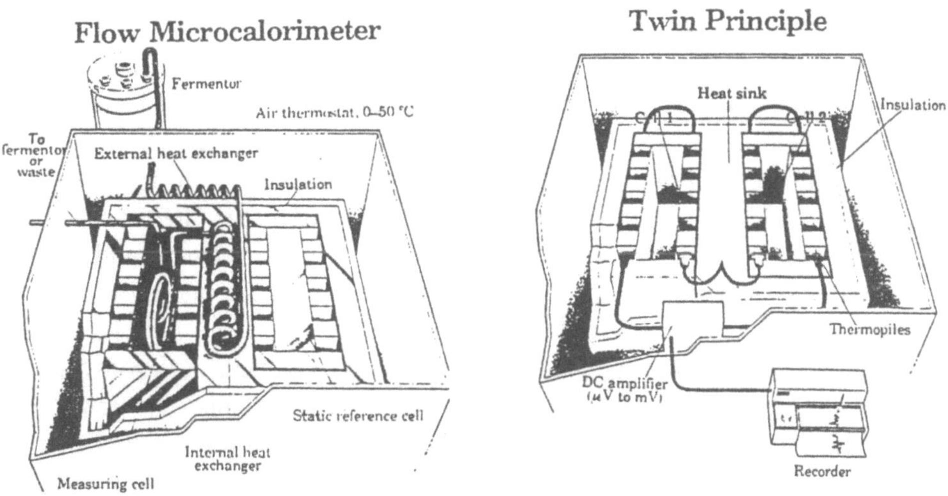

Figure 5.1: Twin principle and flow system of a twin-conduction type microcalorimeter (Courtesy of LKB-Produkter AB, Stockholm).

For steady state or for slow processes, such as those taking place during microbial growth, V is directly proportional to the heat evolved in the reaction vessel. Under these conditions, the microcalorimeter acts as a power meter and can be used for kinetic studies; the recorded power-time curve is thus directly related to the kinetics of the process.

5.2.2 Principal arrangements of microcalorimeters

A calorimeter can be in single or twin arrangement. The single arrangement is mechanically simpler, but the twin calorimeter has some attractive advantages, in particular for microcalorimetry, where twin calorimeters are mostly used: in one vessel, the process to be monitored takes place, while the other vessel acts as a control or reference system to correct for environmental disturbances, which are presumed to influence both vessels identically (see

Figure 5.1).

A second type of arrangement distinguishes between batch and flow microcalorimeters. As the name implies, batch calorimetry refers to a method where the reaction generally takes place in a closed vessel. Batch microcalorimeters can be used for bacterial growth studies, but because providing sufficient oxygen for aerobic cultures in batch microcalorimeters is difficult, most studies deal with anaerobic cultures. Some complications may arise when two or more reacting chemical compounds are brought together in an ampoule: the ampoule usually contains a relatively large gas phase, so that part of the reaction heat of the system will be absorbed by evaporation. Another disadvantage of batch calorimeters is the inaccessibility of the vessels for additional measurements. On the other hand, batch microcalorimeters are suitable for reactions ranging from instantaneous to very slow processes.

It is also possible to bring reaction components together by a continuous-flow or a stopped-flow technique (Figure 5.1). The absence of a gas phase makes flow microcalorimeters ideal for mixing and dilution experiments. Also, flow microcalorimetry allows the measurement of heat production of a bacterial culture grown outside the apparatus (as in a fermentor); it offers a sensitive, reproducible and quantitative analysis of the metabolic reactions under strictly defined culture conditions.

5.2.3 Thermodynamic considerations

The heat output accompanying biological reactions depends only on the initial and the final state of the system; it is independent of the pathway of the reaction, no matter how complex the intermediate steps be.

In biological systems, reactions can be considered to take place in solutions at both constant volume and pressure, so that the concepts of energy and enthalpy become interchangeable, that is: the change in energy content accompanying a reaction is the same as the enthalpy change of the reaction (ΔH). If only expansion work is done, ΔH corresponds to the experimentally measured heat evolution (Q) during a reaction in a closed system at constant pressure.

To perform energy balance calculations of microbial growth, initial and final stages have to be defined. An initial stage can be characterized by a specific substrate and a known amount of cells; the final stage can be selected as the moment when growth ceases and is to be defined by the elemental composition of the cells at this stage to establish a growth reaction.

Battley [6] proposed the following growth reaction for the aerobic growth of *Saccharomyces cerevisiae* (Hansen) on glucose:

$$\begin{aligned} & C_6H_{12}O_{6(aq)} + 3.84\ O_{2(aq)} + 0.29\ NH_{3(aq)} \\ \rightarrow\ & 4.09\ CO_{2(aq)} + 4.73\ H_2O + 1.95\ CH_{1.72}O_{0.44}N_{20.15} \end{aligned} \tag{5.3}$$

Subsequently, Battley [7] measured calorimetrically the enthalpy change (ΔH_{met}) corresponding to this growth reaction and found a value of -2004 kJ per mole glucose; this is close to the theoretically derived value of -1926 kJ per mole corresponding to the catabolic reaction whereby 68 % of the glucose-C is converted to $CO_2 - C$, but less than maximum heat output as computed for the total combustion of glucose, which yields 2834 kJ/mole. Thus, some of the potentially available energy was absorbed somewhere in the biomass, or, if one looks at it the other way round, the efficiency with which enthalpy is absorbed in bacterial growth can be calculated as

$$\%\ enthalpy\ conservation = \frac{\Delta H_{nb} - \Delta H_b}{\Delta H_{nb}} \times 100 \tag{5.4}$$

where ΔH_{nb} represents the enthalpy changes of the reactions that do not involve biomass formation, and ΔH_b represents the enthalpy changes of the reactions that do involve biomass formation [7]. If cellular material is the only product of the growth, this efficiency is a direct measure of enthalpy retained in biomass.

The heat quantity that is produced by the culture at constant pressure and temperature equals the enthalpy change of the total metabolism including the formation of secondary metabolites (H_{met}). The growth reaction can be divided into an exergonic catabolic and an endergonic anabolic reaction [12].

$$Q = \Delta H_{met} = Q_{cat} + Q_{an} \tag{5.5}$$

where Q_{cat} is the heat evolution corresponding to the catabolic part, and Q_{an} is the heat evolution corresponding to the anabolic part.

The catabolic reactions can be described by the following relationship:

$$\begin{gathered} energy\ source \longrightarrow products\ of\ catabolism \\ \Delta H = \Delta H_{cat} \end{gathered} \tag{5.6}$$

When the energy sources and the products of catabolism are known molecules (for example, glucose, organic acids, CO_2 or H_2O), ΔH_{cat} can be computed from the enthalpy of formation of these compounds, which can be found in tables of thermodynamic data (the standard energy of formation of an organic compound is the change in enthalpy accompanying the formation of the compound at 298 K and 1 atm out of elements at equilibrium under these standard conditions).

Anabolism is more difficult to investigate than catabolism. A method to estimate the enthalpy changes corresponding to biosynthesis (ΔH_{an}) is based on the heat of combustion of entire cells. For cells grown on glucose and NH_4OH, the anabolic process can be summarized by

$$C(H_2O) + z\ NH_4OH \quad \underset{\Delta H\,=\,\Delta H_{an}}{\longrightarrow} \quad CH_xO_yN_z + n\ H_2O \qquad (5.7)$$

Different stoichiometries can be obtained for this relation depending on the elementary analysis of the cell.

The heat of formation of cell material can be calculated from the elementary analysis and from the heat of combustion of the microorganisms. However, marked differences have been reported regarding the heat of formation of cell material. Using literature data, Belaich [12] computed enthalpies of formation of microbial cells ranging from -1.21 kJ per g dry weight to -17.02 kJ per g dry weight. In these calculations, Belaich used the heat of combustion of -15.48 kJ per g biomass proposed by Sedlaczk (1964, in [12]) and the value of -22.59 kJ per g biomass reported by Prochazka *et al.* [73]. Lamprecht [54] obtained from literature data a mean value of -19.33 kJ per g biomass for the heat of combustion of various yeasts grown under different growth conditions.

The enthalpy changes ΔH_{an} corresponding to the synthesis of 1 g cell material from glucose and NH_4OH are strongly influenced by the discrepancies between the reported heats of combustion: they vary from -9.12 kJ to +4.39 kJ per g of synthesized cell dry weight [12].

However, the anabolic enthalpy change is often negligible compared to the catabolic one [11][66]. Experiments performed by Erickson and coworkers [27] with complex media or with a carbohydrate as carbon and energy source showed a small anabolic enthalpy change due to the relative similarity between substrate (carbohydrate) and product (biomass). Gustafsson [38] also illustrated the minor contribution of the anabolic enthalpy change by using a carbohydrate as carbon and energy source for the growth of *Saccharomyces cerevisiae*. Using data of Larsson *et al.* [55], she found that the enthalpy change of the anabolic reaction was -28 kJ per mole glucose, or only 1.5 % of the enthalpy change of the total growth reaction (-1908 kJ per mole glucose).

5.2.4 Applications of microcalorimetry

5.2.4.1 Microcalorimetry and Microbial Growth Studies

The earlier power-time curves were derived from studies of bacteria with an unknown metabolism and grown on complex media. The experimental growth conditions were rather ill-defined and the resulting curves were often complex (Boivinet 1964 in [12]). It became evident that growth conditions must be clearly defined to allow quantitative interpretations of heat evolutions accompanying bacterial growth.

The first quantitative study of the energetics of energy-limited bacterial growth was reported by Forrest *et al.* [31] who showed that the rate of heat production by a growing culture of *Streptococcus faecalis* is identical with the rate of biomass increase. Simultaneous increases in both the biosynthesis of new cellular material and in heat evolution have also been shown during most of the growth period for bacteria and yeasts (Boivinet 1964 in [12],[11][23][29][33][40][43]). However, inconsistencies have been reported. A varying specific rate of heat production (dQ/dt per unit biomass) may reflect a change in metabolic activity during growth in batch conditions. For example, when growing *Saccharomyces cerevisiae* aerobically with glucose as the sole carbon and energy source, Fiechter *et al.* [30] found a respiro-fermentative metabolism. This metabolism is characterized by a fluctuating fermentative activity resulting in a changing respiratory coefficient and specific heat production rate during the presumably fermentative phase [15][37].

In case of a sequential utilization of different substrates, power-time curves of more than one phase may be obtained. Organisms may also produce (and subsequently consume) some intermediary products: ethanol, for example, is produced when *Saccharomyces cerevisiae* is grown aerobically on glucose [54], or acetate may be produced by *Escherichia coli* when grown on glucose [28]. Heat measurements must be combined with measurements of other growth parameters, such as cell concentration, substrate concentration or extracellular products, to elucidate the processes of metabolism.

Blomberg and co-workers [14] studied the influence of the physiological stage (defined as a point on the power-time curve) of *Saccharomyces cerevisiae* on tolerance to a sudden increase in the environmental salt concentration. Large changes occurred in the cells within a short time and the physiological stage of the culture appeared to be of utmost importance for the capacity of the cells to tolerate osmotic dehydration, the more tolerant cells being those from the early-respiratory phase.

5.2.4.2 Microcalorimetry and soil microbial ecology

Several authors [36][47][56][65][80][81] have suggested the use of microcalorimetry for measuring metabolic activity in soil.

After amending soils with glucose, Sparling [80][81] found an immediate increase in microbial activity as indicated by increased heat production and respiration rates. The increased heat production rate was stable during 6 hours. The microcalorimetric values correlated well with the rate of respiration and also, although to a lesser extent, with biomass, ATP content and amylase activity.

Kimura and Takahashi [47] studied microbial degradation of glucose in a brown-acid andosol in a conduction-type calorimeter. They obtained linear relationships both between heat evolution and the amount of glucose degraded and between heat evolution and viable cell counts. The heat evolution per unit glucose was 1287 kJ/mole. The average heat production per cell was 5.58 10^{-8} J per cell.

Gustafsson and Gustafsson [36] established a system for toxicity testing using microcalorimetry. The rates of heat production from river sediment samples were monitored after samples had been treated with different concentrations of a quaternary ammonium salt. A good agreement was obtained between heat production and respiration rates.

Likewise, an improved stoichiometry of the growth reaction was verified for the anaerobic respiratory metabolism of sulphate-reducing bacteria [81][82] and an understanding was obtained for the successive uses of available extracellular and of intermediately produced electron acceptors by anaerobic dissimilatory nitrate-reducing bacteria [75][77]. A theoretical approach for the microbial heat evolution as an analytical tool for checking the consistency of biomass and substrate data is given by Luong and Volesky [59]. By simultaneous application of microcalorimetry and respirometry in conjunction with a thermodynamic interpretation of experimental calorimetric-respirometric ratios, Gnaiger and Kemp [35] quantified oxidative and anaerobic components of complex metabolic systems.

Microbial heat evolution has also been put forward as process parameter in fermentations [13][59]. Computerized process control schemes that are used in fermentations are often limited by the number of relevant process variables that can be measured on-line and used as continuous input for the bioprocess computer models; the heat production rate of the biological process can easily be measured quantitatively in industrial-size fermentors, however. This information could be of use provided that suitable models are available to relate the heat production rate of the cell culture to other relevant process variables, such as rates of substrate consumption [55][59] of growth and of oxygen uptake.

5.2.4.3 Other applications of microcalorimetry

Apart from the above applications, microcalorimetry may be useful as an analytical tool for biotechnical, medical or environmental control.

In the food industry, microbiological assay revolves around the screening of products to detect microbes. In this context, microcalorimetry has been applied to the analysis of vacuum-packed food [68].

Microcalorimetry has also been used in the medical sector, especially for the identification of bacteria and for antibiotic sensitivity testing. Johansson *et al.* [44], Newell [68] and López *et al.* [57] analyzed bacterial curves for species identification. At least 200 cultures of clinically important organisms, predominantly *Enterobacteriaceae*, were investigated microcalorimetrically, and almost all organisms were identified [68]. For the use of power-time curves as microbial finger-prints, particular attention must be paid to the preparation of inocula and the culture conditions.

In the field of antibiotic sensitivity testing, Beezer and Chowdhry [10] examined the effect of various antibiotics on 2 strains of *Escherichia coli* grown in a semi-defined medium. The exponential-phase heat output of the bacteria changed upon the addition of antibiotics, the exact change depending on type and concentration of the antibiotics. Flow microcalorimetry has been used by Beezer *et al.* [9] in a semi-automated method of antibiotic sensitivity testing in infected urine samples.

Jolicoeur and Beaubien [45] reported several applications of microcalorimetry on toxicity studies in chemostat-grown cultures. Enthalpy changes for chemostat-grown cultures vary with varying dilution rates [16]. Addition of toxicants will decrease the heat production according to type and concentration of the toxicant.

5.3 Survival of *Pseudomonas fluorescens* Inocula: Influence of the Physiological Growth Stage

Certain *Pseudomonas* strains, particularly *Pseudomonas fluorescens* and *Pseudomonas putida* (fluorescent pseudomonas), are adapted to colonize plant roots and promote plant growth. This growth promotion may result (i) from the production of siderophores (high-affinity iron-chelators) and the resulting iron deprivation of deleterious rhizosphere microorganisms [25][45][49], (ii) from the production of antibiotics that are toxic to soil-borne plant pathogens [86], and (iii) by increasing the availability of phosphorus to the plants by solubilization of phosphorus [69].

Inoculation with *Pseudomonas* did not always result in significant yield increases, however. There are many reasons for this inconsistency, including non-suitability of the soil type [91], lack of iron availability and the nature of the soil microflora [50].

Part of our initial goal was to assess the influence of soil conditioning and incubation temperatures on the survival in soil of *Pseudomonas fluorescens* inocula. It has been shown that the temperature regimes did have an influence on the survival of the inocula, but that there existed, moreover, a non-trivial experimental variability between different inocula, which was high enough to justify a closer investigation into its sources [89]. We thus set out to differentiate between inocula in terms of some physiological criteria. The use of microcalorimetry with respect to such a differentiation will be evaluated and Section 5.3.2 will deal with the influence of the growth stage of *Pseudomonas fluorescens* cells (determined microcalorimetrically) on its survival in soil and in sterilized soil extracts after inoculation. It will be shown that a sizeable part of the experimental error that inoculation trials generally face can be attributed to differences in physiological growth stage between inocula; the use of microcalorimetry to select inocula of the same growth stage markedly reduces the survival variability and may thus increase the probability of success in inoculation applications. It suggests that differences in the physiological stage of the inoculum may be a key factor contributing to the variability in inoculation experiments. This source of variation could then, presumably, be eliminated by standardization of inocula in terms of some physiological criteria.

5.3.1 Microcalorimetry and bacterial growth

Four reasons prompted us to focus on microcalorimetry to study inoculum survival: (i) microcalorimetry measures the total heat production of a bacterial culture, which depends not only on the amount of microbial biomass present but also on the amount of substrate that they metabolize and the efficiency with which they metabolize the C source; hence, the total heat production is a measure for the activity of the microorganisms; (ii) by providing data to compute energy balances and efficiencies, microcalorimetry yields more information than the more conventional monitoring methods for bacterial cultures; (iii) microcalorimetry monitors bacterial cultures continuously and non-destructively; and (iv) microcalorimetric output is reproducible.

To test the hypothesis that microcalorimetry is an appropriate method to study the influence of physiological stages on inoculum survival, we first performed microcalorimetric studies of batch cultures to establish a correlation between heat production and bacterial numbers. The influence of the inoculum density on the heat evolution after inoculation was determined in a next step. Thirdly, we sampled from the culture medium at three different growth stages, using the microcalorimetric output as a control variable to determine the moment of sampling. We then assessed the influence of the growth stage at the moment of sampling on the lag time, growth rate and bacterial numbers (when growth had stabilized) of the subsequent growth of *Pseudomonas fluorescens* in the culture medium.

5.3.1.1 Materials and methods

Wild-type strains of *Pseudomonas fluorescens* 88W1 were grown in a shake culture in a one liter Erlenmeyer flask containing 600 mL King medium B at 28 °C. Bacterial numbers were determined every hour by dilution plating on King medium B spread plates.

The heat production rate (power output) of the cultures was continuously monitored with a twin-conduction type flow microcalorimeter (BioActivity Monitor LKB 2277), the description of which can be found in Monk and Wadsö [61][62]. The voltage signal was recorded by a two-channel potentiometric recorder (LKB 2210-002, 1000 mV). Calibration was performed with an external 300 μW current. The water bath temperature of the microcalorimeter was kept at 25 °C. The culture medium was transferred from the Erlenmeyer flask through the calorimeter by means of a peristaltic pump (LKB 2132 MicroPerpex) at a flow rate of 40 mL/h. After calorimetric measurement, the culture medium was discarded. The transfer time from the Erlenmeyer flask to the waste container was 2 minutes. To avoid contamination, all tubing was thoroughly washed before each run with (i) *NaClO* 10 % (20 min), (ii) sterile distilled water (5 min), (iii) ethanol 95 % (20 min), and (iv) sterile distilled water (5 min) at a flow rate of 40 mL/h.

Culture samples were taken at the outlet tubings of the calorimeter and inoculated into 2 separate batches of 250 mL Erlenmeyer flasks containing 100 mL King medium B. Sampling times were half-way on the upward slope of the power-time curve (5 hours after inoculation; early-exponential growth stage), at its maximum (8 hours after inoculation; exponential growth stage) and 5 hours after the peak (10 hours after inoculation; late-exponential growth stage) (see Figure 5.2). This resulted in 12 flask preparations (2 replicates × 2 batches × 3 sampling times). Through appropriate dilution, approximately the same inoculum size could be obtained for each individual batch. For each of the 12 cultures, bacterial numbers were monitored by hourly dilution plating of aliquots.

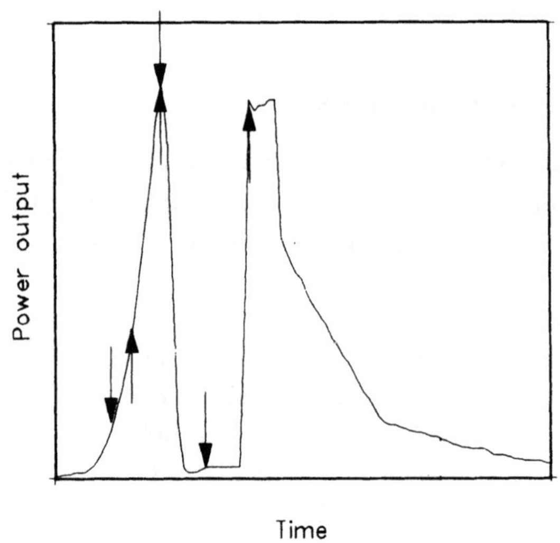

Figure 5.2: Sampling times of *Pseudomonas fluorescens* 88W1 inocula indicated on the power-time curve of the culture. *Downward arrows* (*from left to right*): early-exponential, exponential, and late-exponential growth stages (sampling as in this section); *Upward arrows* (*from left to right*): early-exponential, exponential, early-stationary stages (sampling as in Section 5.3.2).

5.3.1.2 Results and discussion

A strong correlation between heat production and bacterial numbers was borne out by a correlation coefficient of 0.98 (R^2). The average heat output for the formation of one single cell (the slope of the heat production-bacterial numbers curve) was 5.89 10^{-10} J/cell ($R^2 = 0.98$) (results not shown). This is lower than the value that Kimura and Takahashi [47] reported for soil microbes after addition of glucose to soil (5.53 10^{-8} J/cell). For pure cultures, Prat (1969) and Lamprecht *et al.* (1971) (both in [47]) obtained values of 7.50 10^{-9} and 1.46 10^{-8} J/cell. This suggests that *Pseudomonas fluorescens* develops higher energy efficiencies (lower energy losses per cell) on King medium B at 28 °C than the bacteria in the cited examples.

The inoculum size markedly affected the length of the heat production lag: a tenfold increase in the inoculum size reduced the lag time by 3.5 to 4 hours. Neither the specific growth rate nor the apparent growth rate, which is the slope of the heat production curve after logarithmic transformation, was significantly altered (results not shown). These results agree with observations by Shida *et al.* [78] who reported an effect of inoculum size on the duration of the lag time of bacterial growth in liquid culture, but not on the specific growth rate.

The similar influences of inoculum size on the lag times of both heat production and growth and the (equally similar) absence of influence on both the apparent and specific growth rates show, once again, the correlation between heat production and bacterial numbers.

Table 5.1 shows that the growth lag was markedly affected by the growth stage of the inoculum: the duration of the lag time of a culture inoculated with an inoculum in the late-exponential phase was 3 hours, compared with only 1.5 hours for an early-exponential and even 0.0 hours for an exponential-phase inoculum. The final number of bacteria was approximately 5 10^9 cfu/mL for all growth stages.

Since there was no difference between the composition of the preculture medium (from which the inoculum was taken) and the culture medium, it is not surprising that the lag time of a culture from an exponential-phase inoculum tended to zero, since no adjustment of enzymatic activity was necessary. Bacteria from the late-exponential phase, however, might have altered part of their enzymatic system to produce, for example, reserve materials due to nutritional or other imbalances of the culture medium [85]; they would then need a certain time to readjust their enzymatic activity to the fresh medium. The difference in the lag times for cultures inoculated with an in-

Growth stage	Early-exponential		Exponential		Late-exponential	
Replicate	1	2	1	2	1	2
Inoculum size	2.8 10^4	2.9 10^4	7.5 10^5	1.1 10^6	1.4 10^6	1.2 10^6
Lag time	1.5 (b)	1.5 (b)	0.0 (c)	0.0 (c)	3.0 (a)	3.0 (a)
Growth rate	0.40 (b)	0.39 (b)	0.42 (ab)	0.42 (ab)	0.46 (a)	0.55 (a)

Table 5.1: Specific growth rates [1/h] and lag times [h] for *Pseudomonas fluorescens* 88W1 inoculated (inoculum size in [cfu]) into fresh King medium B at three different growth stages. Data followed by the same letter (x) are not significantly different ($p \leq 0.05$).

oculum taken after 5 or 8 hours growth can be due to differences in growth stage and inoculum density; the inocula from the 5-hours-old precultures contained 2.8 10^4 cfu/mL, while the inoculum densities of the 8-hours-old precultures were about 10^6 cfu/mL.

The specific growth rate of a Pseudomonas culture from an early-exponential phase inoculum was significantly lower than that of a culture from a late-exponential phase inoculum; the specific growth rate of a culture inoculated with an exponential-stage inoculum lay in between (without differing significantly from either one).

5.3.1.3 Conclusion

The growth stage of the bacterial cultures at the moment of sampling had an important impact on the duration of the growth lag after inoculation into fresh culture medium. The two replicate batches yielded similar data, indicating that microcalorimetry can –as we hypothesized– be used as a control variable for reproducible sampling of bacterial cultures and that it is sufficiently sensitive to demonstrate inoculum differences hitherto unaccounted for.

5.3.2 Survival in soil

The previous section serves as a preamble to proceed with an assessment of the influence of the growth stage of *Pseudomonas fluorescens* inocula on their subsequent survival in soil. It will be shown that the hypothesis that the use of microcalorimetry can contribute to a reduction of the experimental variability in inoculation trials is correct.

5.3.2.1 Materials and methods

Soil samples were collected from the upper 30 cm of a cultivated loamy sand soil (Spodosol) in Meerle (Belgium) (organic C content: 3.1 %, pH($CaCl_2$, 10 mM): 5.3). Samples were sieved over a 4 mm sieve, stored at 5 °C in polyethene bags and conditioned at 15 °C for the last three weeks prior to the experiments. Soil moisture was 18 % at all times (water holding capacity 21.4 %). Reported soil weights refer to oven-dry weight unless mentioned otherwise.

Neomycin-resistant Tn5 insertion mutants of *Pseudomonas fluorescens* were grown at 28 °C in a 10 liter fermentor (Biolafitte) containing 6 liters synthetic malate medium (malate 5 g/L; NH_4Cl 1 g/L; KH_2PO_4 0.6 g/L; K_2HPO_4 2 g/L; $MgSO_4.7H_2O$ 0.2 g/L; $NaCl$ 0.1 g/L; $MnSO_4.H_2O$ 10 μg/L; $CaCl_2.H_2O$ 5 μg/L; $Fe.EDTA$ 23.375 μg/L; $Na_2MoO_4.H_2O$ 2 μg/L). pH was adjusted to 6.8 with KOH. A sufficient O_2 supply was ensured by aerating with CO_2-free air, the culture was stirred at 100 rpm and the temperature was kept at 28 °C. Precultures were grown aerobically in 100 mL of the same medium with mechanical stirring during 16 hours.

Heat production rate was measured as before (Section 5.3.1). For the experiment studying survival in soil, culture samples were taken halfway on the first upward slope of the microcalorimetric power-time curve (early-exponential growth stage), at the first peak (exponential growth stage) and at the second peak, coming approximately 5 hours after the first peak (early-stationary growth stage) (Figure 5.2). Each sample was centrifuged at 10,000 rpm (10 min, 4 °C). After decantation of the supernatant, the pellet was re-suspended in tap water and the procedure repeated once. The three inocula were diluted so as to obtain comparable inoculum densities and 500 μL aliquots were added to 110 mL plastic beakers containing 12.2 g of (wet) soil each (corresponding to 10 g oven-dry soil). For each growth stage, 60 beakers were inoculated. On days 1, 2, 3, 4, and 7 after inoculation and from then on every week for 7 weeks, four samples were taken from each treatment (destructive sampling) and shaken for 1 hour on a rotary shaker after addition of 95 mL TAE buffer (Trizma base 4.84 g/L, Na_2EDTA 744 mg/L, and $NaCla$ 7.5 g/L) and 10 g gravel. Then serial tenfold dilutions in PBS (phosphate buffer saline, pH 6.8) were plated in triplicate on King medium B with 200 μg neomycin/mL and 100 μg/mL for bacterial counts. The experiment was set up in triplicate.

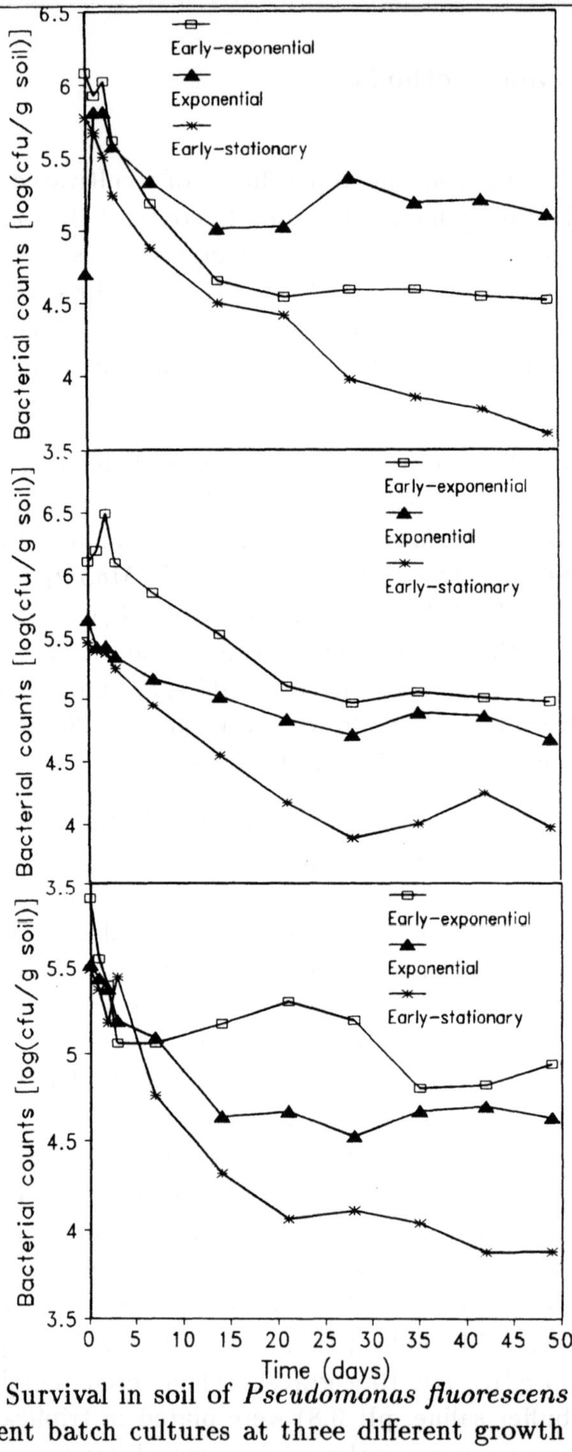

Figure 5.3: Survival in soil of *Pseudomonas fluorescens* 88W1 inocula from three different batch cultures at three different growth stages. *Top*: batch A; *Middle*: batch B; *Bottom*: batch C. Inoculation on day 0.

5.3.2.2 Results and discussion

Figure 5.3 shows the evolution of bacterial numbers in the soil for inocula from three different batch cultures (A, B, and C) from each of the three different growth stages. Bacterial numbers decreased exponentially after inoculation into the soil (except for an increase in bacterial numbers during the first day for batch A of the exponential growth stage and during the first 2 days for batch B of the early-exponential growth stage). The equation

$$\tilde{N}_t = \tilde{N}_e + (\tilde{N}_0 - \tilde{N}_e) \times (1 - R)^t \qquad (5.8)$$

with

$\tilde{N}_t$:	Logarithm of bacterial numbers at time t	[log(cfu/g soil)]
$\tilde{N}_e$:	Logarithm of bacterial numbers at stabilization level	[log(cfu/g soil)]
$\tilde{N}_0$:	Logarithm of initial bacterial numbers	[log(cfu/g soil)]
R :	Daily reduction factor	
t :	Time	[days]

produced a good fit (after omittance of the observations at inoculation day for the treatments with an increase in bacterial numbers right after inoculation). The resulting values of R, $\tilde{N}_e$, and $\tilde{N}_0 - \tilde{N}_e$, and of the variance accounted for by the regression procedure are given in Table 5.2. A high percentage of the variance (more than 99 %) was accounted for by the regression. Regardless of the batch, the same trends were obtained for all three different physiological stages, although the exact values differed sometimes.

Early-exponential and exponential growth stage inocula tended to survive better in soil. The stabilization levels ($\tilde{N}_e$) for the pseudomonas derived from a stationary-phase inoculum (values between 5.94 10^3 and 1.64 10^4 cfu per gram soil) were significantly lower ($p \leq 0.05$) than for the inocula from the earlier growth stages (values ranging from 3.29 10^4 to 1.55 10^5 cfu per gram soil). Even after 3 months, different stabilization levels were recorded for bacteria from different growth stages (data not shown). For the early-exponential growth stage, a mean value of 2.43 ($\pm$0.46) 10^4 cfu per g soil was obtained; for the exponential growth stage, the mean stabilization level was 3.63 ($\pm$0.71) 10^4 cfu per g soil, and it was significantly lower for early-stationary growth stage bacteria, namely 4.02 ($\pm$0.86) 10^3. There was no clear correlation between inoculum size and stabilization level.

The values for $\tilde{N}_0 - \tilde{N}_e$ were also significantly influenced by the growth stage of the inoculum ($p \leq 0.05$). Inocula from the exponential phase showed

Growth stage	Batch	Inoc. dens. [cfu/g soil]	R	$\tilde{N}_e$	$\tilde{N}_0 - \tilde{N}_e$
Early-exp.	A	1.2 10^6	0.1141 ± 0.0122 (b)	4.52 ± 0.42 (b)	1.56 (b)
Exponential	A	5.1 10^5	0.1727 ± 0.0454 (a)	5.19 ± 0.25 (a)	0.54 (a)
Early-stat.	A	5.9 10^5	0.0734 ± 0.0044 (c)	3.77 ± 0.17 (c)	1.84 (c)
Early-exp.	B	1.3 10^6	0.1094 ± 0.0212 (b)	5.14 ± 0.18 (a)	1.29 (b)
Exponential	B	4.4 10^5	0.1545 ± 0.0439 (a)	4.89 ± 0.18 (a)	0.70 (a)
Early-stat.	B	3.9 10^5	0.0850 ± 0.0191 (b)	4.22 ± 0.44 (b)	1.18 (b)
Early-exp.	C	8.1 10^5	0.1530 ± 0.0629 (a)	4.91 ± 0.30 (a)	1.00 (a)
Exponential	C	3.3 10^5	0.1275 ± 0.0283 (b)	4.64 ± 0.30 (a)	0.87 (a)
Early-stat.	C	3.1 10^5	0.0890 ± 0.0087 (c)	3.99 ± 0.25 (b)	1.50 (b)

Table 5.2: Survival in soil of *Pseudomonas fluorescens* 88W1 inocula at three different growth stages, fitted with the non-linear regression model (5.8). Data followed by the same letter (x) are not significantly different ($p \leq 0.05$).

a significantly lower value (between 0.54 and 0.87) than inocula from the other two growth stages (for 2 out of the 3 batches, this difference was significant). The lower value for exponential-phase inocula indicates better survival rates, since the survival rate is equal to

$$\begin{aligned} \frac{\textit{Number of bacteria at stabilization}}{\textit{Initial number of bacteria}} \times 100 &= \frac{N_e}{N_0} \times 100 \\ &= e^{(\ln N_e - \ln N_0)} \times 100 \\ &= e^{-(\tilde{N}_0 - \tilde{N}_e)} \times 100 \qquad (5.9) \end{aligned}$$

The early-exponential and the early-stationary phases produced a wider range of values for $\tilde{N}_0 - \tilde{N}_e$ (from 1.00 to 1.56 for the early-exponential and from 1.18 to 1.84 for the early-stationary phase). Values for the early-exponential phase were significantly lower ($p \leq 0.05$) than for the early-stationary phase inocula in 2 out of the 3 batches.

The lower daily reduction factors that were obtained for the stationary-phase inocula (0.089 to 0.073) may be due to the lower inoculum densities. The exponential-stage inoculum showed the highest daily reduction factors (between 0.127 and 0.173).

The results for the exponential growth stage of all three batches suggest that higher inoculation levels result in both higher stabilization levels and higher survival rates in soil (Table 5.2). The differences are small, however.

Yet, similar results were obtained by Postma *et al.* [72] who obtained significantly higher stabilization and survival levels at higher inoculum densities of *Rhizobium leguminosarum*, and by Postgate and Hunter [70], who reported that denser populations were killed less than sparse populations. Presumably, more cells will reach protective niches at higher inoculum densities [72]. However, there was no clear correlation between inoculum density and bacterial numbers at the end of the experiment for the other two growth stages, which corresponds to results reported by Crozat *et al.* [21] and by Corman *et al.* [20]; they found that the inoculum density did not affect the final population size of *Bradyrhizobium japonicum* strain in a variety of soils.

5.3.2.3 Conclusion

Bacterial survival rates in soil were markedly influenced by the physiological stage of the inocula. Moreover, the selection of inocula with equivalent microcalorimetric power outputs reduced the variability in survival rates after inoculation. These results confirm our hypothesis that microcalorimetry can contribute to a reduction of the experimental error in inoculation trials. These results were confirmed by similar data obtained for *Azospirillum brasilense* inocula [90]. Inoculation of *Pseudomonas fluorescens* of different physiological growth stages around an artificial root [60] resulted, for both immediate and delayed exudation, in higher numbers around the root of exponential growth stage bacteria (as compared to cells of later growth stages) [90].

At this point, it may be anticipated that *Pseudomonas fluorescens* will be shown (in Section 5.4) to produce citrate between both peaks of the power-time curve. The second peak of the power-time curve coincides more or less with the start of a period of (net) citrate consumption by the cells (which were grown on 5 g malate/L, as in this section) and with a second peak in O_2 consumption. Simultaneously, the growth rate decreased; this decrease in growth rate possibly counteracts the successful establishment of the bacteria in soil.

5.4 Microcalorimetry and Bacterial Growth Phenomena

A bacterial culture at a given physiological stage would show a constant rate of heat production per unit biomass (specific heat production rate) [38]. But varying heat productions during culture growth were also reported [84], suggesting changes in the metabolic activity of the bacteria. To understand the underlying mechanisms for this varying heat production rate, a thorough analysis of the energy flows is necessary. Therefore, other growth parameters (such as rate of biomass production, carbon consumption rate and ATP content) have to be monitored simultaneously.

The aim of the present study is to explore in some more detail the links between the heat production rate and a number of other growth parameters for *Pseudomonas fluorescens* and to relate this to the difference in survival ability of inocula sampled at different times on the power-time curve. Although we are not here in the position to examine all of the fundamental physiological growth processes that are involved in the phenomena observed in this or in the previous sections, the results reported here will show that heat production rate is strongly linked with many other growth parameters and that the combination of microcalorimetry with *conventional* analyses opens broad perspectives for the understanding of many metabolic phenomena, including the different survival rates in soils of inocula sampled at different physiological stages.

5.4.1 Materials and methods

A neomycin-resistant Tn5 insertion mutant of strain 88W1 of *Pseudomonas fluorescens* was grown in a 10 L fermentor (Biolafitte) filled with 6 L synthetic malate medium as described previously (Section 5.3.1). Every hour we measured: (i) bacterial numbers (dilution plating of appropriate dilutions on Kings medium B), (ii) cell dry weight (centrifugation of bacterial suspension at 10,000 rpm for 10 min, resuspension of pellet with tap water, centrifugation of washed pellet at 10,000 rpm for 10 min, lyophilization of the pellet), (iii) CO_2 concentration in the headspace by gas chromatography, (iv) C lost as CO_2 (capture of exhaust CO_2 in 1 N $NaOH$ traps, titration with HCl), (v) the amount of malate and citrate in the medium (by High Performance Ion Chromatography Exclusion, using a Dionex Qic chromatograph with a CMMS-1 suppressor and an HPICE-AS1 column. HCl (0.01 N) was used as the eluent at a flow of 0.8 mL/min. The regenerant was a 0.05 N solution of tetrabutyl ammonium hydroxide (Janssen Chimica) at a flow rate of 2 mL/min (see also [76]), (vi) cell protein concentration [μg/g] [58]; using serum bovine albumin as standard), (vii) total C content of the culture [g C/L] (digestion of 2 mL of culture sample with 10 mL of a digestion mixture following the procedure of Amato [3]), (viii) ATP [nmol/mg] [92][93], (ix) O_2 consumption rate [mmole/(g h)] (measured polarographically with a YSI-Biological Oxygen Monitor). Heat production was measured with a twin-conduction type flow microcalorimeter as described previously.

5.4.2 Results and discussion

5.4.2.1 Growth parameters

Results from the measurements of power and other growth parameters of *Pseudomonas fluorescens* are shown in Figure 5.4. The power-time curve was double-peaked. The first maximum (409 μW/mL) was attained after 5 h of growth; a subsequent sharp decline and a very low power output of 9 μW/mL for 7 h was followed by a second peak (426 μW/mL). From then on, the power decreased continuously.

The heat measurements, in combination with other analyses (discussed below), indicated a continuous change in the metabolism of *Pseudomonas fluorescens.* Pearson correlation coefficients between heat production and the other growth parameters (except for O_2 consumption and cell ATP content) were significant ($p < 0.05$), as shown before (Vandenhove *et al.* [89]). Markedly different growth phases could be determined according to power output: up to the first peak, between both peaks, from the second peak on until all malate was metabolized and, finally, from then on until the end of the experiment.

Bacterial numbers, cell dry weight, protein concentration, and malate consumption increased slowly until the first peak of the power-time curve and continued to increase at an accelerated rate between both peaks (Figure 5.4). From the second peak on, the parameters remained more or less constant, apart from bacterial numbers, which showed a distinct decline. Differences in protein content with growth rate –if growth rate increases, protein content decreases– as found by Stouthamer and Bettenhausen [83] were not found for *Pseudomonas fluorescens.* The mean protein concentration was 50.3 % on dry matter basis (data not shown).

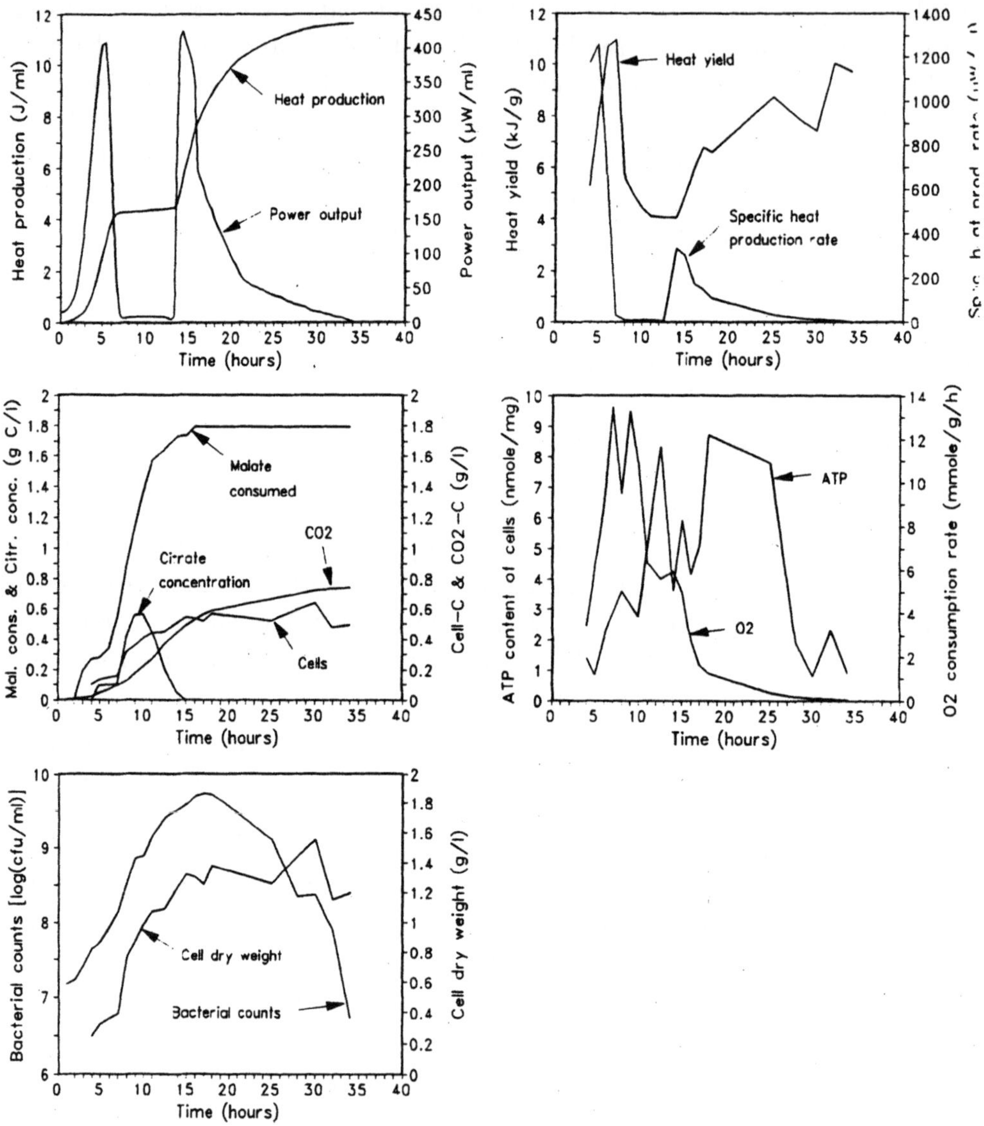

Figure 5.4: Power output and other growth parameters for growth of *Pseudomonas fluorescens* 88W1 on synthetic malate medium containing 10 g malate per L.

Only about half of the malate present in the medium at inoculation was consumed: the *Pseudomonas fluorescens* strain used can only metabolize L-malate, as was found by preliminary testing procedures (API 50 OA, results not shown), whereas the D form (approximately racemic mixture) was used in the experiment. Hence, all consumable malate was metabolized after 16 h (Figure 5.4), when the power-time curve showed a sharp decrease.

During growth of *Pseudomonas fluorescens* citrate is produced. The highest concentration of citrate (1.636 g/L) was found after 10 h (Figure 5.4). At that moment, 41 % of the malate carbon was converted into citrate carbon. All citrate had disappeared from the medium after 15 h of growth, just after the second maximum of the power-time curve.

The evolutions of the specific growth rates on cell-dry-weight basis (μ_{CDW}) and on bacterial-numbers basis ($\mu_{\#}$) are shown in Table 5.3; the evolution of the heat yield is shown in Figure 5.4. The heat yield (ΔQ_X) is obtained by normalization of the heat production rate (dQ/dt) to the rate of biomass formation (dX/dt), or $\Delta Q_X = dQ/dX$, which is independent of the amount of biomass and of the growth rate since $dX/dt = \mu X$. Between both peaks of the power-time curve, when citrate was formed, μ_{CDW} was 0.297 1/h and heat yield averaged around -9 kJ/g. After the second peak of the power-time curve, when all L-malate and citrate was consumed, growth on dry weight basis almost ceased. The values that we obtained are similar to values reported by Blomberg *et al.* [14]. They obtained a specific growth rate of 0.29 1/h during the *respirofermentative* phase for *Saccharomyces cerevisiae* grown on a synthetic glucose medium. During this period, the average heat yield was -9.6 kJ per g biomass formed and considerable amounts of glucose were fermented to ethanol. When all glucose had been metabolized, cells started to respire ethanol; the corresponding heat yield was -57 kJ per g biomass and μ_{CDW} was 0.03 1/h. *Pseudomonas fluorescens* produced citrate between both peaks of the power-time curve (Figure 5.4), and although this phase cannot be called *respirofermentative* but should be regarded as exhibiting overflow metabolism, the agreement is striking.

The higher values of $\mu_{\#}$ as compared to μ_{CDW} (Table 5.3) indicate that cell size decreased for *Pseudomonas fluorescens* until malate depletion. For *Pseudomonas fluorescens*, the maximum individual cell dry weight is reached near the first peak of the power-time curve (between 5 and 6 fg/cell). The decline stabilizes between 0.2 and 0.3 fg/cell at the plateau (second peak) of the power-time curve. The reduction in volume of the bacterial cells from the first peak of the power-time curve on was also verified microscopically (qualitatively). The reason for this reduction remains unclear since

	Growth period *Before first peak*	 *Between both peaks of power-time curve*	 *After second peak*
μ_{CDW}	0.047 ± 0.013	0.297 ± 0.009	-0.004 ± 0.000 (18h-25h)
$\mu_{\#}$	0.362 ± 0.026	0.460 ± 0.060	0.270 ± 0.054 (14h-18h) -0.390 ± 0.075 (18h-34h)

Table 5.3: Specific growth rates on cell-dry-weight basis μ_{CDW} [1/h] and on bacterial-numbers basis $\mu_{\#}$ [1/h] for *Pseudomonas fluorescens* grown on synthetic malate media.

malate was still present in the medium. Maybe the concentration of malate in the medium became too low to permit sufficient diffusion from the culture medium into the cells. The individual cell weight increased again after malate exhaustion; the remaining bacteria probably fed on the dead ones, a phenomenon that Harrison (1960, in [70]) referred to as *cannibalism*.

It is not well known why *Pseudomonas fluorescens* has a period of accelerated growth and very low heat production rate between both peaks. The reduced heat production rate can partly be explained by the formation of citrate. Indeed, full respiration of malate would yield:

$$C_4H_6O_5 + 3\ O_2 \rightarrow 4\ CO_2 + 3\ H_2O \tag{5.10}$$

resulting in a heat production of 1328 kJ per mole malate. Incomplete oxidation of malate to citrate, as in

$$2\ C_4H_6O_5 + 1.5\ O_2 \rightarrow 2\ CO_2 + 2\ H_2O + C_6H_8O_7 \tag{5.11}$$

releases 350 kJ per mole of malate, which is only one fourth of the heat production per mole of malate completely oxidized. This can only partly explain the low heat production rate, however.

It is not clear why $\mu_{\#}$ was higher than μ_{CDW}. This phenomenon often occurs when substrate becomes limiting: the decrease in cell size results in a higher surface-to-volume ratio, facilitating substrate uptake; in this case, however, the medium contained high amounts of carbon, suggesting that malate transport was not limiting. Nevertheless, it does explain why the specific heat production rate (the rate of heat production per unit biomass) remained high until the first peak of the power-time curve (Figure 5.4): under constant environmental conditions, a steady state metabolism results in a constant specific heat production rate, as in continuous cultures [40][45]; cell division entails a high metabolic activity [48], resulting in higher specific heat production and O_2 consumption rates.

The ATP content of the *Pseudomonas fluorescens* cells increased between both peaks of the power-time curve (Figure 5.4). The ATP values obtained are well in line with the values presented by Karl [46] which ranged from 2.9 to 13.8 nmole ATP per mg cell for cells grown aerobically. Simultaneously, O_2 consumption increased and citrate was produced (Figure 5.4). Neijssel and Tempest [67] reported an increased O_2 consumption rate accompanied by an increased ATP content of the cells during overflow metabolism of *Klebsiella aerogenes*; variations in the ATP pool during exponential growth have also been mentioned by Chapman *et al.* (1971), Holms *et al.* (1972) and Gadkari and Stolp (1975) (all in [51]). When all malate was metabolized (after 16 h), ATP increased abruptly, after which the concentration dropped again. According to Forrest [33], the decrease in growth rate at the end of exponential growth means that ATP is no longer required to drive growth at a high rate. Growth may then become energetically uncoupled and the ATP concentration in the organism may increase [32]. Since no energy source (malate or citrate) was available toward the end of the experiment but energy was still needed for maintenance, the ATP pool decreased again.

It is widely accepted that the heat production per mole of oxygen consumption by a microbial culture in aerobic conditions (the calorimetric-respirometric ratio or CRR) is insensitive to biomass yield and constant at about 440 to 470 kJ/mole [13][19][35][74]. The anabolic enthalpy change mostly approximates zero for heterotrophic growth, especially if the degree of reduction of the substrate is close to that of sugars; this is in line with our calculations below. Oxygen consumption, like enthalpy change, originates mainly from catabolism. Hence, the biomass yield can be anticipated not to interfere significantly with the CRR. However, if the catabolic pathway is not completely oxidative, such as in a phase of fermentative or overflow metabolism, oxygen consumption is low relative to heat production. The CRR will then be more (sometimes even much more) negative than -470 kJ per mole oxygen. In contrast, ATP accumulation will cause a CRR less negative than the *normal* -440 to -470 kJ per mole oxygen, since these values assume a complete ATP turnover and resulting constant ATP and ADP concentrations: because if energy is not fully turned over but conserved in ATP production and accumulation, less heat will be produced.

CRR was, in our experiments, greatly affected by the growth phase of *Pseudomonas fluorescens*: during growth of *Pseudomonas fluorescens* in our batch experiments, CRRs ranged from -887 ± 36 kJ/mole until the first peak to -3 ± 0 kJ/mole between both peaks, and from the second peak on, -254 ± 14 kJ per mole O_2 consumed were dissipated as heat (this averages about -127 kJ/mole over the whole experiment). Brettel *et al.* [17][18] reported -850 kJ per mole oxygen for the growth of yeasts on glucose when ethanol was produced. Blomberg *et al.* [14] obtained CRRs ranging from -532 to -4717 kJ per mole oxygen for growth of *Saccharomyces cerevisiae* on glucose medium. Gnaiger and Kemp [35] obtained values as low as -800 kJ per mole oxygen for growth of aerobic mammalian cells.

5.4.3 Conclusions

At this point of our discussion, it seems appropriate to present a more comprehensive view on our results. Heat production was shown to be significantly correlated with most other growth parameters, corroborating the results of previous experiments [89]; the heat production rate reflected changes in metabolic activity. The heat yield, a parameter normalized for biomass and growth rate, still varied during growth, indicating the presence of metabolic shifts. Before the first peak, growth was slow and only little malate was metabolized, but O_2 uptake was increasing. Between both peaks, power output was low notwithstanding the increased O_2 consumption and the high growth rate; this suggests a highly energy-efficient metabolism and perhaps a strong coupling of anabolism with catabolism. In the middle of the *trough*, 12 percent of the initially available energy was conserved in biomass (*Pseudomonas fluorescens*) and only 6 percent was dissipated as heat. In comparison, as much as 10 percent of the energy input was dissipated as heat after 18 h of growth and 22 percent was conserved in the biomass.

It is striking that the very low heat production rate for *Pseudomonas fluorescens* coincided with the production of citrate. Most probably, citrate production should be regarded as resulting from overflow metabolism, as citrate production was proportional to the initial malate concentration in the medium. The production of citrate can partly explain the lower heat production rate: the oxidation of malate to citrate yields only 26 percent of the heat of combustion of malate. Malate depletion caused the heat production rate to drop markedly.

It is thus clear that the shape of the power-time curve is related to the type of metabolic activity and can be related to different physiological states of the bacterial culture. Bacteria sampled from different phases of the power-time curve and subsequently inoculated into soil or soil extracts will, therefore, show differences in survival ability. The results presented here may thus contribute to a clarification of many observations discussed in previous articles [89][91] were we found that, for both *Pseudomonas fluorescens* and *Azospirillum brasilense*, bacteria of the early-exponential and exponential growth stage (sampled at the slope and the first peak of the power-time curve) survived better in soil and in sterilized soil extracts than did bacteria from the early stationary and stationary growth phases (sampled at the second peak and after 24 h growth).

Reasons for a better survival of *Pseudomonas fluorescens* cells sampled at the slope and at the first peak of the power-time curve, may be the following, given the preceding results. First, cells at the slope and at the first peak of the power-time curve are metabollically very active as shown by the high heat yield and O_2 consumption rate (Figure 5.4). Neijssel and Tempest [67] found that metabolically very active cells of different species of bacteria survived better in soil. Second, the maximum individual cell dry weight is reached near the first peak of the power-time curve. Postma *et al.* [72] found a net increase in cell number upon inoculation of large (2.42 μm) *rhizobial* cells, accompanied by a reduction in cell length (1.74 μm). After inoculation of starved bacteria, which were already shorter (1.65 μm) no such increase was detected. A decrease in cell length after introduction of well-fed bacteria into soil was also reported by Crozat *et al.* [21] and by Postma *et al.* [71]. A rapid reduction in size by fragmentation may play a role in the survival mechanisms of bacteria in nutrient-poor conditions [48][64]. Third, bacteria sampled at the second peak of the power-time curve were almost deprived from a carbon source and may have started to adapt themselves to starvation conditions. According to Postma *et al.* [72] starved bacteria survive less well than bacteria taken from affluent growth conditions.

5.5 Bibliography

1. M.J. Acea and M. Alexander 1988. Growth and survival of bacteria introduced into carbon amended soil. *Soil Biology and Biochemistry*, 20, 703-709

2. M.J. Acea, C.R. Moore and M. Alexander 1988. Survival and growth of bacteria introduced into soil. *Soil Biology and Biochemistry*, 20, 509-515

3. M. Amato 1983. Determination of carbon ^{12}C and ^{14}C in plant and soil. *Soil Biology and Biochemistry*, 15, 611-612

4. P.A.H.M. Bakker, J.G. Lamers, A.W. Bakker, J.D. Marugg, P.J. Weisbeek and B. Schippers 1986. The role of siderophores in potato tuber yield increase by *Pseudomonas putida* in a short rotation of potato. *Netherlands Journal of Plant Pathology*, 92, 249-256

5. P.A.H.M. Bakker, A.W. Bakker, J.D. Marugg, P.J. Weisbeek and B. Schippers 1987. Bioassay for studying the role of siderophores in potato growth stimulation by *Pseudomonas* spp in short potato rotations. *Soil Biology and Biochemistry*, 19, 443-449

6. E.H. Battley 1960. Growth reaction equations for *Saccharomyces cerevisiae. Physiologia Plantarum*, 13, 192-203

7. E.H. Battley 1960. Enthalpy changes accompanying the growth of *Saccharomyces cerevisiae* (Hansen). *Physiologia Plantarum*, 13, 628-640

8. J.O. Becker and J.R. Cook 1980. Role of siderophores in suppression of *Pythium* species and production of increased growth response of wheat by fluorescent *Pseudomonas. Phytopathology*, 78, 778-782

9. A.E. Beezer, K.A. Bettelheim, R.D. Newell and J. Stevens 1974. The diagnosis of bacteriuria by flow microcalorimetry. *Science Tools*, 21, 13-15

10. A.E. Beezer and Z. Chowdhry 1980. Microcalorimetrical investigation of drugs. In: A.E. Beezer (Ed.), *Biological Microcalorimetry*, 195-246, Academic Press, London

11. J.P. Belaich, J.C. Senez and M. Murgier 1968. Microcalorimetric study of glucose permeation in microbial cells. *Journal of Bacteriology*, 95, 1750-1757

12. J.P. Belaich 1980. Growth and metabolism in bacteria. In: A.E. Beezer (Ed.), *Biological Microcalorimetry*, 1-42, Academic Press, London

13. B. Birou, I.W. Marison and U. von Stockar 1987. Calorimetric investigation of aerobic fermentations. *Biotechnology and Bioengineering*, 30, 650-660

14. A. Blomberg, C. Larsson and L. Gustafsson 1988. Microcalorimetric monitoring of growth of *Saccharomyces cerevisiae*: Osmotolerance in relation to physiological state. *Journal of Bacteriology*, 170, 4562-4568

15. R. Brettel, I. Lamprecht and B. Schaarschmidt 1980. Microbial investigation of the metabolism of yeasts: VII. Flow microcalorimetry of aerobic batch cultures. *Radiation and Environmental Biophysics*, 18, 301-309

16. R. Brettel, I. Lamprecht and B. Schaarschmidt 1981. Microbial investigation of the metabolism of yeasts. Growth in chemostat cultures on glucose. *European Journal of Applied Microbiology and Biotechnology*, 11, 205-211

17. R. Brettel, I. Lamprecht and B. Schaarschmidt 1981. Microcalorimetric investigations of the metabolism of yeasts: Growth in batch and chemostat cultures on ethanol medium. *European Journal of Applied Microbiology and Biotechnology*, 11, 212-215

18. T.J. Burr and A. Caesar 1984. Beneficial plant bacteria. *CRC Critical Revue on Plant Science*, 2, 1-20

19. C.L. Cooney, D.I. Wang and R.I. Mateles 1968. Measurement of heat evolution and correlation with oxygen consumption during microbial growth. *Biotechnology and Bioengineering*, 11, 269-281

20. A. Corman, Y. Crozat and J.C. Cleyet-Marel 1987. Modelling of survival kinetics of some *Bradyrhizobium japonicum* strains in soil. *Biology and Fertility of Soils*, 4, 79-84

21. Y. Crozat, J.C. Cleyet-Marel and A. Corman 1987. Use of the fluorescent antibody technique to characterize equilibrium survival concentrations of *Bradyrhizobium japonicum* strains in soil. *Biology and Fertility of Soils*, 4, 85-90,

22. K.G. Davies and R. Whitbread 1989. Factors affecting the colonization of a root system by fluorescent pseudomonas: The effect of water, temperature and soil microflora. *Plant and Soil*, 116, 247-256

23. Z. Dermoun and J.P. Belaich 1979. Microcalorimetric study of *Escherichia coli* aerobic growth: Kinetic and experimental enthalpy associated with growth on succinic acid. *Journal of Bacteriology*, 140, 377-380

24. M. Dupler and R. Baker 1984. Survival of *Pseudomonas putida*, a biological control agent, in soil. *Phytopathology*, 74, 195-200

25. Y. Elad and R. Baker 1985. The role of competition for iron and carbon in suppression of chlamydospore germination of *Fusarium* spp. by *Pseudomonas* spp. *Phytopathology*, 75, 1053-1056

26. Y. Elad, I. Chet and R. Baker 1987. Increased growth response of plants induced by rhizobacteria antagonistic to soilborne pathogenic fungi. *Plant and Soil*, 98, 325-330

27. L.E. Erickson, I.G. Minkevich and V.H. Eroshin 1978. Application of mass and energy balance. *Biotechnology and Bioengineering*, 20, 1595-1621

28. R. Eriksson and I. Wadsö 1971. Design and testing of a flow microcalorimeter for studies of aerobic bacterial growth. In: *Proceedings 1st Biophysics Congress*, Vol. IV, 319-327, Medical Academy, Vienna

29. R. Eriksson and T. Holme 1973. The use of microcalorimetry as an analytical tool for microbial processes. *Biotechnology and Bioengineering Symposia*, 4, 581-590

30. A. Fiechter, G.F. Fuhrmann and O. Käppeli 1981. Regulation of glucose metabolism in growing yeast cells. *Advances in Microbiological Physiology*, 22, 123-183

31. W.W. Forrest, D.J. Walker and M.F. Hopgood 1961. Enthalpy changes associated with the lactic fermentation of glucose. *Journal of Bacteriology*, 82, 685-690

32. W.W. Forrest and D.J. Walker 1965. Synthesis of reserve materials for endogenous metabolism in *Streptococcus faecalis*. *Journal of Bacteriology*, 89, 1448-1452

33. W.W. Forrest 1972. Microcalorimetry. In: J.R. Morris and D.W. Ribbons (Eds.), *Methods in Microbiology*, Vol. 6B, 285-318, Academic Press, London

34. F.P. Geels and B. Schippers 1983. Reduction of yield depression in high frequency potato cropping soil after seed tuber treatments with antagonistic fluorescent *Pseudomonas* spp. *Phytopathology*, 108, 207-214

35. E. Gnaiger and R.B. Kemp 1990. Anaerobic metabolism in mammalian cells: Information from the ratio of calorimetric heat flux and respirometric oxygen flux. *Biotechnology and Bioengineering*, 1016, 328-332

36. K. Gustafsson and L. Gustafsson 1983. Heat changes, respiratory activities, ATP pools and metabolic potentials in natural sediment samples treated with a quaternary amine. *Oikos*, 4, 64-72

37. L. Gustafsson 1987. Microcalorimetry as a tool in microbiology and microbial ecology. In: M.A. Sleigh (Ed.), *Microbes in the Sea*, 167-181, Ellis Horwood Ltd., Chichester

38. L. Gustafsson 1991. Microbiological microcalorimetry. *Thermochimica Acta*, 193, 145-171

39. Y. Ishikawa, Y. Nonoyama and M. Shoda 1981. Microcalorimetric study of aerobic growth of *Escherichia coli* in batch culture. *Biotechnology and Bioengineering*, 23, 2825-2836

40. Y. Ishikawa and M. Shoda 1983. Calorimetric analysis of *Escherichia coli* in continuous culture. *Biotechnology and Bioengineering*, 25, 1817-1827

41. A. Iswandi, P. Bossier, J. Vandenabeele and W. Verstraete 1987. Relation between soil microbial activity and the effect of seed inoculation with the rhizopseudomona strain 7NSK2 on plant growth. *Biology and Fertility of Soils*, 3, 147-151

42. A. Iswandi, P. Bossier, J. Vandenabeelc and W. Verstraete 1987. Effect of seed inoculation with the rhizopseudomona strain 7NSKS on the root microbiota of maize (*Zea mays*) and barley (*Hordeum vulgare*). *Biology and Fertility of Soils*, 3, 153-158

43. S. Itoh and K. Takahashi 1984. Calorimetric studies of microbial growth: Kinetic analysis of growth thermograms observed for bakery yeast at various temperatures. *Agricultural Biological Chemistry*, 48, 271-275

44. A. Johansson, C.E. Nord and T. Wadström 1975. Comparative evaluation of bacterial growth patterns based on microcalorimetric data. *Science Tools*, 22, 19-21

45. C. Jolicoeur and A. Beaubien 1986. Microcalorimetric studies of microbiological methods and inhibition bases for in vitro toxicity evaluation. In: G. Bitton and B.J. Dutka (Eds.), *Toxicity testing using microorganisms*, Vol. I, 115-151, CRC Press, Florida

46. D.M. Karl 1980. Cellular nucleotide measurement and application in microbial ecology. *Microbiological Reviews*, 44(4), 739-796

47. J. Kimura and K. Takahashi 1985. Calorimetric studies of soil microbes: Quantitative relation between heat evolution during microbial degradation of glucose and changes in microbial activity in soil. *Journal of General Microbiology*, 131, 3083-3089

48. S. Kjelleberg, B.A. Humphrey and K.C. Marshall 1983. Initial phases of starvation and activity of bacteria at surfaces. *Applied and Environmental Microbiology*, 46(5), 978-984

49. J.W. Kloepper, J. Luong, M. Teintze and M.N. Schroth 1980. Enhanced plant growth by siderophores produced by plant growth-promoting rhizobacteria. *Nature*, 286, 885-886

50. J.W. Kloepper, R. Lifshitz and M.N. Schrot 1988. *Pseudomonas* inoculants to benefit plant production. *ISI Atlas of Science: Animal and Plant Science*, 86, 60-64

51. C.J. Knowles 1980. Microbial metabolic regulation by adenine nucleotide pools. In: B.A. Haddock and W.A. Hamilton (Eds.), *Microbial Energetics*, 27th Symposium of the Society for General Microbiology, 241-283, Academic Press, London

52. J. Kubát, B. Novák and M. Ružička 1980. Modelling the growth of bacteria during glucose decomposition in soil under different temperatures. *Plant and Soil*, 55, 77-84

53. B.S. Kundu and A.C. Gaur 1979. Rice response to inoculation with N_2-fixing and P-solubilizing microorganisms. *Plant and Soil*, 79, 227-234

54. I. Lamprecht 1980. Growth and metabolism in yeasts. In: A.E. Beezer (Ed.), *Biological Microcalorimetry*, 43-112, Academic Press, London

55. C. Larsson, A. Blomberg and L. Gustafsson 1991. Use of microcalorimetric monitoring in establishing continuous energy balances and in continuous determination of substrate and product concentrations of batch-grown *Saccharomyces cerevisiae*. *Biotechnology and Bioengineering*, 38, 447-458

56. K. Ljungholm, B. Norén and I. Wadsö 1979. Use of microcalorimetry for the characterization of microbial activity in soil. *Oikos*, 33, 15-23

57. D. López, M. Veñas, J.G. Lorén and J. Bermúdez 1987. Analysis of microcalorimetric curves for bacterial identification. *Canadian Journal of Microbiology*, 33, 6-11

58. O.L. Lowry, N.J. Rosebrough, A.L. Farr and R.J. Randall 1951. Protein measurement with the folin phenol reagent. *Journal of Biological Chemistry*, 193, 265-275

59. J.H. Luong and B. Volesky 1983. Heat evolution during the microbial process: Estimation, measurements and applications. *Advances in Biochemical Engineering/Biotechnology*, 28, 1-40

60. R. Martens 1982. Apparatus to study the quantitative relationship between root exudates and microbial populations in the rhizosphere. *Soil Biology and Biochemistry*, 14, 316-317

61. P. Monk and I. Wadsö 1968. A flow microreaction calorimeter. *Acta Chemica Scandinavica*, 22, 1842-1852

62. P. Monk and I. Wadsö 1969. Flow microcalorimetry as an analytical tool in biochemical and related areas. *Acta Chemica Scandinavica*, 23, 29-36

63. J.K. Moore, H.D. Braymer and A.D. Larson 1983. Isolation of a *Pseudomona* spp. which utilizes the phosphonate herbicide glyphosate. *Applied and Environmental Microbiology*, 46, 316-320

64. R.Y. Morita 1988. Bioavailability of energy and its relationship to growth and starvation in nature. *Canadian Journal of Microbiology*, 34, 436-440

65. U. Mortensen, B. Norén and I. Wadsö 1973. Microcalorimetry in the study of the activity of microorganisms. *Bulletin of the Ecological Research Community (Stockholm)*, 17, 189-197

66. M. Murgier and J.P. Belaich 1971. Microcalorimetric determination of the affinity of *Saccharomyces cerevisiae* for some carbohydrate growth substrates. *Journal of Bacteriology*, 105, 573-579

67. O.M. Neijsel and D.W. Tempest 1975. The regulation of carbohydrate metabolism in *Klebsiella aerogenes* NCTC 418 organisms, growing in chemostat culture. *Archives of Microbiology*, 106, 251-258

68. R.D. Newell 1980. The identification and characterization of microorganisms by microcalorimetry. In: A.E.Beezer (Ed.), *Biological Microcalorimetry*, 187-194, Academic Press, London

69. A. Ortuño, A. Hernansáez, J. Noguera, V. Morales and P. Armero 1978. Acción solubilizadora de fosforo por *Aspergillus niger y Pseudomonas fluorescens. Anales de Edafología y Agrobiología*, 37, 1311-1361

70. J.R. Postgate and J.R. Hunter 1962. The survival of starved bacteria. *Journal of General Microbiology*, 2, 233-263

71. J. Postma, J.D. van Elsas, J.M. Govaert and J.A. van Veen 1988. The dynamics of *Rhizobium leguminosarum* biovar *trifolii* introduced into soil as determined by immunofluorescence and selective plating technique. *FEMS-Microbiology Ecology*, 53, 251-260

72. J. Postma, C.H. Hok-A-Hin and J.H. Oude Vossehaar 1990. Influence of the inoculum density on the growth and survival of *Rhizobium leguminosarum* biovar *trifolii* introduced into sterile and non-sterile loamy sand and silt loam. *FEMS Microbiology Ecology*, 73, 49-58

73. G.J. Prochazka, W.J. Payne and W.R. Mayberry 1970. Calorific content of certain bacteria and fungi. *Journal of Bacteriology*, 104, 646-649

74. T.W. Randolph, I.W. Marison, D.E. Martens and U. von Stockar 1990. Calorimetric control of fed-batch fermentation. *Biotechnology and Bioengineering*, 36, 678-684

75. H.E. Reiling and H. Zuber 1983. Heat production and energy during growth of a prototrophic denitrifying strain of *Bacillus stearothermophilus. Archives of Microbiology*, 136, 243-253

76. R.D. Rocklin, R.W. Slingsby and C.A. Pohl 1986. Separation and detection of carboxylic acids by ion chromatography. *Journal of Liquid Chromatography*, 9, 757-775

77. M.O. Samuelsson, P. Cadez and L. Gustafsson 1988. Heat production by the denitrifying bacterium *Pseudomonas fluorescens* and the dissimilatory ammonium-producing bacterium *Pseudomonas putrefaciens* during anaerobic growth with nitrate as the electron acceptor. *Applied and Environmental Microbiology*, 54, 2220-2225

78. T. Shida, K. Komagata and I. Mitsugi 1975. Reduction of lagtime in bacterial growth: Effect of inoculum size and nutrients. *Journal of General and Applied Microbiology*, 21, 75-86

79. J.L. Sinclair and M. Alexander 1984. Role of resistance to starvation in bacterial survival in sewage and lake water. *Applied and Environmental Microbiology*, 48, 410-415

80. G.P. Sparling 1981. Heat output of the soil biomass. *Soil Biology and Biochemistry*, 13, 373-376

81. G.P. Sparling 1981. Microcalorimetry and other methods to assess biomass and activity in soil. *Soil Biology and Biochemistry*, 13, 93-98

82. L.H. Stevenson and M.D. Socolofsky 1966. Cyst formation and poly-β-hydroxybutyric acid accumulation in *Azotobacter*. *Journal of Bacteriology*, 19, 304-310

83. A.H. Stouthamer and C. Bettenhausen 1973. Utilization of energy for growth and maintenance in continuous and batch cultures of microorganisms. *Biochimica Biophysica Acta*, 301, 53-70

84. P.J. Stoward 1962. Thermodynamics of microbial growth. *Nature*, 194, 977-978

85. S. Tal and Y. Okon 1985. Production of the reserve poly-β-hydroxybutyrate and its function in *Azospirillum brasilense* cd. *Canadian Journal of Microbiology*, 31, 608-613

86. L.S. Thomashow and D.M. Weller 1988. Role of the phenazine antibiotic from *Pseudomonas fluorescens* in biological control of *Gaeumannomyces graminis* var. *tritici*. *Journal of Bacteriology*, 170, 3499-3508

87. A.S. Traore, C.H. Hatchikian, J.P. Belaich and J. Le Gall 1981. Microcalorimetric studies of the growth of sulfate-reducing bacteria: Energetics of *Desulfovibrio vulgaris* growth. *Journal of Bacteriology*, 145, 191-199

88. A.S. Traore, C.H. Hatchikian, J. Le Gall and J.P. Belaich 1982. Microcalorimetric studies of the growth of sulfate-reducing bacteria: Comparison of the growth parameters of some *Desulfovibrio* species. *Journal of Bacteriology*, 149, 606-611

89. H. Vandenhove, R. Merckx, H. Wilmots and K. Vlassak 1991. Survival of *Pseudomonas fluorescens* inocula of different physiological growth stages in soil. *Soil Biology and Biochemistry*, 23, 1133-1142

90. H. Vandenhove 1992. *Microcalorimetric Characterization of Bacterial Inocula.* PhD thesis, Katholieke Universiteit Leuven, Belgium

91. H. Vandenhove, R. Merckx, M. Van Steenbergen and K. Vlassak 1993. Microcalorimetric characterization, physiological stages and survival ability of *Azospirillum brasilense*. *Soil Biology and Biochemistry*, 25, 513-519

92. J.D. Van Elsas, A.F. Dijkstra, J.M. Govaert and J.A. van Veen 1986. Survival of *Pseudomonas fluorescens* and *Bacillus subtilis* introduced into two soils of different textures in field microplots. *FEMS Microbiology Ecology*, 48, 410-415

93. W. Verstraete, F. van de Werf, F. Kucnerowicz, F. Ilaiwi, L.M.J. Verstraeten and K. Vlassak 1983. Specific measurement of soil microbial ATP. *Soil Biology and Biochemistry*, 15, 391-396

94. L.M.J. Verstraeten, K. de Coninck, K. Vlassak, W. Verstraete, H. van de Werf and M. Ilaiwi 1983. ATP content of soils estimated by two contrasting extraction methods. *Soil Biology and Biochemistry*, 15, 397-402

Part II

Model based Control

Chapter 6

On-line Data Acquisition

P.A. WILLEMS and J.P. OTTOY
Universiteit Gent
Vakgroep Toegepaste Wiskunde, Biometrie en Procesregeling
Coupure Links 653 B-9000 Gent (BELGIUM)

ABSTRACT – *The several steps needed for bioprocess control are considered from the point of the hardware. This includes such topics as measuring, signal conditioning, data conversion, controllers, actuator interfacing and galvanic isolation.*

KEYWORDS – *Measurements, Data Conversion, Data Transmission, Galvanic Isolation, Computer, Bioprocess Control*

J.F.M. Van Impe et al. (eds.), Advanced Instrumentation, Data Interpretation, and Control of Biotechnological Processes, 161-190.

6.1 Introduction

In this chapter we will deal with the necessary hardware for the practical realization of bioprocess control. Although many principles are possible, we will limit the scope of the subject here to the control strategies which use an electronic digital controller.

In order to be able to control a biological process, the data generated by the process have to be transported to the controller. The controller output has then to be fed back into the system under control. Of course, in the process there is a need (i) for signal conditioning to improve the signal quality, (ii) for data conversion between the several steps, and (iii) for data transmission.

The hardware structure of a process controlling device based on analog signals is roughly schematized in Figure 6.1 [10].

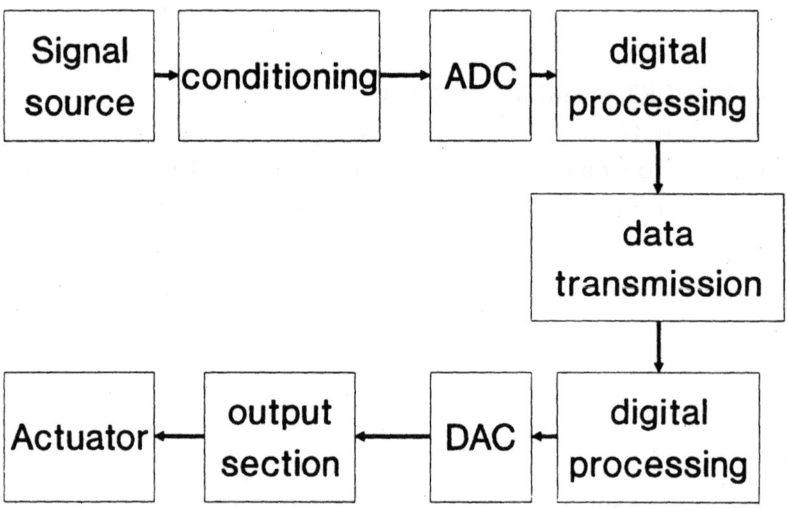

Figure 6.1: Hardware structure.

The actual sensors produce some signal, called the source. In a first step the signal has to be conditioned to make the raw sensor signal suited for the subsequent steps. The signal is then discretized using an analog to digital (A/D) convertor. The digital signal can then be processed by a process computer or a programmable logic controller (PLC). Once the control action has been determined, the output of the intelligent controller has to be made suitable for driving an actuator. This can be done using a digital to analog

(D/A) convertor and some power electronics. In this process there is of course need for data transport, be it in analog or digital form.

However, not all control strategies need to rely upon continuous signals. In many cases a number of discrete steps suffice for the controller to work.

Another, often overlooked topic is the need for galvanic separation of some critical steps in the signal path.

6.2 Measuring Principles

6.2.1 Introduction

The raw signal is generated by some sensor. This can be, e.g., a temperature sensor, a pH electrode, or a dissolved oxygen (DO) electrode. Each of these sensors generates a tiny electrical signal which subsequently needs to be conditioned.

However, most of the signals are not only a function of what the sensor is intended for, but of a number of other external factors as well. Often the sensor does not even measure what it tells it is measuring.

A pH electrode does not measure pH, but the potential over some glass membrane, which is influenced by pH as well as by a number of other factors.

An even more striking example of the fact that the sensor does not measure what it is sold for can be found in the dissolved oxygen sensor. There are besides the DO concentration plenty of factors influencing the ultimate output of the DO electrode such as the atmospheric pressure, the salt content, the tangential velocity of the liquid (stirring speed), fouling of the membrane, the temperature and more.

From those examples it should be clear that the sensors do not always measure what they are intended for and that this is an important element in data interpretation. However, the raw sensor signal can also have other shortcomings.

Mostly the sensor signals are very weak so that they are quite susceptible to noise. For instance, a pH electrode gives a signal which varies only with some 59 mV per pH unit over an internal resistance of the electrode in the order of 250 MΩ. A DO electrode gives a current in the order of nA. A PT100 temperature sensor gives only a tiny change in electrical resistance.

6.2.2 Low level electrical measurements

Basically all necessary measurements become, one way or another, electrical measurements. In most cases the raw signals are (very) low level signals. For instance, a voltage measurement is considered to be a low level one if $V_s < 1\ \mu\text{A} \cdot R_s$, in which V_s stands for the required voltage sensitivity in volts, R_s the resistance of the source in ohms and 1 μA is the likely current interference present if no special care is taken to minimize its effects. Suppose one wants a pH measurement with a precision of 0.02 pH units using a pH electrode with an internal resistance of 250 MΩ. Then one finds:

$$V_s = 0.02\ \text{pH} \cdot 0.059\ \text{V/pH} = 1.18\ \text{mV} << 1\ \mu\text{A} \cdot 250\ \text{M}\Omega = 250\ \text{V}$$

So a pH measurement is clearly a very low level electrical measurement.

There are five different commonly occurring low level measurements. These are: (i) measuring a low voltage ($< 1\ \mu$V); (ii) measuring a low current ($< 1\ \mu$A); (iii) measuring a low resistance (< 100 mΩ); (iv) measuring a high resistance (> 1 GΩ); (v) measuring a voltage from a source with a high impedance [5].

6.2.2.1 Measuring low voltage

For measuring a low voltage the major problems are the *offset voltage* (a voltage which is present even if no signals are applied) and the *noise voltage*. The offset voltage is likely of thermal origin. To avoid it, one has to make sure that all connections are at the same temperature and that they are all of the same metal (crimped Cu-Cu connections). Alternatively one can wait until there is temperature equilibrium (leave the apparatus on for several hours before continuing) and then adjust the offset.

The noise voltage is likely due to magnetic interferences. If this is the case one has to try to remove or shield the measurement from magnetic fields. Also one can arrange the leads as twisted pairs in order to reduce the area enclosed. This should be done both for the signal cables as well as for the cables carrying large currents.

6.2.2.2 Measuring low current

The low current measurements also suffer from *offset currents* and *noise currents*. The offset can be due to insulator leakage. Some insulators are in fact better than others, e.g., an epoxy board has a volume resistivity of

$10^9\ \Omega$ to $10^{13}\ \Omega$, whereas PTFE has a volume resistivity of more than 10^{18} Ω. Moreover, a conducting water film is much more likely to occur on an epoxy board than on a teflon one. So one has to choose the right insulator and make sure that it is always very clean. There can also be a meter bias current, which can be avoided by choosing an appropriate meter such as a picoamperemeter or an electrometer.

The noise current can be due to electrostatic coupling, which can be counteracted by shielding. Also one has to avoid high voltage and movement nearby. In addition, vibration and deformation can give rise to noise currents due to triboelectric and piezoelectric effects. To avoid such problems one can use low noise cables and insulate them from vibration.

6.2.2.3 Measuring low resistance

The three main problems with low resistance measurements are: (i) *offset resistance*, (ii) *drift* in readings and (iii) *noise*. The offset resistance is due to the resistance of the connecting wires and can be overcome by using the 4 wire method (Kelvin connections) shown in Figure 6.2.

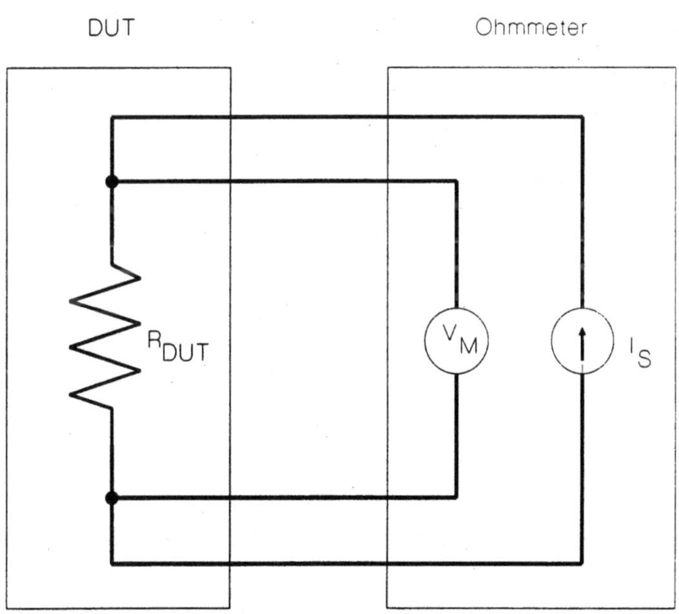

Figure 6.2: Measuring resistance using Kelvin connections.

In this method a well calibrated current runs across the resistance, while one measures the corresponding potential over the resistance.

The drift in readings can be caused by the warming up of the device under test (DUT). To overcome this, one can use several single shot measurements. This effectively limits the time a current flows through the DUT and thus the warming up.

The noise is most likely of magnetic origin. To cope with it one can take measures similar to those in the case of low voltage measurements.

6.2.2.4 Measuring high resistance

The resistance of the voltmeter can be *too low* compared to the DUT. To overcome this, one can source a voltage and measure the corresponding current. The other possible errors are similar to those discussed next and have similar remedies.

6.2.2.5 Measuring voltage from high R source

The readings can be *too low* or *noisy*.

In the case of too low readings probable problems are the presence of a non-negligible shunt resistance or a non-zero offset current. Noisy readings are most likely introduced by electrostatic coupling.

The shunt resistance, which appears as a resistance parallel to the meter, can be due to the insulation material whose resistivity is not high enough. This is illustrated in Figure 6.3. In that case, one has to look out for insulation materials with a very high volume resistivity such as PTFE, Kel-F or sapphire. Also polyethylene and polystyrene have a high volume resistivity.

Also a water film on the apparatus can cause a shunt resistance. It should be kept in mind that a water film can be formed on many insulator surfaces such as polystyrene, polyethylene, ceramic, nylon, epoxy or phenolic insulators. PTFE, sapphire and Kel-F are very good in this respect.

Another way to overcome the shunt resistance is by guarding. This consists of totally surrounding the input leads by a conductor at essentially the same potential as the input, but driven by a low impedance source. Although this does not influence the resistance, it does influence the leak currents.

In the case of noisy readings due to electrostatic coupling, one should avoid high voltage nearby. Care should be taken of movement also. Another way is shielding of the apparatus.

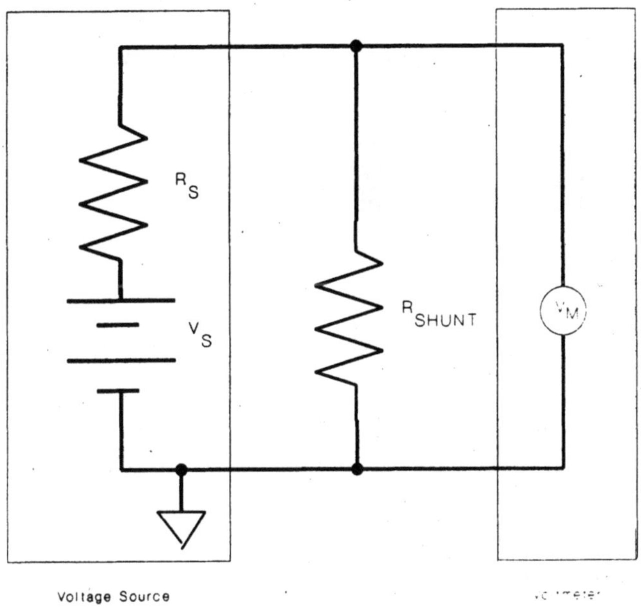

Figure 6.3: Shunt resistance influencing meter readout.

6.2.3 Pulses

Some measurements do not generate analog quantities but a number of pulses. In such cases the signal is rather immune to noise. This is a very elegant way of transporting signals. Basically the interfacing consists in counting the pulses (possibly per unit of time).

Often the pulses themselves are generated by optocouplers installed on some moving parts. The optocoupler consists of a light generating element, most often a light emitting diode (LED) and a light sensitive element, most often a phototransistor. The resistance of the phototransistor depends on the light which reaches it. They are often used in positioning systems and in tachometers. A well known example of measuring through pulses generated by an optocoupler can be found in ordinary computer mice. Figure 6.4 gives a view on the way a mouse operates.

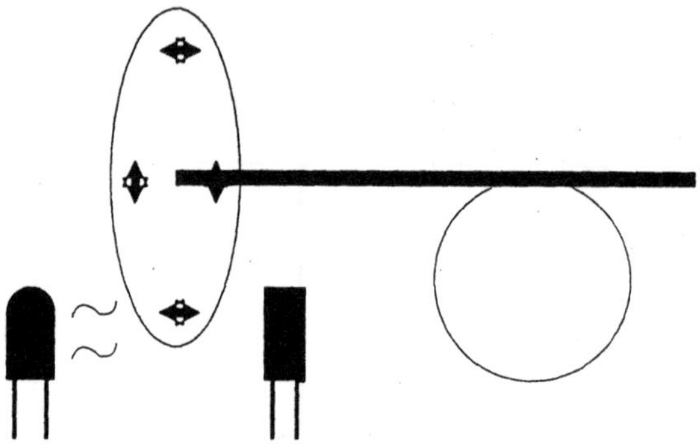

Figure 6.4: Internal structure of computer mouse.

The ball makes a wheel turn. In the wheel there are a number of slits. The number of slits which passes between the LED and the photodiode is detected as a number of generated pulses.

Pulses are used for checking certain processes such as the pulses to activate the stepper motors of, e.g., automatic burettes.

It should also be pointed out that there are very efficient methods for converting a frequency to a voltage as well as converting a voltage to a frequency. In fact some analog to digital convertors even use a voltage to frequency conversion as an internal intermediate step [1].

6.3 Signal Conditioning

In most cases the raw sensor signal is not suited for further handling. It will have picked up some noise, which has to be filtered out. Also the signal levels will not be appropriate for direct handling by the biocontroller.

In the signal conditioning step the signal is made suitable for further handling and it is also improved in quality.

There are several requirements in signal conditioning. A first requirement is the matching of the impedances of the sensor and the subsequent electronics. Another requirement is the removal of unwanted components in the signal (but without removing the essential components). Possibly, the level of the signal has to be shifted. This shifting can be necessary to remove a constant high level of the signal, leaving the changing, lower level for subsequent handling.

Signal conditioning for analog signals is mostly done on basis of OpAmps [7][8]. An OpAmp, which stands for Operational Amplifier, is an electronic device which was originally developed for use in analog computers. It can amplify signals, add them up, subtract them, and it can be used to construct many sorts of filters. Nowadays, OpAmps have a wide number of applications, including signal conditioning.

Concerning filtering it should be emphasized that, although in some cases digital filtering is possible, analog filtering is in most cases indispensable because of the aliasing phenomena which occur in digital filtering if the signal still would contain frequencies above the Nyquist frequency, being the half of the sampling frequency. Analog filtering is immune to such phenomena as aliasing, and is therefore necessary to remove the high frequency components, even if the signal has to undergo digital filtering afterwards.

6.4 Data Conversion

6.4.1 Types of signals

In general terms one can distinguish between three major types of signals. The *real phenomena* are in analogous form. This means that the signal or its measured value can have at any time any value within a range.

This signal is most often transformed to some intermediate *sampled signal*. A sampled signal can have any value within a range, but is only defined at certain time instances. Alternatively, it can only change values at certain time instances.

The next step is the *digital signal*, which is a discrete signal both towards the time as towards the possible values.

6.4.2 Sample and hold

Often the raw sensor signal just varies too fast for discretizing. In order to be able to use such a signal, one has to take from time to time the actual value of the signal. This is accomplished by a sample and hold strategy, for which BIFET OpAmps are the most appropriate. Sample and hold consists of electronically taking an instant value of the signal (sample) and then keep this value for some time to this signal level (hold). While this voltage level is kept constant, it is clear that subsequent conversions, which require the signal to be constant over some time interval, can be easily effected. Of course, sample and hold is only valuable if the signal varies only over a limited range or does not oscillate with a big amplitude around some real value. If the signal would change too much, then the sample and hold will return meaningless instant values of the signal.

Sample and hold can be seen as a capacitor which is loaded over some resistor each time a switch is closed (Figure 6.5). While the switch is open, the voltage over the capacitor remains constant.

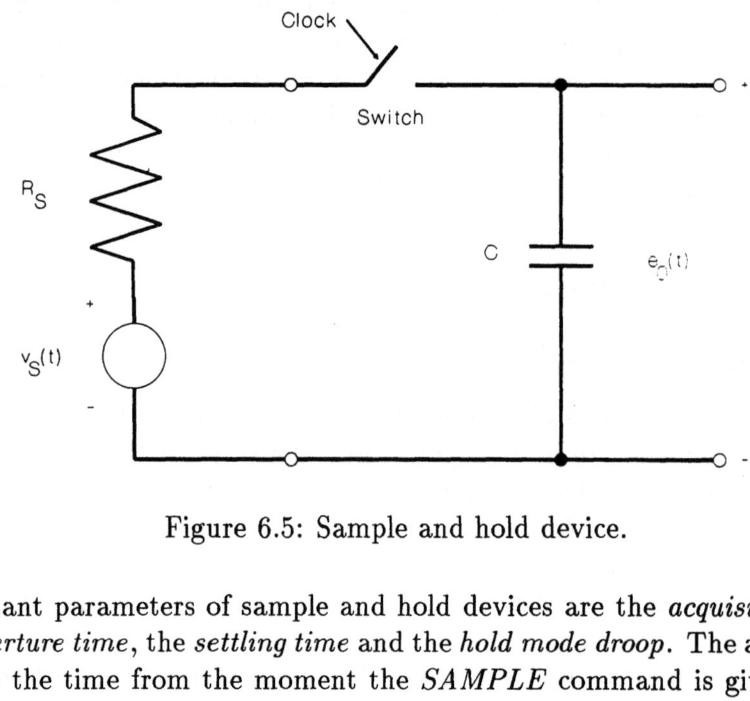

Figure 6.5: Sample and hold device.

Important parameters of sample and hold devices are the *acquisition time*, the *aperture time*, the *settling time* and the *hold mode droop*. The acquisition time is the time from the moment the *SAMPLE* command is given to the time instant the output is within a specified error band around the input signal. The aperture time is the time interval during which the *HOLD* command is given and the time instant the sampler remains open. The settling time is the time required for the transient oscillation to settle to within a certain percent of the full scale. The hold mode droop is the slight decrease in output voltage due to leakage currents [3].

6.4.3 A/D conversion

Many signals are in some analog form. In order to be able to use them in a digital controller they have to be transformed to digital signals. This can be done using A/D convertors [9]. In this section only linear A/D convertors will be considered. The linear A/D convertor has its steps ideally of equal width, in contrast to the non-linear A/D convertor, which has a specified non-linear transfer function between the nominal midstep values and the corresponding step widths.

The most important characteristics of an A/D convertor are its *speed* and its *resolution*. The speed is expressed in samples per second. It ranges from a few samples per second to several millions samples per second. The higher the speed, the more expensive the device. For bioprocess control rather slow A/D convertors are adequate in most cases. Another important speed consideration is the sustained speed. In fact, some devices can work at a very high speed, but only for a very limited time (sometimes only a few ms at maximum speed). One should also be aware of the fact that in most cases, if one samples at a high speed, the accuracy is strongly negatively influenced.

The resolution of an A/D convertor is often expressed in *bits*. This means that the analog signal, which is transformed in the A/D convertor into a digital signal, is represented by the number of specified bits. The more bits, the more accurate the signal can be represented. In fact the number of possible states is given by $number\ of\ states = 2^{number\ of\ bits}$. The relationship between an analog signal and its digital representation is represented in Figure 6.6 for the ideal case.

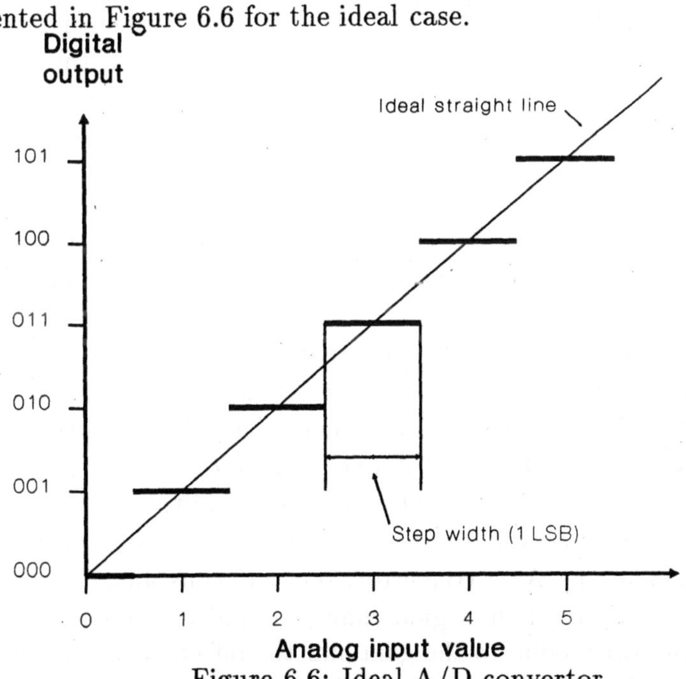

Figure 6.6: Ideal A/D convertor.

The step between two adjacent digital levels and the according change in interpretation of the analog signal is called 1 LSD (Least Significant Digit). This is also the unit in which most errors are expressed.

The resolution should be chosen so that it matches the performance of the measuring circuit. If the resolution would be chosen much coarser, one would loose information. If the resolution would be chosen too fine, this would add an excessive cost to the discretizing system and, moreover, one would measure all kinds of artifacts.

One should also consider what is the most important: the exact actual value or the exact average value when matching the performance of the measuring circuit and the resolution of the A/D convertor. Consider a given A/D convertor which can determine signal levels $\ldots, 20, 21, 22, \ldots$ Consider also a signal which is in reality constantly 20.75. In case there is no noise, the A/D convertor will continuously show the value 21, which is in fact the value most closely to the real value. The mean value is also 21, which differs from the expected value of 20.75. Now suppose there is some noise. If the noise is small, then the same arguments as before will apply. However as the noise increases, the A/D convertor will from time to time find an actual value which is closer to 20 than to 21. If this is the case, the value 20 will be issued. This actual value is not so close to the real value of the signal. However, the mean value will be much better than before. If the noise still increases, then the average will not improve any further but the actual values will deviate more and more from the ideal value. Possibly, also the average will drift from its ideal value.

An important issue is the generation of possible errors during the A/D conversion. The quantization error (Figure 6.7), which is due to the fact that most analog values have to be rounded to be transformed to digital values, is inevitable.

It is maximally 1/2 LSD. As it is inherent to the device used, it can be made smaller by substituting the A/D convertor used for another one which has more bits of resolution (but also a higher price). If the signal does not use the full range of the A/D convertor, one can also change the conditioning of the signal, by amplifying it, possibly following a shift.

There are a number of *static* conversion errors [4], which occur even if the signal, applied to the A/D convertor, remains constant. The *dynamic* errors however only apply if the signal changes rapidly. Amongst the static errors one can find errors due to the gain and the offset, which are not fully adjusted [9]. Also there are errors of non-linearity, errors of non-monotonic behavior, missing codes and the rollover error. The linearity error (Figure

6.8) is the difference between the actual step value and the nominal step value after adjusting offset and gain.

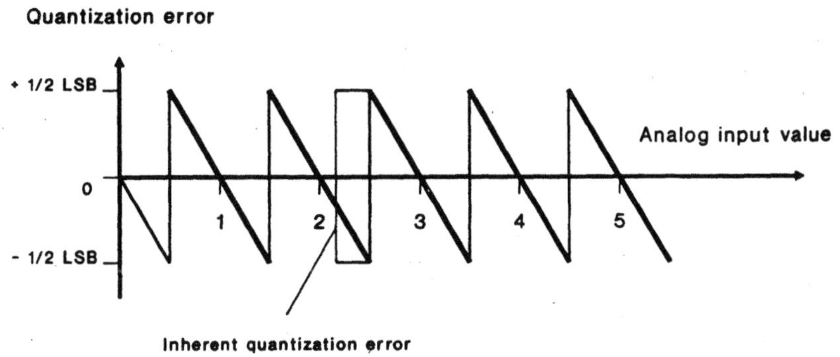

Figure 6.7: Quantization error.

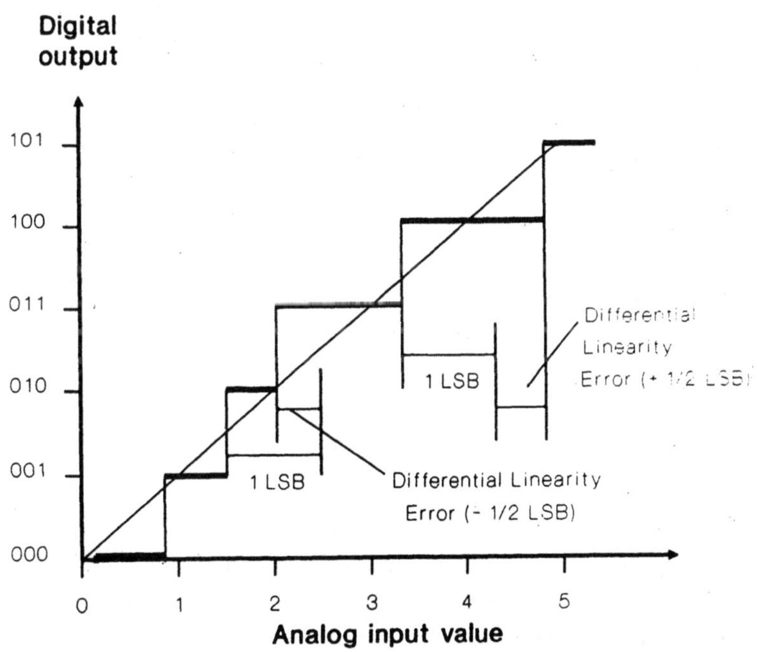

Figure 6.8: Errors of non-linearity.

In the case of non-monotonic behavior (Figure 6.9), some values which are higher on the input scale are actually measured as lower on the output scale.

If there are missing codes (Figure 6.10), then some output values, which are expected to be present, are never generated, even if the signal is such that this output value should be generated.

The rollover error is the difference in output readings if the same analog input would be switched between positive and negative values.

In the dynamic performance one has to deal with the conversion rate or the conversion time, which is the delay between the issuing of the command to perform a conversion and the complete conversion. This gives an upper limit to the number of A/D conversions that can be effected per unit of time.

The most important of course is the total error. As this error has contributions of the offset error, the gain error, the linearity error and the quantization error (which is for every A/D convertor maximally 1/2 LSB), the maximal total error is certainly more than 1/2 LSB.

An A/D convertor typically gets a voltage signal of zero to a few (2 to 10) volts and needs only a low current μA or lower as input.

6.4.4 D/A conversion

D/A conversion is just the opposite of A/D conversion. A digital signal is available and this is transformed to an analog signal (voltage or current). If the D/A convertors produce current signals, in general, greater speeds are possible.

In general, the digital to analog convertor (DAC) will produce an output that is [1]:

$$X = k \cdot A \cdot B$$

where X is the output voltage or current, k a constant (often 1), A the analog reference current or voltage, and B the binary applied word.

Any D/A convertor needs a reference voltage or reference current. On some D/A convertors the reference is inside the convertor module, on others the reference is external. If the reference is external, the D/A is called a multiplying D/A convertor. In many models the internal reference source is brought outside and must be connected to the reference input. Those models have the convenience of an internal reference and the possibility of using an external reference.

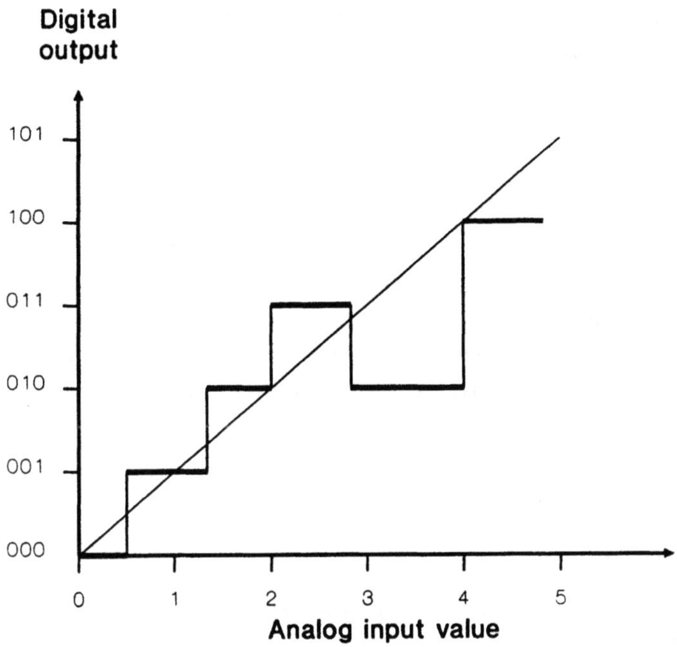

Figure 6.9: Errors of non-monotonicity.

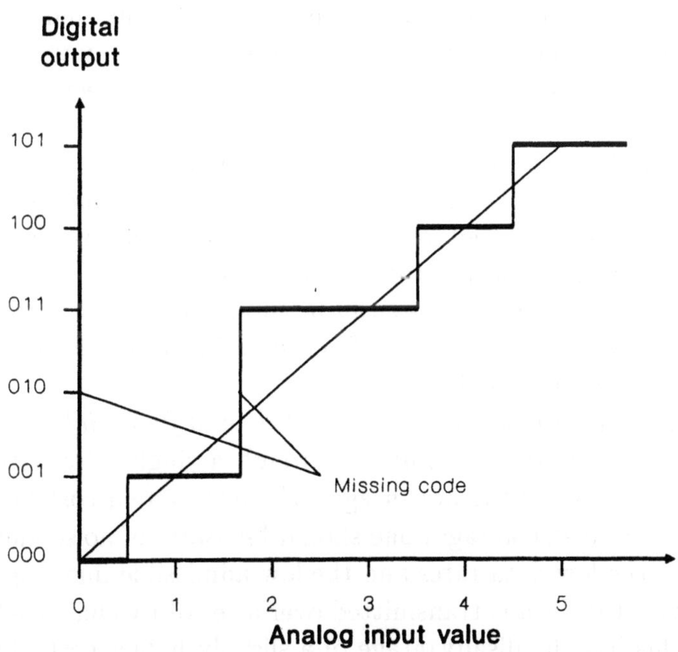

Figure 6.10: Missing codes.

The output characteristics of a D/A convertor are illustrated in Figure 6.11.

As can be seen from this Figure the step response of a D/A convertor needs a certain time before it reaches and remains in the specified error band. This time t_{sd} is called the digital settling time. The time before the onset of the change of the signal is called the delay time [4].

Another important parameter is the maximum rate of change of the analog signal when a change of the digital input causes a large step in the analog value. This parameter is called the slew rate.

Sometimes undesirable transients in the analog output occur following a code change at the digital input. Those are called glitches (Figure 6.12) and have the glitch area (time integral of the analog value of the glitch transient) as most important parameter.

6.5 Data Transmission

6.5.1 Introduction

There are several strategies to transport the signals or the data from the point where they are generated or captured to the point where they are needed. The raw signal itself can be transported, although this is often error prone as the signal is (very) low level. There are all kinds of transport media including mechanical, hydraulic, acoustic, electromagnetic and electrical/electronic means. In this section only electrical/electronic ways will be considered.

One can distinguish between the various types of signals being transmitted. One possibility is the transmission of *analog signals*. Those are proportional to the measured variable and can be both voltage signals or current signals. Another way is the use of *digital signals*. Those can be sent over serial or parallel lines, or they can be modulated using a carrier signal. Still another way is the transport of pulse trains which have a frequency proportional to the measured variable.

The data transport can be *single ended* or *differential*. In the single ended case, the data are transmitted through a single wire with reference to the ground. This has the advantages of a low system cost and is simple to implement. As disadvantages one should be aware of noise and crosstalk, ground shifts, the low data rates and the low admissible line lengths. In the differential case the data is transmitted over a set of two balanced transmission lines. This has the disadvantage of a slightly higher cost. The benefits are a high common mode voltage rejection, improved speed, and the possibility to drive longer line lengths.

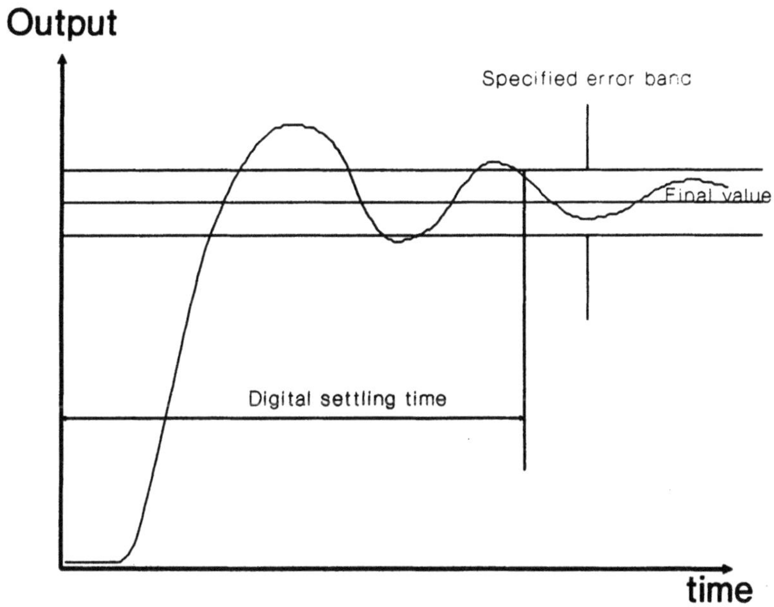

Figure 6.11: Analog output characteristics of D/A convertor.

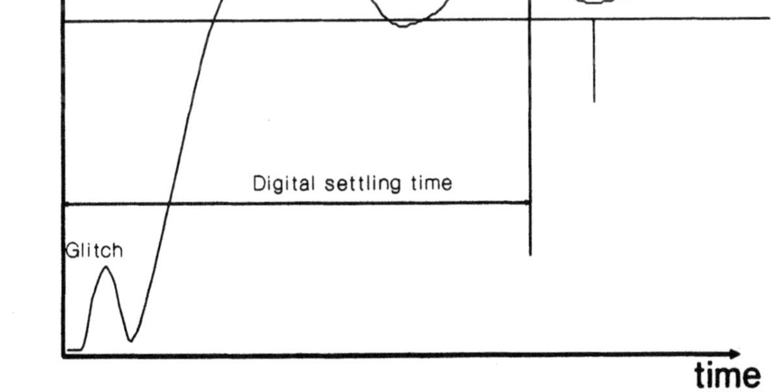

Figure 6.12: Analog output characteristics of D/A convertor with glitches.

Also one has to consider the data transport speed, the throughput, the reaction time, and the maximum length of the transmission channels as well as flexibility and cost. Other points are the degree of integration, reliability, possibility of galvanic separation, and the possibility of exchange during operation.

6.5.2 Analog data transmission

The raw electrical signals from the measuring device can be used to transport the signal in an analog way, but only over short distances as this is error prone due to their weak nature.

After amplification the signals can also be transported as voltage or current signals. Amongst those the 0-20 mA and the 4-20 mA current loops are widely used. The latter has some advantages over the former, notably the possibility to distinguish between an interrupted circuit and a minimal signal. Also as the current is never zero, the remote apparatus can get its power from the current loop. Both types are very immune to noise.

In the current loop there is always a current source and a current regulating device (which can be the same). There can be any number of current sensing devices in series in the loop, provided their total electrical resistance is lower than some predefined value, often 600 Ω (Figure 6.13).

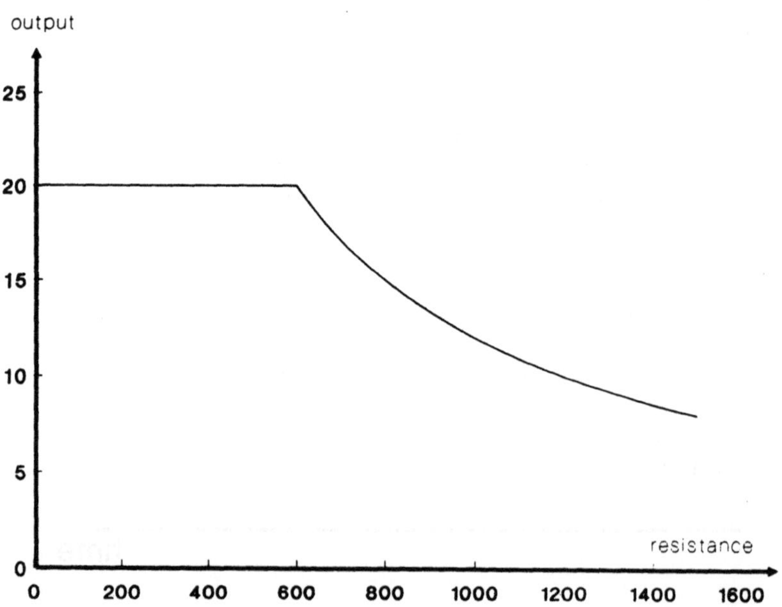

Figure 6.13: Influence of external resistance to 20 mA signal.

If the resistance would be higher, then the maximum output voltage of the source would be insufficient to generate the required current. If for some reason the current loop is in short circuit, this can not harm the apparatus, as the current will have its normal operating value. In some cases it is even mandatory to short circuit a current loop if no current sensing devices are present.

Voltage signals, such as a 0-1 V, a 0-5 V, or a 0-10 V signal, are connected to the sensing devices in such way that all sensing devices are in parallel to each other. The total electrical resistance should however be higher than some predefined value such as 100 kΩ. If the resistance would be lower, the generated voltage could drop. In case of a short circuit, the voltage would drop in most cases to zero. In a few devices, however, short circuits on the voltage output permanently damage the device. Voltage signals are more susceptible to noise than current loops.

Current signals can be transformed to voltage signals by simply inserting an appropriate resistor in the circuit and using the voltage generated across this resistance. The transformation of voltage signals to current signals is a bit more complicated and often needs an additional external power supply.

The advantage of analog signals is that the sensed information is in principle present at any single point in time. However, some devices reconstruct the analog signal from a digital source, in which case the previous argument does not necessarily hold.

This way of data transmission has also some disadvantages. There can be only one signal transported over a current loop, which most often consists of two wires. The maximum precision of the signal is limited. Analog data handling is more expensive than digital data handling. If galvanic separation is required, this is more costly than in the case of digital or pulse signals. Also the signal has a direct current (DC) component (in fact the signal is mainly DC), which can have some drawbacks in certain circumstances: there is, e.g., a need for some (intermediate) conversion if galvanic isolation is required.

6.5.3 Digital data transmission

In the case of digital data transmission, the data is transported at two different signal levels. If some noise disturbs the signals, this has no influence at all as long as the signals can correctly be classified according to their category.

The signals can be of the DC type or they can be modulated using a carrier, in which case the DC component becomes negligible. This is beneficial as it allows the signal to be handled using transformators and the risk of a signal being out of range diminishes. Also the modulated signal carries in itself a clock signal, thus eliminating the need for another connection. The modulated signal is also better distinguishable from noise signals, which enhances the error detection possibilities.

An important point is that, in general, not all of the information is available at the same time instant. This gives rise to the need of synchronization between the sender and the receiver. The loss or the misinterpretation of a single bit can completely invalidate a complete measurement. There are several ways for error detection such as parity (making sure that the number of ones in a transmitted byte is always even or always odd), sum checks, cyclic redundancy check (CRC) and redundancy (sending the same message several times). A sum check is a specific check developed when groups of digits are summed, usually without regard to overflow, and that sum checked against a previously computed sum (checksum), to verify that the message has not been changed [11].

Also one should consider the speed of the digital link, which is expressed in bits per second. It should be stressed however that the real throughput of actual information is lower than the physical speed, due to the overhead such as start bits, stop bits, error detection schemes and so on. Digital signals can also be multiplexed, which stands for the concurrent transmission of more than one information stream on a single channel [11]. This way several signals can be transported over the same set of physical wires.

6.5.3.1 Networks

There are quite a lot of different hardware and software networks. They vary in such characteristics as their speed and reaction time; the number of nodes; the maximum total length; flexibility in installation; linking with the application; coding of data; price; ...

The choice of the network is mainly dependent on the application. In general one could distinguish amongst 4 levels of application [2]. A *sensor network* is used on the sensor and actuator level. Typical examples are ASI (`Actuator Sensor Interface`), I^2C (`Inter IC bus`) and `Interbus-S`. One level up one can find the *field networks*, such as `DIN-meßbus` which are used on the process level. The higher levels are the *process/cell networks* which are used for the operational management, and the *factory networks*, which are used throughout the plant. The lower levels tend to have less data transport, over short distances. These levels have mostly a time critical character. On the other hand the higher levels have larger amounts of data to transport, over longer distances, but their time dependency is not so critical. There are no networks which are affordable and useful at all levels of the hierarchy.

Independent of the level of the application, several network topologies are possible such as the *bus*, the *ring*, the *star* and the *tree* topology [6]. In the bus structure all devices are connected with the bus. The nodes can send and receive on the same connection. The bus remains functional even if not all nodes are functional. The bus is a linear topology. The ring structure passes through all nodes. Begin and end are connected to each other, and are often not distinguishable from other nodes. Each connection connects one sender to one receiver. The ring is only functional if all nodes are functional. In the star topology all the nodes are connected to the same central point. The tree can have branches on random locations. It is however not suited for high speeds neither for high distances.

In most cases arbitrage, determining which sender is allowed to send during a certain time interval, is necessary. This can be done in several ways. Only *master/slave*, *token bus* and *token ring* networks will be considered here as they are most common. As for master/slave, only the master has the right to send. However, the master can send the slave a message that it is allowed to send during some short time interval. In the token bus a special message, the token, is exchanged between the connected nodes. The node which has the token is the one allowed to send. A disadvantage is the long access time; an advantage the fact that each node will have to wait only a maximum predictable amount of time. In the token ring the token is passed from node to node. Each node passes all messages which are not meant for it unchanged. A defective node however impairs the complete network. In some cases the token may get lost, which totally impairs the operation of the token bus or token ring network.

6.5.3.2 Parallel communication

In parallel communication, bytes are demultiplexed over several wires. Synchronization is required between the lines, as the data is only valid if all bits, which belong together, reach the receiver synchronously. This diminishes the maximum cable length as the transport over the cable tends to disrupt the synchronization. Also as there are several conductors, the solution tends to be rather expensive if one would nevertheless try to connect over long distances. A typical example of parallel communication can be found in the centronics printer interface. Another example is the Small Computer Systems Interface (SCSI).

6.5.3.3 Serial communication

In serial communication the individual bits of data are consecutively sent over a single data channel. As an example consider the 232 interface, which is a way of single ended, digital, serial data transmission.

Often when some apparatus is sold, it is specified *standard 232 interface* present. This can be a very misleading term. One could think that if two devices each have a 232 interface, they can be connected without any problems. In this overview it is illustrated that one should be very careful when dealing with a 232 interface. In fact there is more than one 232 standard, notably the EIA-232 (or RS-232) standard, which was introduced in 1962 in an attempt to standardize the interface between Data Terminal Equipment (DTE) and Data Communications Equipment (DCE), or said in simpler terms between a modem and a terminal. There was an important revision of the standard in 1969, the RS-232-C. There was another revision in 1986, the RS-232-D and this one has again been superseded by the EIA/TIA-232-E standard, which brings it in line with the CCITT V24 and V28 standards.

Two devices which are based on the same 232 norm can still cause problems when one is trying to communicate between them. There are different types of *connectors* (DB 25, DB 9, OEM layout, ...) and the actual connections can sometimes vary within one type of connector.

Even if the connections are right, then still some problems remain. The actual way of transmission can be done according to many different schemes. The *speed* of transmission, in bits per second, can have values such as 110, 150, 300, 600, 1200, 2400, 4800, 9600 and 19200. In some equipment even non-standard values are used such as 38400 or 115200 bps. Moreover, the number of bits per byte can have the values 5, 6, 7, or 8. There is always 1 startbit per byte, but the number of stopbits can vary from 1 to 2. In some cases there can be a parity bit (used for error detection). This parity can be odd or even.

When one has considered all the foregoing, then one should continue considering the *protocol* of data transmission. The data can be sent as soon as they are generated. In this case the remote controller should be ready to accept the data at any time. This can be accomplished on a PC by using a polling or an interrupt driven strategy, the first being the most easy to implement, the latter the more reliable one in case other actions from the controller are required. Also one can use some special hardware (a FIFO buffer). There can be some protocol between the two devices, defining when data can be sent, e.g., the XON/XOFF protocol or some hardware protocol such as RTS/CTS (Request to send/ Clear to send). This is often only possible to a very limited extent, because the data generating devices often have a very limited data storage capacity.

Finally, one should also consider what the meaning is of the individual bytes and decode the relevant information within a stream of non-relevant information.

As the 232 interface is single ended, it has also the limitations and drawbacks of single ended systems. The RS-422 and the RS-485 are differential digital transmission line standards which improve on the EIA-232 limitations. The advantages are: a higher data rate (up to 10 Mbps); a longer line length (up to 1200 m); less noise sensitive. With RS-422 there is always one driver and up to 10 receivers. The RS-485 is meant to be an upgrade and a more flexible version of RS-422. RS-485 allows for multiple (up to 32 according to the standard, more in some non-standard configurations) drivers and receivers on a single line. As for the 232, neither the RS-422 nor the RS-485 specifies the full system. In fact the standard only specifies the electrical characteristics. All the other relevant parameters, such as protocols, timing and connector choice, are left to be defined. Whether the data are serial or parallel is not specified by the standard. As a result, one should be even more conscious when using these standards than when using the 232 standard.

6.5.4 Pulse trains

In a number of cases the data are transmitted using some standardized pulses. The frequency of those pulses is directly proportional to the measured value. This is only possible if the measured signal is rather constant (no high frequency nor medium frequency components). The advantages are that those systems are rather immune to noise, can transport over long distances without data deterioration, and that over any relevant time interval the data are present.

The pulses can simply be counted, or alternatively they can be converted to analog signals using frequency to voltage convertors.

6.5.5 Internal data transport

In many cases the controller can be a PC. The peripheral signals should reach the memory (RAM) or the central processing unit (CPU) before they can be handled. There are various ways to accomplish this.

In some cases the data generating device (e.g., an A/D card) has to the CPU the characteristics of a memory location. On that location the CPU can always find the latest disponible value of the measurement. This is not necessarily the most recent value, as there can be some delay in the A/D card itself. In other cases this latest value is present on a port address. Port addresses are a special set of locations reserved for input/output (I/O) operations. In both cases the CPU should copy this value to some other reserved memory location if this value will be needed in the future as the next value will overwrite the previous one.

As an example, consider the RS-232-C interface as found in IBM PC compatible computers and configured as COM1. The received character always appears at the same port address, notably $03F8 [13]. Once the next character is fully received, the previous one is lost on that location.

In order to move the data generated to some other memory location, one can use several strategies. The simplest one is *polling*. The CPU continuously monitors the location. If necessary it moves the new data. While polling the CPU can only execute simple additional tasks in order to be fast enough not to loose any data.

Another strategy is to use some hardware or software *protocol*, which allows the remote device to change the data only at specified instants.

Still another strategy is the *interrupt* driven strategy. In this case the remote device changes the data at any moment. However, the activity of the device is monitored and the CPU is warned each time a relevant event occurs. At that moment the CPU ceases its activities, handles the interrupt (moving the data), and then resumes its previous activities. This has the advantage that the processor can do almost every possible action without the risk of loosing data. The disadvantages are the more complex programming and the increased overhead. Another disadvantage is that the interrupt driven system is only beneficial for data transports which do not generate much data (up to a few thousands interrupts per second are possible).

Yet another strategy is using Direct Memory Access (DMA). In this strategy the DMA chip of the computer will move the data into the final memory locations without intervention of the CPU. This is often done using a process called *cycle stealing.* During one of the cycles the DMA places the data on the bus, whereas the rest of the time the CPU is the busmaster. The benefits are the very high throughput and the fact that the CPU can continue its own activities. Some drawbacks are the fact that often the devices can not have a sustained throughput, but in fact give the data in chunks. Also, often, the CPU is not aware of the changed data and if the reserved place is too small, the problems of handling the data remain. Devices on basis of DMA data handling are often more expensive than others.

One can also provide a special memory which accepts the data, and provides the data in the same order to the main unit when it is asked for it. This type of memory is called a FIFO (first in first out) buffer. The advantage of this is that the data generation and the data handling are independent.

6.6 Controlling Devices

There are several possible intelligent controllers. Amongst those one can find *computers*, *programmable logic controllers* (PLC), *microcontrollers*, as well as *specific hardware*.

Whereas the first two can exist as independent devices, the last one is most often integrated in some bigger unit.

6.6.1 Computers

Many types of computers are suited for some kind of control. The possibilities of a computer for control not only depend on the computer itself, but also on the operating system. An example can be found in the `IBM` compatible PC. If such a machine is working under `DOS` with a given hardware configuration, the throughput of controlling actions is most of the time much higher than the same machine operating under `MS Windows`.

Computers often need to be equipped with additional hardware, such as A/D or D/A cards. Often those cards have their own processor, so they can operate for some time without intervention of the host computer. There are even add-on cards which can autonomously control some parameters with a moderately complex control algorithm, such as PID.

Advantages of computers are their availability and large installed base, the availability of software tools, the flexibility, and the possibility of implementing very complex algorithms.

The drawbacks of common computers are the possibility of endless loops due to programming errors, which can delay the controlling action too long;

the lack of robustness; the limited environmental range. However special hardware and operating systems exist, which totally invalidate this list of drawbacks, but add to the cost.

Computers seem to be the best choice in non-critical applications and in research. If fact, most of the process control on lab scale and on pilot plants is done by using the IBM PC compatible computers equipped with additional I/O cards. Using special real time operating systems (RTOS), they can also be used in critical, demanding applications.

6.6.2 PLC

In a traditional PLC a program is executed as a big loop from the first line to the last line and this is continuously repeated. As there are no possibilities for skipping some lines or jumping back (all lines are evaluated in turn, even the non-used ones), the cycle of the program is guaranteed to take always the same amount of time. It is evident that it is impossible for a traditional PLC to be trapped in some endless subloop. This feature adds strongly to the overall reliability of a PLC based system. Moreover, the hardware of PLC's is often of a more rugged design than the hardware of a typical PC. On the other hand, PLC's provide much lower flexibility than computers and they are also not suited for implementation of complex control algorithms.

New generations of PLC's have emerged which have possibilities for jumps as well as possibilities for implementing elaborate control algorithms, making the distinction between computers and PLC's a bit fuzzy.

Because of their reliability PLC's are the preferred choice for full scale plants such as wastewater treatment plants or production facilities.

6.6.3 Microcontrollers

Microcontrollers are integrated circuits, very comparable to the CPU of a computer. They have however special features to enhance their applicability in controllers with a minimum of peripheral electronics. Microcontrollers exist which contain the necessary hardware for interfacing with a 232 serial line. Some also contain incorporated A/D and/or D/A circuits. A watchdog may also be included, preventing the processor to stay in a tight endless loop.

A typical example of the application of microcontrollers is the RODTOX [14], an apparatus for quick determination of the biological oxygen demand (BOD) as well as the toxicity of wastewaters. It uses the 8052-AH-Basic, a microcontroller with 3 serial ports, as central unit. The three ports are used for the communication with a printer, a screen, and for an internal communication. The programs necessary for the operation of the RODTOX are stored in external memory and are interpreted with the incorporated Basic interpreter. A major drawback of the 8051/8052 series of microcontrollers is its limited capabilities [12]. In the RODTOX the limit of the capabilities is almost reached.

6.7 Actuator Interfacing

In order to be able to control some bioprocess, some actuator needs to be activated. This can be a pump, a valve, a piston, a heater, a cooling element, an engine, a motor, ... Most often these actuators have power requirements exceeding the power source capabilities of the controller, hence needing power driver circuits.

There are basically two ways of controlling an actuator: (i) it can be controlled via an *on/off control*, or (ii) it can be controlled with some –on the average– *continuous control*. This second type includes –besides the voltage proportional and the current proportional controls– the controls effected by pulses.

There exist convertors which transform some electrical signal to a non-electrical controlling signal such as a hydraulic or pneumatic signal. In this text, however, only purely electrical ways of actuator interfacing will be discussed.

For actuator interfacing, one can use such devices as power transistors (especially MOSFET transistors), special integrated circuits, electronic switches, as well as mechanical relays.

One should consider several aspects when selecting the load device. First of all, it should be determined whether the load has a resistive, a capacitive, or an inductive nature. Each of these requires a different way of handling. The purely resistive load can be dealt with by simply examining voltage and current, considering heat dissipation and duty cycle, and adding some safety margins. Note however that sometimes the resistance can be temperature dependent, which makes that the current or the voltage can change with time. For instance, this is particularly true if one turns on a lamp. There is a rapid initial current surge due to the low resistivity of the cold filament, reducing with time as the filament reaches its operating temperature. Although the phenomenon is of a very short duration, it has to be taken into consideration. One could make a power supply which can deliver a high peak current, or one could limit the current.

For capacitive loads there is always a very high initial current, which needs to be limited. However capacitive loads are rare.

In the case of inductive loads, such as solenoids or motors, energy can be passed from the source or switch to the load and vice versa. The returned energy has to be dissipated without damaging the devices. For solenoids found in relays and electronic valves one should keep in mind that the inductance changes as the mechanical movement occurs. Also in most cases, more energy is required to activate solenoids than to hold them in their final position.

In determining the interface circuit requirements one should not only consider operating voltage and current at continuous duty, but also the transient conditions occurring during switching. The switching speed is also of major importance, although in bioprocess control mostly slow speeds will be adequate as in the process itself only relatively slow variations occur.

Very often the actuator interfacing is done by switching. This is even the case in such applications where a continuous control seems to be used. As an example one can take a DC (direct current) motor. One might think that in contrast to a stepper motor (which is by its nature always controlled with pulses) a DC motor is controlled by varying the voltage level. However, a DC motor is most often controlled by a technique called pulse width modulation (PWM). Hereby pulses of varying time duration are created and sent to the motor, giving a very good control on the motor speed. PWM is so important that many microcontrollers have incorporated features for PWM. This is for instance the case in the 8052 series. Other examples include lamps and heating devices, which are most often also controlled by rapid switching. In some cases this switching can disturb some very sensitive measurements. If this is the case, one should try a continuous control of the otherwise switched element. However, continuous control is often much more expensive.

6.8 Galvanic Isolation

In bioprocess control one often needs to interconnect a multitude of instruments, be it sensors, (intelligent) controllers, or actuators. The devices are connected to each other in several ways. They are connected through their power supply, through the measuring parts (possibly via the liquid), through the data transport channels, and via the ground.

Quite often there are undesired electrical currents flowing via one or more of those ways. Especially the so-called ground loops are feared in bioprocess control. This is illustrated in Figure 6.14.

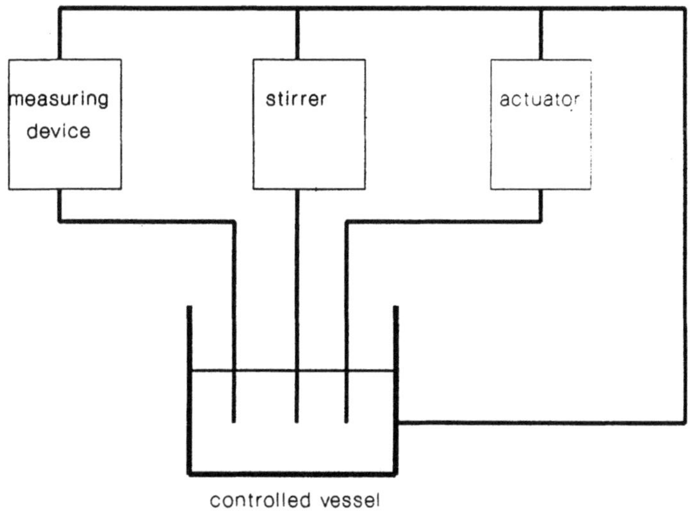

Figure 6.14: Ground loops.

To cope with these problems one should focus attention to two major points. First, all grounding should be done to one unique ground point, thus avoiding the possibility of ground loops. Second, whenever appropriate, one should provide galvanic isolation. The galvanic isolation between two points lets pass the information (be it digital information, pulses, or analog information such as the magnitude of the current) but isolates the two devices electrically from each other, thus reducing the mentioned the problems for the major extent.

Galvanic isolation has also other benefits. It adds to the safety. Also in case of failures, the failure will in most cases not extend to both sides of the galvanic isolation.

Most often galvanic isolation is obtained by using *transformers* (only possible if the signal does not contain a DC component or if it is modulated), or by using *optocouplers* (for pulses and digital information).

6.9 Bibliography

1. J. Carr 1980. *Microcomputer interfacing handbook: A/D & D/A*, Tab
2. H. Hakkesteegt and K. Bangma 1993. *Intelligente Sensornetwerken*, CME
3. B. Kuo 1980. *Digital control systems*, Holt, Rinehart and Wilson Inc.
4. Texas Instruments 1992. *Lineair Circuits Data Book part 2*
5. Keithley 1992. *Low Level Measurements Handbook*
6. R. Meyers (Ed.) 1989. *Encyclopedia of telecommunications*, Academic Press
7. *Operationele versterkers, deel 1*, RB elektronica, mei 1994
8. *Operationele versterkers, deel 2*, RB elektronica, juni 1994
9. G. Peltz 1990. AD-/DA- omzetters, *Elektuur*
10. S. Ramsdale, D. Robinson, L. Goudge, T. Ardley, E. Davies, D. Cox, B. Barrett, R. Asmussen, A. Webber,G. Arnold, D. Cotton, J. Holland, M. Maytum, B. Mullet, C. Wray 1994. *Seminar Reference Book, Linear Mixed Signal Design Seminar 1994*, Texas Instruments
11. C. Sippl 1985. *Macmillan Dictionary of data communications*, The Macmillan Press
12. J. Steeman 1992. Werken met microcontrollers uit de MCS-51-familie, *Elektuur*
13. B. Teliki 1994. Applied I/O for the PC. *Electronics world + wireless world*, 100, 452-458
14. P. Willems and L. Van Vooren. *Chemische en mikrobiologische monitoring van de waterkwaliteit, deel I, Algemene metodes*, Verhandelingen van de Faculteit Landbouwkundige en Toegepaste Biologische Wetenschappen, Universiteit Gent, 32, 5/1-12, PUO, 2-12-92

Chapter 7

Data Processing for System Identification

P. VAN OVERSCHEE
Katholieke Universiteit Leuven
Departement Elektrotechniek ESAT-SISTA
Kardinaal Mercierlaan 94 B-3001 Leuven (BELGIUM)
email: peter.vanoverschee@esat.kuleuven.ac.be

ABSTRACT – *The quality of black box models highly depends on the characteristics of the signals applied to the identification algorithms. In this chapter it is showed how measured data can and should be pre-processed before applying linear system identification techniques. We discuss how to remove disturbances from the data (e.g., spikes, drifts, offsets, differences in input and output power) and how to compensate for long time delays and non-linearities. We describe simple but effective techniques to remove these degrading effects from the measured signals. Thus the quality of the identified model can be significantly improved.*

KEYWORDS – *Identification, Processing, Filtering, Delayed Systems*

7.1 Introduction

Today, a variety of black box identification methods for linear systems is available (see for instance [2][4]). These methods have been implemented in commercially available packages [3][5]. The basic black box identification setup is illustrated in Figure 7.1.

J.F.M. Van Impe et al. (eds.), Advanced Instrumentation, Data Interpretation, and Control of Biotechnological Processes, 191-209.

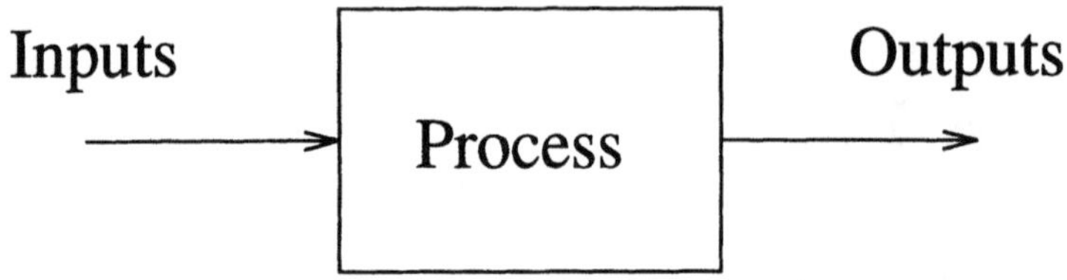

Figure 7.1: The typical black box identification setup. Inputs (u_k) and outputs (y_k) are measured from the process. They are then used to identify a model of the plant.

A common point between all of the methods is that they start from measured data. These data have to be conditioned to make them suitable for black box identification methods. In this chapter, a short overview of simple preprocessing methods is given. Even though simple in nature, these methods can have a significant effect on the end result. The major part of this chapter has been borrowed and/or adapted from [1] to which the reader is referred for a nice overview of practical issues of black box system identification. The methods described in this chapter have been implemented in a graphical environment in [6].

The quality of the identified model heavily depends on the characteristics of the signals offered to the identification algorithm. Disturbances with a bad influence on the quality of the model are for instance: spikes, drifts, offsets and significant differences in power of the various input and output signals. Almost all industrial data sets will be corrupted by this type of *noise*, and it is thus advisable to try to reduce this as much as possible. How to cope with these disturbances is discussed in Section 7.2, 7.3 and 7.5.

Another problem encountered in industrial practice is the presence of, often relatively long, time delays in the measured transfers. If the signals applied to the identification algorithms are not compensated for these time delays, the delays have to be estimated by the algorithms. This implies that many extra parameters, corresponding to the time delays, have to be used for the modeling. This has a deteriorating influence on the quality of the models ultimately obtained. Therefore, it is preferred to estimate the time delay separately from the parameter identification and to correct the signals for the estimated time delay as much as possible. How to estimate delays is described in Section 7.4.

Finally, a diagnostic test is discribed for detecting non-linearities in process models and for examining possibilities for the linearization of the process. This is the topic of Section 7.6.

7.2 Trend Determination and Correction

The choice made with respect to the inputs of the system, does in general, not include all process variables that influence the selected outputs. The inputs that have not been selected for the modeling may however contribute to changes found at the output. During the modeling phase, these disturbances are considered as colored output noise. Often the characteristics of these contributions are known and in many cases the effects are slow variations of the outputs, which are called trends. These trends have a bad effect on the quality of the models. Some examples of trends in industry are:

- A temperature measurement that is influenced by the ambient temperature. When the measurement campaign takes more than one day, a periodic disturbance (a trend) can appear in the data. This trend is due to the day-night temperature cycle.
- A glass furnace melting quartz sand to glass. The batch of sand from which the furnace is fed has to be changed regularly. Every time a new batch is introduced, sand with a different composition is poured into the furnace. Typically this leads to a ramp-like disturbance at the output of the process. The disturbance is a ramp since at the moment of change over from one batch to another, the furnace still contains sand of the old batch (with the old composition). The effects of the new composition become gradually visible while the two types of sand mix in the furnace.

In general, only part of a period to a few periods of the trend signal will be present in the measured data interval. Typically, black box identification techniques rely on the fact that the disturbance v_k is uncorrelated to (for instance) the inputs u_k. When there are N measurements, this means that:

$$\frac{1}{N}\sum_{k=1}^{N} u_k v_k \simeq 0$$

Now, if v_k is a large disturbance with only one or two periods in the considered time interval, this condition will not be satisfied. Furthermore, the sum will even not be small. On the other hand, faster disturbances will have less influence on the estimated model, since they average out faster.

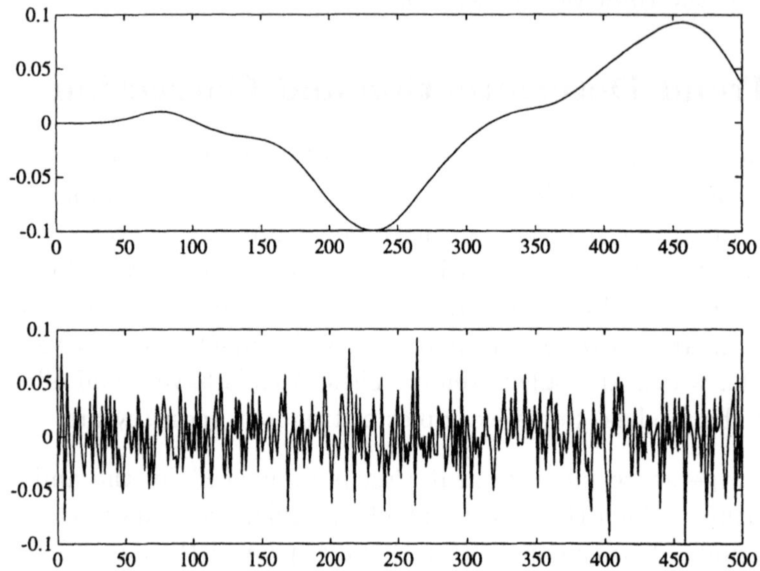

Figure 7.2: Top: a low frequency disturbance which is defined as a trend. Bottom: a high frequency disturbance.

As an example of this effect: consider two classes of disturbances with the same maximum amplitude (around 0.1). The first set of disturbances has a low frequency content, while the second set has a frequency content which is higher (relative to the interval under consideration). Figure 7.2 shows a member of each disturbance set (v_k). Checking the expected value of $\sum_{k=1}^{N} u_k v_k$ shows that the quantity is about 5 times larger for the first class than for the second class.

A possible approach to separate a trend from a signal is based on the determination of the trend by filtering the signal with a low pass filter and subtracting the trend from the original signal. However, filtering of the signal with a low pass filter introduces a phase shift between the original signal and the trend signal obtained. This is illustrated in Figure 7.3. As can be seen from Figure 7.4, subtraction of this shifted trend from the original signal leaves undesired signal components in the detrended signal. To overcome this problem, the signal can be filtered twice: once with a strictly causal low pass filter (time running forward) and a second time with the corresponding strictly anti causal low pass filter (time running backward). The average of the sum of the two filtered signals will not be shifted any more and can therefore be subtracted from the original signal (see Figures 7.3 and 7.4). It should be noted that due to the low frequency character of the filters, care has to be taken of the initial conditions of the filters. The transients take a long time to die out. It is thus best to estimate the initial state of the filters and to use this estimate to start up the filters. Finally, note that most software packages support zero phase shift filtering.

The filters should be designed in such a way that the trends are determined as accurately as possible and that no relevant information in the signal outputs will be lost. In practice the filter design will usually be a compromise between these two demands. There are a couple of things worth noting about this *detrending* scheme:

- Since the data have to be filtered in a causal and anti-causal direction, it is not possible to apply this technique to on-line processing.

- Instead of low pass filtering, one could also apply a high pass filter to the data. This would imply that the signal at the output of the filter is the detrended signal we want to obtain (we do not have to subtract the filtered and original signal any more, so the phase shift is not important). The filter will introduce a phase shift at the inputs *and* outputs, but since this shift is the same for both signals, it will not significantly influence the quality of the identified model. However, in practice the inputs are typically not disturbed by trend disturbances. This implies that only the output signals have to be detrended. If only a phase shift is introduced at the outputs, this shift will be identified as part of the model. So, often it is better to avoid phase shifts. This is exactly what the scheme above achieves.

After the trend correction one should also do an offset correction. This is necessary to linearize the signals around the operating point, since often, in an industrial environment, the offset levels are very high, while the variations around the offsets are small. For instance in an industrial glass oven, the nominal temperature would be 1600 °C, while the variations around this working point would be of the order of 1 °C. Neglecting to remove the offsets leads to a totally distorted model, that has a DC gain equal to the offset of the output divided by the offset of the input. This fraction will in general (unless the system is linear over the whole working area) not be equal to the local (around the working point) DC gain of the process.

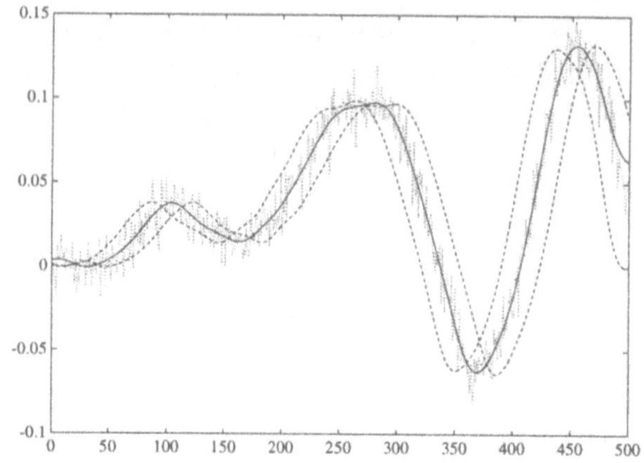

Figure 7.3: The dotted line shows the original signal with a clear sinusoidal trend in it. The dashed lines show the forward and backward filtered signals. The full line shows the trend that was determined as the mean of the forward and backward filtered signal.

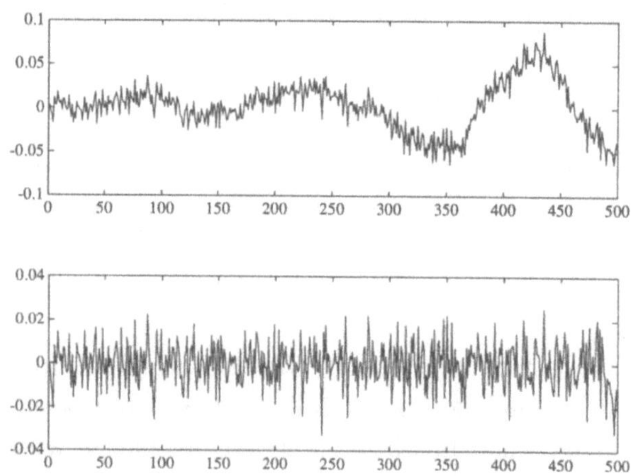

Figure 7.4: The top figure shows the difference between the original signal and the forward filtered signal. Clearly there is an undesired residual signal present. The bottom figure shows the difference between the original signal and the average filtered signal. Now all trend disturbances are removed.

7.3 Peak Shaving

Another problem often met in industrial practice is the disturbance of measured signals with spikes. These noise contributions may be introduced in the sensors and in the often long leads from the sensors to the measuring equipment by switching actions of some high power process machinery. The amplitudes of these spikes are in general very large compared to the signal changes obtained from the process (see also Figure 7.5). Most of the time these spikes last from one sample to tens of samples. If these spikes are not removed from the signals, they may form an important part of the noise energy. As a consequence, the spikes may have a considerable influence on the quality of the resulting model, even though they have no relationship to the process at all. A possible (statistical) way of *shaving* these peaks is the following four step method (Figure 7.5 illustrates the four steps):

1. Clip the signal amplitudes to values never reached by the real process signals (the real process signals are most of the time clearly visible):

$$\tilde{s}_k = f_k^{cl}.s_k$$

with $\tilde{s}_k$ the clipped signal, s_k the measured signal and

$$f_k^{cl} = \begin{cases} s_{max}.[\frac{1}{s_k}] & s_k \geq s_{max} \\ 1 & s_{min} < s_k < s_{max} \\ s_{min}.[\frac{1}{s_k}] & s_k \leq s_{min} \end{cases}$$

2. Compute a trend signal of $\tilde{s}_k$ as described in the previous section. Call this signal $\overline{s}_k$. This signal is plotted in the third figure of Figure 7.5.

3. Compute the standard deviation of the trend-corrected, clipped signal (with s_a the average of $\tilde{s}_k - \overline{s}_k$):

$$\sigma = \sqrt{\sum_{k=1}^{N} [(\tilde{s}_k - \overline{s}_k) - s_a]^2}$$

4. Interpolate all samples of the original signal s_k that are outside a band defined by the trend signal plus and minus α times the standard deviation obtained from the previous step. The permitted signal band is thus given by:

$$x_k = \begin{cases} \overline{s}_k + \alpha\sigma & \text{upper limit} \\ \overline{s}_k - \alpha\sigma & \text{lower limit} \end{cases}$$

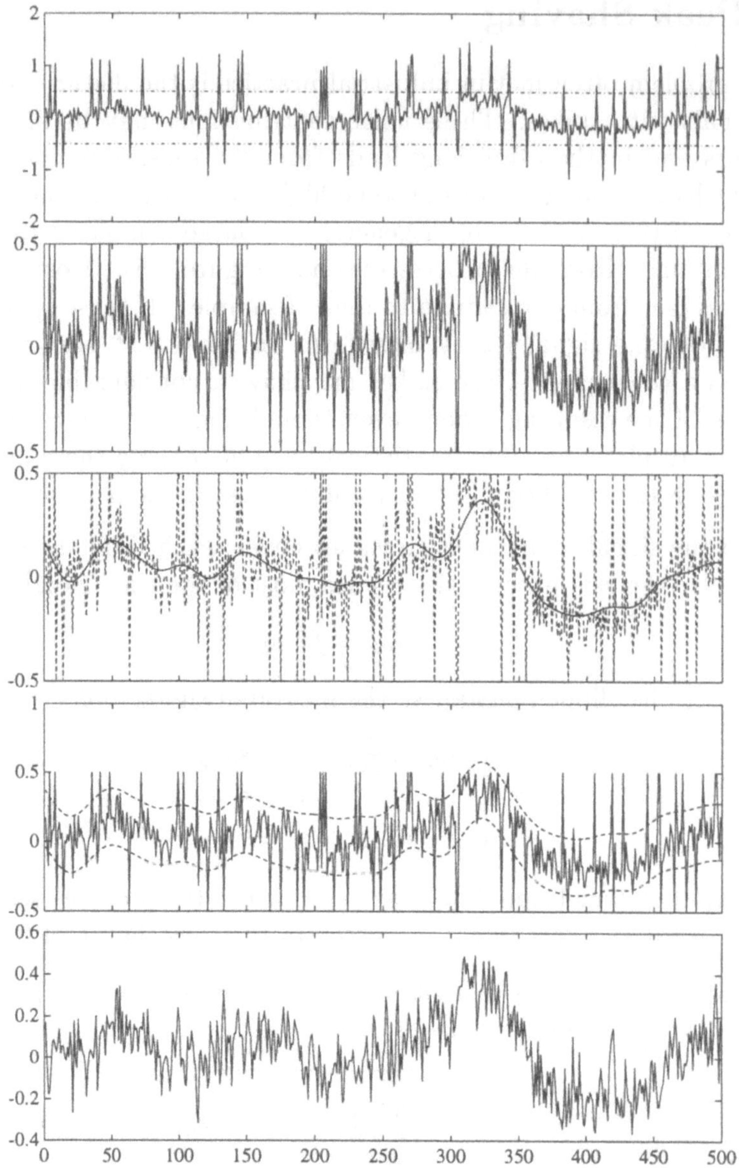

Figure 7.5: The different steps of the peak shaving process. Figure a/ shows the original signal with the peaks. Figure b/ shows the clipped signal $\tilde{s}_k$. Figure c/ shows the trend on top of the signal. Figure d/ shows the 1 σ upper and lower bound. Finally, Figure e/ shows the peak shaved signal. In this example, the interpolation is performed using a second order spline function. Clearly all *spikes* from Figure a/ have been removed in Figure e/.

> All consecutive sample values that are outside the permitted band are replaced by values obtained from a linear interpolation starting from the last accepted value and reaching to the first sample value within the permitted band after the spike.

In the last step, the multiplication factor α has to be chosen such that no process signal values are outside the permitted signal range (typically $\alpha \simeq 3$). Using this signal treatment the influence of the spikes on the model ultimately obtained can be neglected.

7.4 Estimation of Time Delays

In industry, one is often confronted with the problem that the signals to be used for process modeling can only be measured with, sometimes significant, time delays. The measurement of tube dimensions in a glass tube production process may serve as an example (see Figure 7.6). In this case, the sensors used cannot withstand the high temperatures at the place where the shaping of the tube takes place. As a consequence, the tube dimensions can only be measured after the glass tube has reached a temperature to which the sensors can resist. This cooling takes time and this time will appear as an extra time delay in the measured responses. This type of transport delays occurs in many processes.

If measured signals contain time delays, and if the signals are not compensated for these time delays before using them for process identification, the time delays have to be estimated as part of the model. Although they *are* part of the model, the time delays do not contribute to the further description of the dynamic behavior of the process. They just indicate after what time the effect of the inputs becomes noticeable at the outputs. Since these delays introduce extra model parameters to be estimated, it is better to estimate them separately. If the identification is done off-line, the signals can then be compensated for these delays, and a simple model (that only contains the system dynamics) can be estimated from these compensated signals.

Estimation of the delays in process input-output transfer can be done in several ways. A first way is by measuring the step response of the system. Inspection of a step response reveals the number of time samples the process output is not reacting to the input. This number is exactly equal to the time delay. Unfortunately, step response measurements are not always at hand.

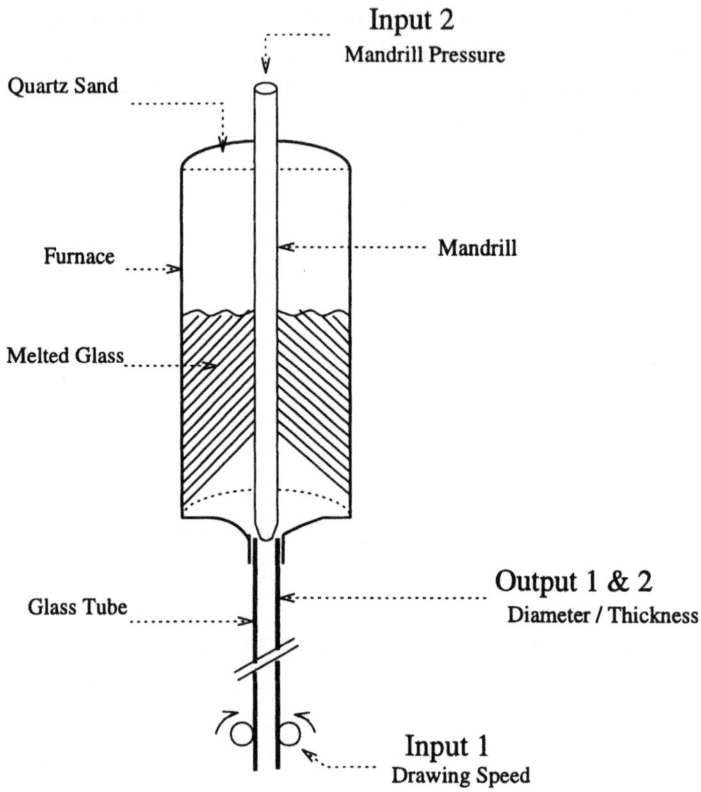

Figure 7.6: A glass tube manufacturing process. To produce glass tubes, quartz sand is fed into the top of the furnace. The sand is melted to glass inside the furnace. The glass tubes are drawn at the bottom. The inputs are drawing speed and mandrill pressure. The drawing speed is the speed at which the tubes are pulled out of the bottom of the machine. The mandrill pressure is the pressure applied to the mandrill at the top of the machine. Through the mandrill, this pressure is applied to the inside of the tubes when they are pulled out of the machine. The important outputs are the geometrical parameters of the tubes: mean diameter and thickness.

On top of that, they are often very noisy, so that it is not immediately clear where the *real* response is starting.

A more *noise-robust* way of estimating time delays is by making use of correlation techniques. One method is based on the analysis of the cross-correlation between process input and output signals. The input signal applied to the process is assumed to be a stationary, white, (inter-channel independent for multi-input multi-output systems), zero mean noise sequence with standard deviation σ:

$$\mathbf{E}[u_k . u_l] = \sigma^2 \delta_{kl}$$

If the input signal is not white, it can be pre-whitened by using the right filters. Furthermore, the input signal is assumed to be uncorrelated with the output noise signal. The process response to an arbitrary input signal may be written as a convolution of the impulse response h_k and the input u_k:

$$y_k = \sum_{i=0}^{\infty} h_i u_{k-i} + v_k \tag{7.1}$$

The cross-correlation between process output and process input is given by:

$$\psi_{yu}(\tau) = \lim_{N \to \infty} \frac{1}{N+1} \sum_{i=0}^{N} y_{i+\tau} u_i$$

Substitution of y_k by (7.1), and using the fact that output noise and inputs are uncorrelated, leads to:

$$\psi_{yu}(\tau) = \sum_{j=0}^{\infty} h_j \psi_{uu}(j - \tau)$$

with ψ_{uu} the auto-correlation function of the input. Since the input is white noise, its auto-correlation is just an impulse at zero. Using this fact, we find:

$$\psi_{yu}(\tau) = \sigma^2 h_\tau$$

Consequently the cross-correlation function obtained will have the shape of the impulse responses in case the input signal is a stationary, white, inter-channel independent, zero mean noise sequence. Otherwise the cross-correlation function is the convolution of the impulse responses with the auto-correlation function of the input signal.

The time delay can be found from the cross-correlation function by looking for the beginning of a sequence of values that differ significantly from zero. The time delay can then be compensated for by shifting the output signal in the right direction until the time delay is zero.

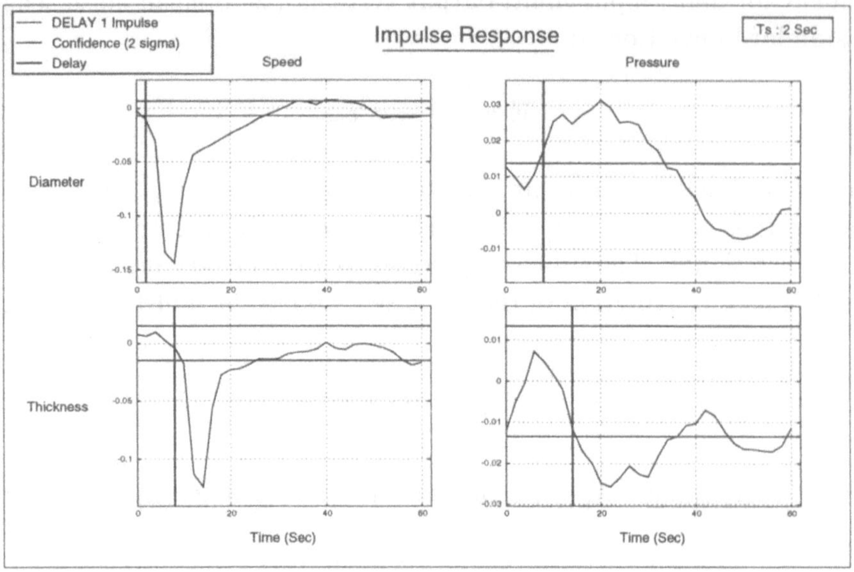

Figure 7.7: Cross-correlation between the inputs and the outputs of the glass tube manufacturing process. The horizontal lines are the confidence bounds, while the vertical lines are the estimated time delays.

As an example we revisit the industrial tube manufacturing process (Figure 7.6). Figure 7.7 shows the cross-correlation between the inputs and outputs. This cross-correlation is similar (within a scaling factor) to the impulse response. The negative correlations are also calculated. Due to the causality of the system, these negative correlations are expected to be zero (the output at time instance 10 can not be influenced by the input at time instance 20). Due to the finite data length and the presence of noise, they are different from zero. However, these negative time lags enable us to estimate confidence bounds on the estimated impulse response.

$$\begin{aligned} \text{upper} &= 3\sigma[\, \psi_{yu}(\tau < 0)\,] \\ \text{lower} &= -3\sigma[\, \psi_{yu}(\tau < 0)\,] \end{aligned}$$

All values outside this confidence region can be considered to be significantly different from zero. These confidence regions provide a helpful tool in estimating the time delays. If the time delay is ambiguous, it is better to underestimate it and then add extra delays during the modeling (try out different delays). This is because an overestimation of the time delay leads to signals that are not causally dependent on each other.

7.5 Filtering

Identification of the dynamic behavior of a process requires quite some experiments to be done. These experiments are needed to acquire the appropriate *a priori* information (e.g., relevant inputs and outputs, sensitivities of the outputs to certain input changes, disturbance characteristics, ...) and for the collection of process data that will be used for the estimation of the model parameters. It is very important that the information obtained from these experiments is reliable, and therefore filtering is required. Filtering has to prohibit aliasing effects and has to reduce the influence of disturbances on the measured signals. However, filtering of the data may give rise to undesirable loss of relevant process data. Therefore the design of appropriate filters is an important step in the configuration of the measuring equipment.

Many of the disturbances entering the recorded process signals do not have their sources in the process. Therefore these disturbances have no relation with the part of the signals relevant for the modeling process. To reduce the influence of disturbances, two measures can be taken:

- Filter out all irrelevant information during the recording of the process data.
- Record extra information of the process in order to use part of that information to increase the ratio between relevant process information and blurring disturbances.

The first action implies a filtering action, generally low pass, applied to the signals, at the time they are recorded. The filters are designed so that they will only pass information in the frequency range that is relevant for the identification of the system dynamics. The filters are also needed to serve as anti-aliasing filters. The design of the filters is mainly governed by the following rules:

- In the pass band, the filters are not allowed to have noticeable influence on the signal characteristics (flat pass band).
- The cut-off frequency of the filters has to be chosen so high that the filters do not introduce any damping at frequencies covered by the process dynamics.
- Disturbances at frequencies beyond the bandwidth of the process have to be removed as much as possible, since this will increase the signal to noise ratio.

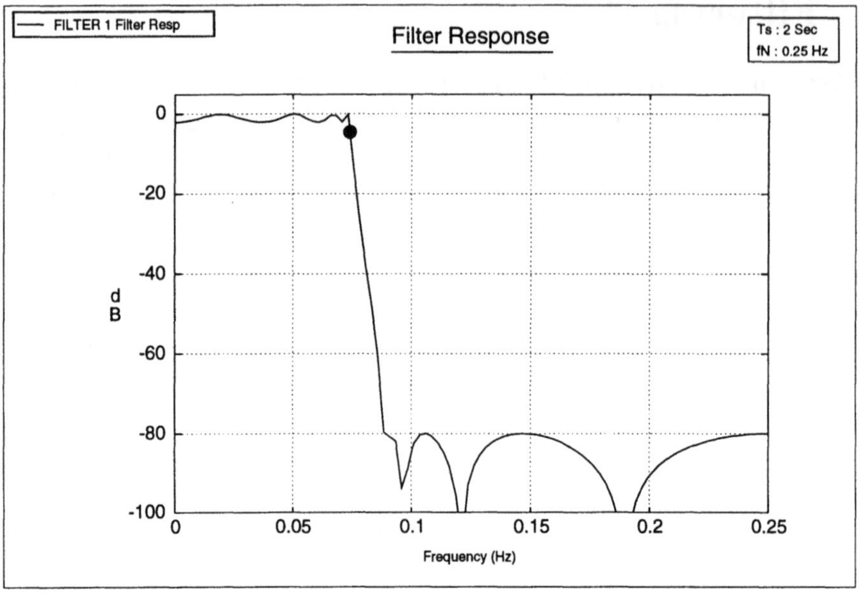

Figure 7.8: An eight order elliptic low pass filter. While such filters are very hard to implement reliably in hardware, they can be easily designed and implemented in the computer.

The second measure involves sampling of the process signals at a frequency higher than the sampling frequency essentially needed for the identification of the process dynamics. The sampling rate is chosen 5 to 10 times the bandwidth of the process. The recorded signals will have much redundancy, that can be used for signal processing (trend removal, peak shaving, estimation of time delays, ...). The basic steps would thus be:

- Sample the system at a high sampling frequency.
- Pre-process the signals.
- The signals recorded at this high sampling frequency will not contain much power relevant at frequencies outside the bandwidth of the system. Only the disturbances can play an important role at these high frequencies. It is highly recommended to filter and decimate (under sample) the signals after the pre-processing, as to bring the sampling frequency down. A major advantage of this procedure is that, once the signals have been recorded in the computer, they can be easily

filtered with an optimally designed filter of high order. These filters are typically impossible to implement in hardware (as capacitors and resistors) but they can be easily designed and implemented in the computer. Figure 7.8 shows an example of a typical filter.

7.6 About Linearity

Most methods for the dynamic black box identification of process models use linear models. No industrial process however is linear. An important step in the modeling of a process is the test whether the amplitude ranges applied to the process are allowable without violating the assumption of linearity in practice. For a process to have a linear transfer the following condition has to be fulfilled (superposition theorem):

$$u_1 \rightarrow y_1 \quad \text{and} \quad u_2 \rightarrow y_2 \quad \Rightarrow \quad \alpha u_1 + \beta u_2 \rightarrow \alpha y_1 + \beta y_2$$

A first test to be done is the test for *steady state linearity* of the process in the surroundings of the working point. If a process has non-linear steady state responses, in most situations linearization is possible with simple linearizing polynomial functions, as will be explained below. A test signal that may be applied separately to each input of the process for testing steady state linearity is the stair-case signal of Figure 7.9.
The time interval τ (equal to 2 on Figure 7.9) between the steps must be chosen in accordance with the response time of the process. For the analysis of the steady state behavior the process must reach its steady state for each step applied. The maximum amplitude of the signal (0.3 on Figure 7.9) has to exceed the amplitude range that has to be covered by the model. Because each separate step in the signal has the same amplitude, also the response to the different steps needs to have the same amplitude if the process is steady state linear. From Figure 7.9, it is clear that the process under consideration here is not linear. The response to the large input amplitudes is clearly saturated.

To test the linearity of the steady state responses a polynomial can be fitted through the steady state amplitudes. Each of the levels of the input stair-case of Figure 7.9 is called u_i. Accordingly, each of the average steady state responses at the output is called y_i. Now, for each i (and there are as many i's as there are stairs in the input signal, say N), we need to have:

$$\beta_\mu u_i^\mu + \beta_{\mu-1} u_i^{\mu-1} + \ldots + \beta_2 u_i^2 + u_i = \alpha_\nu y_i^\nu + \alpha_{\nu-1} y_i^{\nu-1} + \ldots + \alpha_1 y_i + \alpha_0$$

According to this equation the steady state input-output behavior of the system (for this operating point) has to satisfy:

$$\begin{pmatrix} u_1^\mu & u_1^{\mu-1} & \dots & u_1 & -y_1^\nu & \dots & -y_1 & -1 \\ u_2^\mu & u_2^{\mu-1} & \dots & u_2 & -y_2^\nu & \dots & -y_2 & -1 \\ \vdots & \vdots & \vdots & \vdots & \vdots & \vdots & \vdots & \vdots \\ u_N^\mu & u_N^{\mu-1} & \dots & u_N & -y_N^\nu & \dots & -y_N & -1 \end{pmatrix} \begin{pmatrix} \beta_\mu \\ \beta_{\mu-1} \\ \vdots \\ 1 \\ \alpha_\nu \\ \vdots \\ \alpha_0 \end{pmatrix} = P\beta = 0 \tag{7.2}$$

In general, this equation will not hold exactly if the order selected for the polynomials is less than the number of amplitude levels (N) applied to the process. In that case, the polynomial coefficients can be determined through a (total) least squares approximation.

In most practical cases a third order polynomial will be sufficient as a describing curve. For example, a third order polynomial very well describes the beginning of a saturation (as in Figure 7.9).

If non-linearities are found in the steady state behavior of the process, linearizing polynomial functions can be designed to linearize the behavior. Most non-linearities encountered in practice are caused by sensors (e.g., thermocouples) and actuators (e.g., valves). Compensation for these non-linearities can be done by application of a linearizing polynomial function to the corresponding output and/or input signal (see Figure 7.10).

The linearizing functions can be easily found as follows. With the coefficients obtained from Formula (7.2):

$$\begin{aligned} p_u(x) &= \beta_\mu x^\mu + \beta_{\mu-1} x^{\mu-1} + \ldots + \beta_2 x^2 + x \\ p_y(x) &= \alpha_\nu x^\nu + \alpha_{\nu-1} x^{\nu-1} + \ldots + \alpha_1 x + \alpha_0 \end{aligned}$$

we find that (with $p_u^{-1}(x)$ the inverse polynomial function of the polynomial function $p_u(x)$):

$$\begin{aligned} Li &= p_u^{-1}(x) \\ Lo &= p_y(x) \\ u^* &= p_u(u) \\ y^* &= p_y(y) \end{aligned}$$

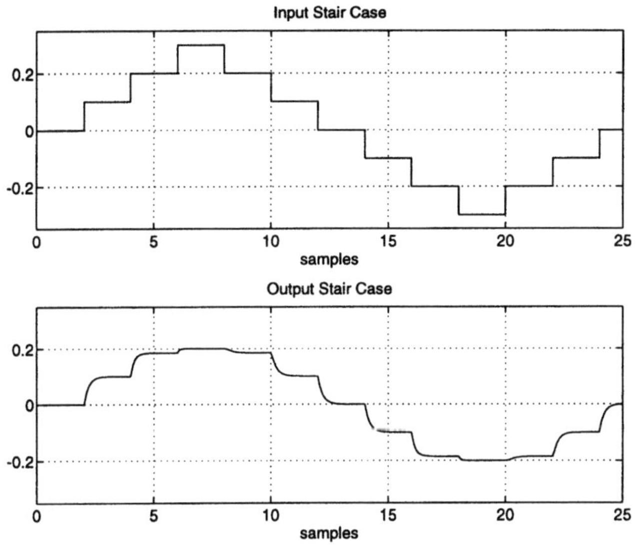

Figure 7.9: An example of a stair-case input and output signal to test for steady state linearity of the process. The output signal is clearly saturated for large input signals.

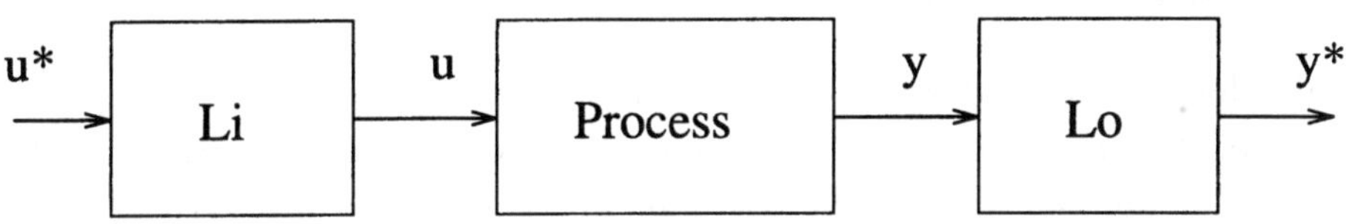

Figure 7.10: Linearizing functions applied to the inputs and outputs of the process.

The corrected inputs (u^*) and outputs (y^*) are shown in Figure 7.11.

Besides the steady state non-linearities, also dynamic non-linearities have to be investigated. A similar test for dynamic non-linearities as the one discussed above is based on the same stair-case test signal, but now as the envelope of an amplitude modulated sinusoidal signal. Application of similar test signals with different frequencies of the carrier wave, gives insight in the non-linearities at different frequencies. The analysis discussed for the steady state behavior can also be done for the various frequency points. However, in practice, linearization of dynamic non-linearities will be more difficult than linearization of static non-linearities.

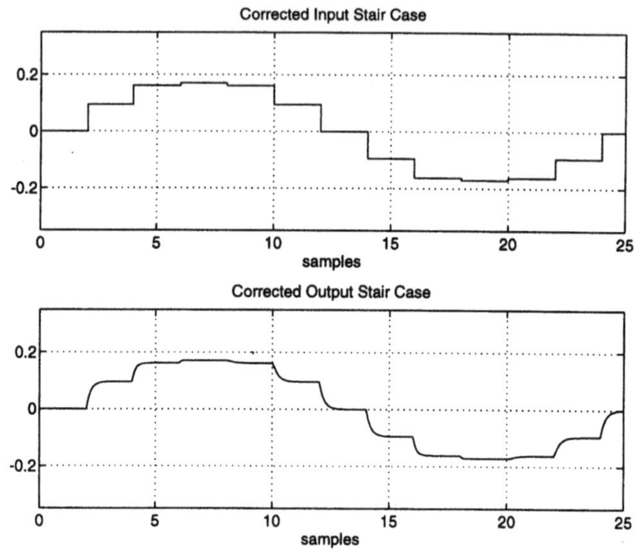

Figure 7.11: The corrected inputs and outputs after linearization. The steady state transfer between these corrected signals is linear.

7.7 Conclusions

In this chapter we have revised some simple techniques to pre-process measured data before using them for system identification. It was described how peak, trend and noise disturbances can be easily removed. We have also treated the estimation and the correction of large time delays. Finally we have provided a simple scheme for compensation of steady state non-linear behavior.

Acknowledgments

Author Peter Van Overschee is a postdoctoral researcher with the Fund for Scientific Research - Flanders.

7.8 Bibliography

1. T. Backx 1987. *Identification of an Industrial Process: A Markov Parameter Approach.* Doctoral Dissertation, Technical University Eindhoven (The Netherlands)

2. L. Ljung 1987. *System Identification: theory for the user.* Prentice-Hall Information and System Sciences Series

3. *Matlab*, The Mathworks Inc., Massachusetts (USA)

4. T. Söderström and P. Stoica 1989. *System identification.* Prentice-Hall International Series in Systems and Control Engineering

5. *Xmath*, Integrated Systems Inc., CA (USA)

6. *Xmath ISID - Build for Interactive System Identification*, Integrated Systems Inc., CA (USA)

Chapter 8

Error Diagnosis and Data Reconciliation Using Linear Conservation Relations

C. HELLINGA*, B. ROMEIN
K.Ch.A.M. LUYBEN and J.J. HEIJNEN
Delft University of Technology
Kluyver Laboratory for Biotechnology
Julianalaan 67 2628 BC Delft (THE NETHERLANDS)
* to whom all correspondence should be addressed

ABSTRACT – *In biotechnological conversion processes, black box process descriptions can easily be derived and are most useful to process measured data. Such descriptions relate the net compound flows to and from the process by means of the conservation laws for chemical elements, charge, enthalpy, ... When several compound flows (conversion rates) are measured, other conversion rates can generally be calculated this way. If a calculated conversion rate is also measured, one has obtained more than one estimate of that rate. By comparing these estimates, measurement errors or errors in the black box description can be detected and sometimes even be located. When the estimates are combined, a more accurate estimate for the conversion rate is obtained. This chapter provides an introduction to these techniques for fault detection and data reconciliation.*

KEYWORDS – *Error Diagnosis, Fault Detection, Data Reconciliation, Conservation Relations*

J.F.M. Van Impe et al. (eds.), Advanced Instrumentation, Data Interpretation, and Control of Biotechnological Processes, 211-244.

8.1 Introduction

Essential concentration levels or reaction rates in biochemical conversions are often difficult to monitor because appropriate sensors are not available, or measurements are too inaccurate. At the same time measurement technology is rapidly developing and in fermentation industry, an increasing amount of process data will become available. It is therefore necessary to examine methods that allow a proper transformation of measured data into *derived values* that are more significant for identifying the process state. A mathematical operation on two independent measurements with limited accuracy will generally lead to a figure that is less accurate than either of the two. Hence, it is important to reduce the inaccuray in the raw data as far as possible. This contribution focusses on the use of linear conservation relations to (i) detect or even diagnose errors in data sets, (ii) improve data accuracy (i.e., suppress stochastic variations), and (iii) calculate non-measured quantities from measured ones.

8.2 Measurement Inaccuracies

8.2.1 Stochastic variations

Every measurement is inaccurate to some extent. Measurements are always corrupted with stochastic deviations or noise. Part of this noise originates from the measurement device itself: the measurement noise. The rest is generated by the process and is called system noise.

$$\textit{total noise} = \textit{measurement noise} + \textit{system noise}$$

Noise is inherent to all physical systems but its level can be reduced by appropriate measures in terms of measurement and process design. Once the data are available they can be filtered to further reduce the stochastic variations. It is important to note that even small variations can lead to considerable inaccuracy in derived values. An example may be useful to illustrate this.

EXAMPLE: oxygen analysis in the gas phase

A paramagnetic analyzer is quite commonly used for oxygen concentration measurements in the exhaust fermentor gas. Typically, such an instrument is accurate up to the last displayed digit, corresponding to 0.01 vol% oxygen.

So, if air from the environment is lead through the instrument, the display could show 20.94-20.96 vol%. This is indicated as an absolute measurement error of 0.01 vol% (relative error: 0.05 %). If the analyzer is connected to the off-gas of a fermentor, the display could show for instance 19.92-19.98 vol%. Assuming that the absolute measurement error is independent of oxygen content in the gas, the system noise results in an absolute error of 0.02 vol%. Measurement noise is related to principal and practical limitations of the measurement device, but maybe also depending on the way the *sensor* is interfaced to the sample, to signal transport, and so on. In this example, measurement noise can be decreased by stopping the air flow through the device before the display is read out, as flowing air influences the torque measurement in such an instrument. By taking the average value of a number of samples, simple *statistical filtering* can be applied in addition. This way, the absolute error can typically be decreased by a factor 10. System noise can for example be due to variations in the gas flow rate. When the oxygen uptake rate by the organisms is constant, a varying flow rate leads to oxygen concentration fluctuations in the effluent gas. In larger scale installations more prominent disturbances are generated by variations in the size distribution of air bubbles rising in the fermentor (even slugs with the size of the reactor diameter can occur in industrial bubble columns). This affects the gas hold-up and the oxygen transfer rate to the liquid phase directly, so that again the oxygen concentration in the exhaust gas fluctuates. ■

8.2.2 Systematic deviations

Filtering can not deal with systematic deviations (bias), that can be due for example to drift or an incorrect calibration of the measurement device. Commonly, bias can be eliminated, for example by re-calibration. However, if it is not detected, it will lead to an erroneous identification of the process state. It is therefore important to have methods that can distinguish bias from noise.

8.2.3 Derived values

Often, measured process values are combined into one figure that is more informative about the state of the process than the individual measured values. An example is the oxygen uptake rate (OUR) of a culture. It is a direct measure for biomass activity, especially when it is expressed per unit biomass dry weight. Under (near) steady state conditions, the OUR equals the oxygen transfer rate (OTR):

$$OTR = \Phi_{air} \cdot (C_{O,in} - C_{O,out}) \tag{8.1}$$

To calculate the (maximum) error in the OTR, some simple error propagation rules, which are included in Appendix A, will be used here. For the above example, it follows from these rules that the error in the concentration difference is $2 \cdot 0.03 = 0.06$ vol%, which is a relative error of 6 %. Assuming an air flow rate of 1 L/min, with a relative error of 0.5 %, the relative error in OTR becomes 6.5 %. A very large number compared to the 0.05 % relative error in the primary oxygen measurements!

Things get even worse if the air flow rate through the fermentor is not constant. Then, the off-gas flow rate must be measured. The air flow rate can vary due to:

- a difference between oxygen uptake by the micro-organisms and their carbon dioxide production rate
- exchange of other volatile/gaseous compounds

Using the same numbers as above, and assuming a relative error in the exhaust flow rate measurement of 0.5 %, the absolute error in the OTR is now calculated according to:

$$\begin{aligned}\Delta OTR &= (\Delta\Phi_{air,in}/\Phi_{air,in} + \Delta C_{O,in}/C_{O,in}) \cdot \Phi_{air,in} \cdot C_{O,in} \\ &+ (\Delta\Phi_{air,out}/\Phi_{air,out} + \Delta C_{O,out}/C_{O,out}) \cdot \Phi_{air,out} \cdot C_{O,out} \quad (8.2)\end{aligned}$$

yielding a value of 0.0026 L oxygen/min. Hence the relative error in OTR is 26 %! It may be clear that considerable errors may appear even after elementary data manipulations.

8.3 Introduction to Gross Error Detection and Data Reconciliation

8.3.1 Gross error detection

To detect an error in a measured quantity, the measured value must be compared with some reference:

1. When only one measurement device is used, physical limits can be put on either the measured value or on its rate of change. Limits to the value can either be absolute or relative to the mean value within a certain time frame.

2. When the quantity is being measured with several independent devices (the quantity is overdetermined), the measured values can be mutually compared.

3. Measured values of other quantities can be used to calculate the one under consideration. The measured value is now compared with an estimated value. The quantity is overdetermined as well. Other measurements in combination with *a priori* knowledge (e.g., a model) serve as the reference.

A test for item (1) is relatively easy to implement. *Hard* physical limits as well as the maximum possible rate of change can be specified rather easily. However, such a test only reveals severe malfunctioning of the measurement device. Drift, calibration errors, and other errors that are often relatively small can not be detected this way. The items (2) and (3) provide methods that are much more sensitive. These methods can indicate *outliers*: values that deviate much more from other measured values within a certain time frame than would be expected from the statistical properties of the stochastic value. This is also termed a *gross error.* For example, a difference between two independent values of 3 times the standard deviation could be a clear test boundary. The chance for such an event to happen when no error is present (and the stochastic values are normally distributed, having approximately the same standard deviation) is only 2 %. In Appendix B a very short introduction to some elementary statistics is provided. A statistical test reveals gross errors rather than systematic deviations. A certain series of gross errors in time can of course indicate a systematic deviation. Therefore, in the remainder of this text the term gross error will prevail.

The key to error detection and diagnosis is that the measured quantity is *overdetermined.* For example, when two thermocouples are used to measure the temperature, the temperature measurement is one time overdetermined. If the two values differ significantly (i.e., in relation to the standard deviation of the measurement method), it is likely that one of the two sensors produces erroneous values.

In a system that is at least one time overdetermined, a gross error can be detected.

If a third thermocouple is added (the temperature is now two times overdetermined), and one sensor yields a value that differs significantly from the other two, it is most likely that the deviating sensor is the erroneous one.

In a multiple overdetermined system, a systematic error can not only be detected, but its source can be identified as well.

In the above example simply more than one sensor was used for the same measurement. If different quantities are measured, which are related to each other via some (set of) mathematical equation(s), they can be used to predict each others value.

Process models can supply the equations between measurements of different process variables to become overdetermined without performing multiple measurements for one variable.

In (bio-)chemical processes, linear equations expressing the elemental conservation laws with respect to the chemical elements (and/or enthalpy, charge, ... [6]), can conveniently be used to provide such relations between the conversion rates of involved reaction compounds. Conversion rates are typically determined by multiplying a measured flow rate with a concentration difference (continuous reactor, chemostat), or by multiplying a concentration difference with a reactor volume and divide it by the elapsed time between the concentration measurements (batch reactor). Combinations are possible as well (fed-batch reactors or transient behavior of a continuous reactor).

8.3.2 Data reconciliation

When there is no evidence for any error, the overdetermination can be used to *improve the accuracy of the data.* The more the system is overdetermined, the higher the gain in accuracy. If a value is overdetermined, one has more than one estimate for its value: the measurement(s) and/or calculated values. All estimates have their own inaccuracy. By taking the average value of these estimates, the standard deviation of this average value is smaller than the smallest standard deviation of the original values (see also Appendix B). In fact, it is irrelevant whether these samples have been gathered with one probe or with various probes, as calculated value or not. This way of improving the data accuracy is termed data reconciliation, or (when using balances for the chemical elements, ...) balancing. When linear conservation relations have been formulated, they can be used for calculating non-measurable conversion rates as well. Of course, it makes sense to first try to find evidence for errors, then (if no evidence was found) balance the measured conversion rates, and next calculate the non-measured rates, taking advantage of the improved accuracy of the measured data.

The remainder of this chapter will deal with the mathematical background of data reconciliation, fault detection/diagnosis and calculating non-measurable rates using conservation relations.

8.4 Gross Error Detection and Data Reconciliation Using One Conservation Relation

8.4.1 Gross error detection

The principle of gross error detection will be demonstrated in this section with a simple example. In Figure 8.1 a reaction vessel is schematically drawn in which microbial growth occurs in batch mode. It is assumed that

- nitrogen is fed to the system by means of ammonia to maintain a constant pH value;
- for each consumed ammonia molecule, 1 proton is released, which is the only pH influencing reaction;
- besides biomass no products are formed containing nitrogen;
- the ammonia concentration in the vessel is constant (at a not necessarily known level);
- evaporation of ammonia can be neglected.

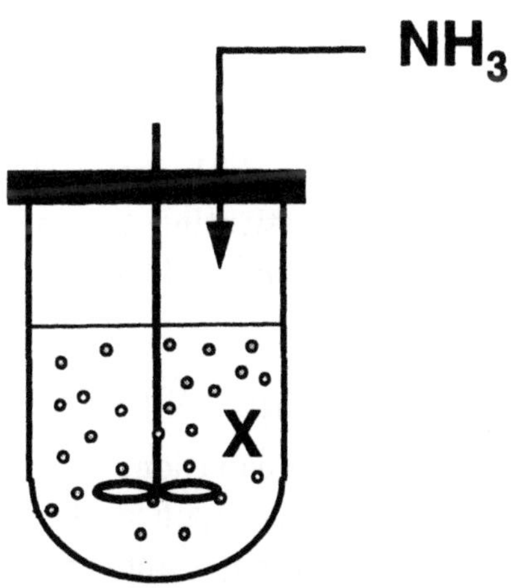

Figure 8.1: Microbial growth in a batch reactor with ammonia addition.

Under these conditions, the nitrogen balance reads as follows:

$$1 \cdot r_{NH3} + 0.15 \cdot r_X = 0 \tag{8.3}$$

It states that the N-supply rate (r_{NH3}) equals the N-uptake rate by the biomass ($0.15 \cdot r_X$). The factor 0.15 originates from our *a priori* knowledge of the elemental composition of biomass ($CH_{1.8}O_{0.5}N_{0.15}$). Note that such a chemical composition formula only indicates the ratio of C, H, O and N inside the biomass. For convenience the number of C-moles is set to 1 in this formula, so that it is said to be on C-mole basis.

Now assume that the ammonia flow to the system is *measured* by multiplying the influent flow rate with its ammonia concentration. As each NH_3 molecule contains one atom N the ammonia flow equals the N flow to the system. Consequently, the biomass production rate can be calculated.

In this example Equation (8.3) is *determined.* The unknown r_X can be calculated from the known r_{NH3}.

> *A system of equations is exactly determined when all unknown variables can be calculated from the known (i.e., measured) variables.*

The accuracy of r_X^c depends on the accuracy of:

- r_{NH3}^m (the NH_3 concentration and the ammonia flow rate)
- the knowledge of the number of atoms N in the two compounds

and on the validity of the above assumptions. The superscripts ^{c} and ^{m} denote calculated and measured values respectively.

When the biomass production rate is measured in addition, Equation (8.3) is *overdetermined.* For now, it will be assumed that the chemical composition formulae are exact, and that the above assumptions hold. It is highly unlikely that after substituting the measured conversion rates into Equation (8.3) this equation will hold exactly (because of the stochastic variations). Instead we read:

$$1 \cdot r_{NH3}^m + 0.15 \cdot r_X^m = \varepsilon \tag{8.4}$$

ε is the residual of the balance equation.

If the absolute value for ε is significantly larger than the weighted sum of the standard deviations for r_{NH3} and r_X ($s_{r}{}_{NH3} + 0.15 \cdot s_{r}{}_X$) then the conclusion could be that r_{NH3} and r_X are not consistent. Before such a consistency test will be formulated more precisely, the subject of balancing experimental data will be dealt with in the next section.

8.4.2 Data reconciliation

The previous example can be elaborated to show how overdetermination of a quantity can be used to decrease the inaccuracy of its value. When no proof for an inconsistency has been found, balancing proceeds as follows. It is demanded that the constraint imposed by Equation (8.3) holds: the sum of N-production and N-consumption must be zero. Both measured conversions can be adjusted to satisfy this equality and the question is how this can be done rationally.

Two correction terms are introduced, denoted by $\hat{\delta}_{NH3}$ and $\hat{\delta}_X$. The diacritical mark ˆ indicates that these correction terms are estimated. The *real* correction terms (that are unknown) are defined as:

$$\delta_{NH3} = r_{NH3} - r^m_{NH3} \quad (8.5)$$

$$\delta_X = r_X - r^m_X \quad (8.6)$$

Combining the Equations (8.3), (8.4), (8.5) and (8.6) gives:

$$1 \cdot \delta_{NH3} + 0.15 \cdot \delta_X = \varepsilon \quad (8.7)$$

ε is known (8.4) and it is demanded that the estimates for δ_{NH3} and δ_X obey in analogy to Equation (8.7):

$$1 \cdot \hat{\delta}_{NH3} + 0.15 \cdot \hat{\delta}_X = \varepsilon \quad (8.8)$$

To solve Equation (8.8) a criterion is needed. It can be proven [5] that the best solution for $\hat{\delta}_{NH3}$ and $\hat{\delta}_X$ is found when the following function J is minimized:

$$J = \hat{\delta}^2_{NH3}/\mathrm{var}(\delta_{NH3}) + \hat{\delta}^2_X/\mathrm{var}(\delta_X) \quad (8.9)$$

δ_i is used to compensate for the stochastic variations in r_i, so that its distribution must be equal to the distribution of r_i. Since the errors in r_i are supposed to be normally distributed (see Figure 8.2), $\mathrm{var}(\delta_i) = \sigma^2_{ri}$, which can be estimated from the measurements (s^2_{ri}, Appendix B). Equation (8.9) states that

> *The more inaccurate the measurement, the larger the correction term may be.*

The solution to the Equations (8.8) and (8.9) is given by [1]:

$$\hat{\delta}_{NH3} = 1 \cdot \mathrm{var}(\delta_{NH3}) \cdot \varepsilon/\mathrm{var}(\varepsilon) \quad (8.10)$$

$$\hat{\delta}_X = 0.15 \cdot \mathrm{var}(\delta_X) \cdot \varepsilon/\mathrm{var}(\varepsilon) \quad (8.11)$$

var(ε) can be estimated from (8.7) recalling the rules (a,b: stochastic variables; k: constant):

$$\text{var}(a+b) = \text{var}(a) + \text{var}(b)$$
$$\text{var}(k \cdot b) = k^2 \cdot \text{var}(b)$$

$$\text{var}(\varepsilon) = 1^2 \cdot \text{var}(\delta_{NH3}) + 0.15^2 \cdot \text{var}(\delta_X) \tag{8.12}$$

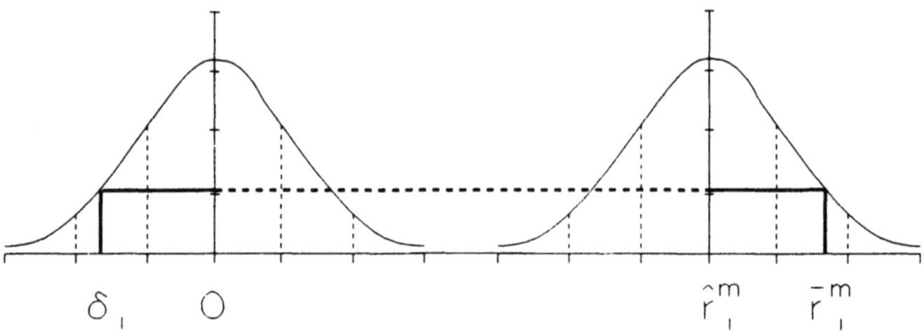

Figure 8.2: Relation between the distributions of the correction term δ and the conversion rate r.

It can easily be verified that Equations (8.10) and (8.11) obey Equation (8.8). So, the estimated values for both conversion rates are:

$$\hat{r}_{NH3} = r^m_{NH3} + \hat{\delta}_{NH3} \tag{8.13}$$
$$\hat{r}_X = r^m_X + \hat{\delta}_X \tag{8.14}$$

It can be shown [11] that the standard deviations of the estimated values for the conversion rates (8.13) and (8.14) are smaller than the standard deviations of the original measurements. *Hence we gain accuracy by balancing our data.* The balanced values obey by definition the balance equations.

Again the prerequisite is stressed that the balance equations must be valid (no compounds may be omitted and the chemical composition of the compounds must be known accurately).

8.4.3 Gross error detection: a statistical criterion

The statistical test that was indicated intuitively in a previous section can now be formulated more precisely. Let us consider the quantity h [8]:

$$h = (\hat{\delta}_{NH3})^2/\mathrm{var}(\delta_{NH_3}) + (\hat{\delta}_X)^2/\mathrm{var}(\delta_X) \quad (8.15)$$

If h is large, the correction terms are large with respect to the standard deviations of the original measurements. Again, intuitively it is reasonable to reject the hypothesis that there are no measurement errors if $h >> 2$ for Equation (8.15). An exact measure can be obtained when it is realized that h possesses a χ-square distribution, because all of its terms have been squared. In a χ-square table a value of 5.02 can be found at the 97.5 % confidence level. For the number of degrees of freedom, one must choose the number of independent equations that formed the set of balance equations (in this example 1).

Hence, if in this example $h > 5.02$, one can conclude with a probability of 97.5 % that there is something wrong. One or both conversion rates can be wrong, but there can also be mistakes in our assumptions. A discussion on the nature of errors can be found in the section on fault diagnosis.

8.5 Gross Error Detection and Data Reconciliation Using Multiple Conservation Relations

8.5.1 Formulation of the equations

Instead of using only the N-balance, the method can take into account all linear conservation relations that are relevant for the system under consideration. Here, the method will be worked out using an ethanol producing yeast fermentation as an example with 4 elemental balances (C, H, O, and N).

To write down the conservation relations, all relevant compounds and their elemental composition must be known. The reaction itself is considered as a black box[1]: no assumptions are made with respect to reaction

[1] For this reason, the overall model is often termed a black box process model as well. This may be confusing as pure black box models are commonly only flexible mathematical structures for describing process input-output relations. Here, constraints based on *a priori* process knowledge are applied to process inputs-outputs so that the model should rather be considered as a (dark) grey box.

mechanisms (kinetics) or yields (stoichiometry). Only compounds appearing or disappearing in the environment of the (bio-)catalysts are taken into account (Figure 8.3).

Figure 8.3: Black box representation of a (bio-)chemical conversion.

The following overall reaction equation is another way to represent the process in relevant terms:

$$(\cdot)C_6H_{12}O_6 + (\cdot)NH_3 + (\cdot)O_2 \rightarrow 1CH_{1.8}O_{0.5}N_{0.2} + (\cdot)H_2O + (\cdot)CO_2 + (\cdot)C_2H_5OH$$

The molecular formulae for glucose, biomass and ethanol can be recognized here. It is assumed that no *a priori* knowledge is available with respect to the reaction rate or the reaction stoichiometry. As the consumed or produced amounts are relative to one another, one of the coefficients may be set to an arbitrary value. Here, the stoichiometric coefficient for biomass has been chosen to equal unity.

The elemental balances can be concisely represented with the aid of the elemental composition matrix and the vector of conversion rates, as shown in Figure 8.4.

This matrix contains exactly the same information as the reaction equation. The C-balance can be recognized from the first matrix row and reads as follows:

$$1 \cdot r_G + 1 \cdot r_C + 1 \cdot r_X + 1 \cdot r_E = 0 \tag{8.16}$$

By convention, rates denoting consumption terms are negative and rates denoting production terms positive. The C-balance is an equation with 4 unknown rates, so if 3 rates are measured, the fourth rate can be calculated, in which case the equation is determined.

Measuring all 4 reaction rates leads to an overdetermined equation, so that fault diagnosis and balancing can be performed.

$$\underbrace{\left(\begin{array}{c|ccccccc} & Gluc & NH_3 & O_2 & CO_2 & Biom & H_2O & Ethanol \\ \hline C & 1 & 0 & 0 & 1 & 1 & 0 & 1 \\ H & 2 & 3 & 0 & 0 & 1.8 & 2 & 3 \\ O & 1 & 0 & 1 & 2 & 0.5 & 1 & 0.5 \\ N & 0 & 1 & 0 & 0 & 0.2 & 0 & 0 \end{array}\right)}_{\text{Elemental composition matrix } \mathbf{E}} \cdot \underbrace{\left(\begin{array}{c} r_G \\ r_N \\ r_O \\ r_C \\ r_X \\ r_W \\ r_E \end{array}\right)}_{\text{Conversion rates vector } \mathbf{r}} = \left(\begin{array}{c} 0 \\ 0 \\ 0 \\ 0 \end{array}\right)$$

Figure 8.4: Representation of 4 elemental balances in terms of the elemental composition matrix and the vector of conversion rates.

8.5.2 Classification of the conversion rates

In case of one equation it is straightforward to see whether the equation is overdetermined or not. For a set of equations it is much more difficult to see whether measured quantities are (multiple) overdetermined and whether non-measured quantities can be calculated (Figure 8.5). This is the central problem in analyzing the potentials of such a system description. In the above example, one could state superficially that there are 4 (balance) equations, and 7 variables (the conversion rates of the relevant compounds), resulting in 3 degrees of freedom. If 3 rates are measured, the other 4 can be calculated. If 4 rates or more are measured the system is one time or multiple overdetermined. A first objection to this reasoning could be that one has to prove first that the relations are not linear dependent (**E** must be of full rank). Another problem is the distribution of the variables over the equations. Consider again the N balance. Thanks to this equation, r_{NH3} and r_X are overdetermined if they are both measured. If these are the only measurements for the system in Figure 8.4, none of the other rates can be calculated. Thus, the system of equations is at the same time overdetermined and underdetermined, which means that one has to make the classification indicated in Figure 8.5 for each individual variable!

Van der Heijden *et al.* [2] developed a mathematical approach to solve the classification problem. It is an integral element of the derivation of generally valid equations for data reconciliation and will be treated in the next section.

BALANCEABLE	NOT BALANCEABLE	measured rates
CALCULABLE	NOT CALCULABLE	non-measured rates

Figure 8.5: Classification of the conversion rates.

8.5.3 The general equations for gross error detection and data reconciliation

To derive generally valid equations for data reconciliation and gross error detection, first the matrix equation indicated in Figure 8.4 is rewritten as:

$$\mathbf{E} \cdot \mathbf{r} = \mathbf{0} \tag{8.17}$$

E is the elemental matrix and **r** the vector of conversion rates. To start with, the conversion rates need to be subdivided into rates that will be measured ($\mathbf{r}^m$) and the other rates that can (possibly) be calculated ($\mathbf{r}^c$). For the given example the following division could be made:

$$\underbrace{\begin{pmatrix} r_G \\ r_N \\ r_O \\ r_C \\ \\ r_X \\ r_H \\ r_E \end{pmatrix}}_{\text{vector } \mathbf{r} \text{ of all rates}} \quad \begin{matrix} \left.\begin{matrix} \\ \\ \\ \\ \end{matrix}\right\} \text{vector } \mathbf{r}^m \text{ of measured rates} \\ \\ \left.\begin{matrix} \\ \\ \\ \end{matrix}\right\} \begin{matrix}\text{vector } \mathbf{r}^c \text{ of non-measured rates} \\ \text{that are possibly calculable}\end{matrix} \end{matrix}$$

Now, matrix **E** can be split into two matrices $\mathbf{E}^c$ and $\mathbf{E}^m$ that correspond to these two vectors, so that Equation (8.17) can be rewritten as:

$$\mathbf{E}^c \cdot \mathbf{r}^c = -\mathbf{E}^m \cdot \mathbf{r}^m \tag{8.18}$$

For the given example, $\mathbf{E}^m$ and $\mathbf{E}^c$ read as follows:

$$\mathbf{E}^m = \begin{pmatrix} 1 & 0 & 0 & 1 \\ 2 & 3 & 0 & 0 \\ 1 & 0 & 2 & 2 \\ 0 & 1 & 0 & 0 \end{pmatrix} \qquad \mathbf{E}^c = \begin{pmatrix} 1 & 0 & 1 \\ 1.8 & 2 & 3 \\ 0.5 & 1 & 0.5 \\ 0.2 & 0 & 0 \end{pmatrix}$$

The vector of rates that are not measured, can be expressed in terms of the measured rates according to:

$$\mathbf{r}^c = -(\mathbf{E}^c)^\# \cdot \mathbf{E}^m \cdot \mathbf{r}^m \tag{8.19}$$

The $^\#$ mark indicates the *pseudo-inverse* of a matrix [7]. The normal inverse is not appropriate here, as $\mathbf{E}^c$ may be not invertible (one is completely free in splitting the vector of conversion rates). The pseudo-inverse always exists, which means that every element of $\mathbf{r}^c$ gets a value with Equation (8.19). Since $\mathbf{r}^c$ contains by definition all elements of $\mathbf{r}$ that are not measured, the calculability of the elements of $\mathbf{r}^c$ must be investigated (see below) before the meaningful elements of $\mathbf{r}^c$ can be selected.

Now, by substituting Equation (8.19) again in Equation (8.18) a non-trivial expression results [2], due to the properties of the pseudo-inverse matrix:

$$\underbrace{\left(\mathbf{E}^m - \mathbf{E}^c \cdot (\mathbf{E}^c)^\# \cdot \mathbf{E}^m\right)}_{\text{redundancy matrix } \mathbf{R}} \cdot \mathbf{r}^m = 0 \tag{8.20}$$

Van der Heijden *et al.* [2] call the resulting matrix expression in the term to the left hand side the *redundancy matrix* $\mathbf{R}$. With $\mathbf{R}$ a concise expression can be obtained that contains only the measured conversion rates:

$$\mathbf{R} \cdot \mathbf{r}^m = \mathbf{0} \tag{8.21}$$

If rank $\mathbf{R} = 0$ no rates can be balanced ($\mathbf{E}^c$ is normally invertible). If rank $\mathbf{R} > 0$, the columns of $\mathbf{R}$ can be inspected to find out *which* rates are balanceable; a column with only zeros corresponds to a measured rate that is not balanceable.

Moreover, matrix $\mathbf{R}$ indicates whether all non-measured rates of $\mathbf{r}^c$ are calculable or not. Let e be the number of independent linear equations in matrix $\mathbf{E}$ (i.e., rank $\mathbf{E}$] and c be the number of elements in vector $\mathbf{r}^c$, then all rates are calculable if rank $\mathbf{R}$ has its maximum value, being $e - c$. If not all elements of $\mathbf{r}^c$ are calculable, a *Singular Value Decomposition* [8] of matrix

$\mathbf{E}^c$ is useful to make a distinction between the calculable and not calculable ones. Figure 8.6 summarizes this analysis.

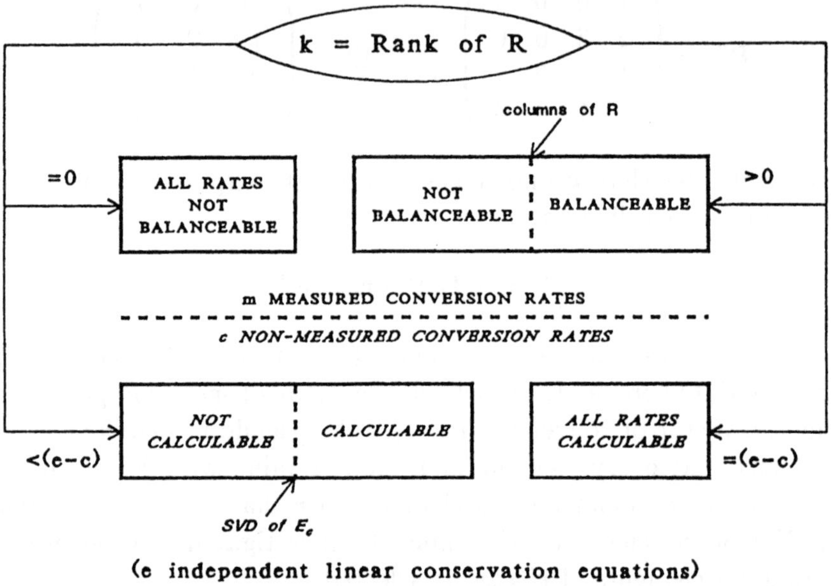

Figure 8.6: The use of matrix $\mathbf{R}$ and matrix $\mathbf{E}^c$ to classify the conversion rates.

This method can be used to investigate *a priori* whether a given set of measurements is appropriate to calculate non-measured conversion rates, or for performing gross error detection and data reconciliation. When a proper measurement set has been chosen for obtaining an overdetermined system of equations, data reconciliation proceeds as follows.

In Equation (8.21) the vector of measured rates is replaced by the vector of (measured and) balanceable rates ($\mathbf{r}^{m'}$). Also, the rows and the columns of matrix $\mathbf{R}$ that are related to the non-balanceable rates are removed. This way, an expression is obtained for the balanceable rates only:

$$\mathbf{R}' \cdot \mathbf{r}^{m'} = \mathbf{0} \tag{8.22}$$

Now, the vector of actual measurements for the balanceable rates $\bar{\mathbf{r}}^{m'}$ can substitute $\mathbf{r}^{m'}$. Due to the presence of noise the product of $\mathbf{R}'$ and $\bar{\mathbf{r}}^{m'}$ will not yield the zero vector, but a vector of residuals ε instead:

$$\mathbf{R}' \cdot \bar{\mathbf{r}}^{m'} = \varepsilon \tag{8.23}$$

The variances and covariances of the elements of ε (these elements are correlated because the elements of the (independent) measurements $\bar{\mathbf{r}}^{m'}$ appear in various equations) are expressed in a covariance matrix $\mathbf{P}_{\boldsymbol{\varepsilon}}$. A covariance matrix contains the variances of a vector in its diagonal elements and the covariances between elements of that vector in the off-diagonal elements.

$$\mathbf{P}_{\boldsymbol{\varepsilon}} = \mathbf{R} \cdot \mathbf{P}_{\tilde{\mathbf{r}}^m} \cdot \mathbf{R}^T \tag{8.24}$$

$\mathbf{P}_{\tilde{\mathbf{r}}^m}$ is the covariance matrix of the balanceable conversion rates. If the measurements are not cross-correlated, $\mathbf{P}_{\tilde{\mathbf{r}}^m}$ is a diagonal matrix containing the (estimated) variances of the measured rates. The χ-square distributed test function h, that was seen already in Equation (8.15), is now calculated for the situation with many balance equations, according to:

$$h = \boldsymbol{\varepsilon}^T \cdot \mathbf{P}_{\boldsymbol{\varepsilon}}^{-1} \cdot \boldsymbol{\varepsilon} \tag{8.25}$$

Note that h can be calculated before balancing has been performed.

h is again compared with tabulated values for the χ-square distribution on the required confidence level. The number of degrees of freedom of the χ-square distribution equals the rank of matrix $\mathbf{R}'$ (i.e., its number of rows if all dependent rows were removed from $\mathbf{R}$).

EXAMPLE

Suppose $\mathbf{R}'$ has two rows, and h is to be tested at the 97.5 % confidence level. Then the tabulated value for the χ-square distribution reads 7.38. If $h \geq 7.38$ the hypothesis that no error is present is rejected. If $h < 7.38$ there is no evidence for any error at this level of confidence, and the equations can be used for balancing. ■

The balanced rates follow directly from:

$$\hat{\mathbf{r}}^m = (\mathbf{I} - \mathbf{P}_{\tilde{\mathbf{r}}^m} \cdot (\mathbf{R}')^T \cdot \mathbf{P}_{\boldsymbol{\varepsilon}}^{-1} \cdot \mathbf{R}') \cdot \tilde{\mathbf{r}}^m \tag{8.26}$$

The variances in the elements of $\hat{\mathbf{r}}^m$ can be found on the diagonal of the covariance matrix $\mathbf{P}_{\hat{\mathbf{r}}^m}$:

$$\mathbf{P}_{\hat{\mathbf{r}}^m} = \mathbf{P}_{\tilde{\mathbf{r}}^m} - \mathbf{P}_{\tilde{\mathbf{r}}^m} \cdot (\mathbf{R}')^T \cdot \mathbf{P}_{\boldsymbol{\varepsilon}}^{-1} \cdot \mathbf{R}' \cdot \mathbf{P}_{\tilde{\mathbf{r}}^m} \tag{8.27}$$

For the best estimation of the value of the calculable rates ($\hat{\mathbf{r}}^c$) the balanced values $\hat{\mathbf{r}}^m$ are used, because they have a smaller standard deviation than the measured rates. This way, the variances of $\hat{\mathbf{r}}^c$ are kept minimal. These variances are again found on the diagonal of the covariance matrix $\mathbf{P}_{\hat{\mathbf{r}}^c}$:

$$\hat{\mathbf{r}}^c = -(\mathbf{E}^c)^{\#} \cdot \mathbf{E}^m \cdot \hat{\mathbf{r}}^m \tag{8.28}$$

$$\mathbf{P}_{\hat{\mathbf{r}}^c} = (\mathbf{E}^c)^{\#} \cdot \mathbf{E}^m \cdot \mathbf{P}_{\hat{\mathbf{r}}^m} \cdot (\mathbf{E}^m)^T \cdot ((\mathbf{E}^c)^{\#})^T \tag{8.29}$$

In Appendix C a computer program (MACROBAL) is presented that can be used for classifying the conversion rates and subsequently for performing data reconciliation and calculating the non-measured rates.

8.6 Gross Error Diagnosis

If the value of h is too large according to the χ-square test on the selected confidence level, an error may be present. Here, some methods will be presented to *locate* the error and estimate its magnitude, if the measurements are multiple overdetermined. The potential error sources that can be distinguished are:

1. At least one of the conversion rates was determined wrongly.
2. Some component(s) was (were) not taken into account in the black box process description.
3. One or more chemical composition formulae are wrong.

In addition, it is important to realize that a high value for h can also be caused by too small estimates for the standard deviations of the conversion rates. If, on the contrary, the specified standard deviations are too large, the error diagnosis system becomes too insensitive. It cannot be stressed enough that the correct (estimates for the) standard deviations must be used. Determining these standard deviations is not always so easy. Measurements must be carried out under real process conditions, as the contributions of both measurement noise and system noise must be taken into account. Moreover, the *measured* quantities are often derived values (such as the conversion rates) so that the standard deviation must be calculated from those of the primary measurements [4].

In this section the vector comparison test is described that uses the *direction* of $\boldsymbol{\varepsilon}$ (the relative size of its elements) to identify the error source. Note that first the statistical test on the *length* of $\boldsymbol{\varepsilon}$ (Equation (8.25)) must be used to indicate the presence of errors. As will be explained, the test can be used for all three error categories mentioned above. The main idea is that hypotheses with respect to possible errors are translated into vectors with the dimension of $\boldsymbol{\varepsilon}$. If the directions of $\boldsymbol{\varepsilon}$ and the compare vector are the same, according to some statistical criterion, the hypothesis that the error expressed by the compare vector is present, is not rejected. To demonstrate this, first the detection of errors in the measured conversion rates will be discussed.

8.6.1 Measurement errors

In order to understand how the direction of ε is related to the conversion rates, the (unrealistic) situation is considered that all rates from Figure 8.4 have been measured and that $\mathbf{E}$ is of full rank. Equation (8.23) can then be rewritten as:

$$\mathbf{E} \cdot \bar{\mathbf{r}}^m = \varepsilon \tag{8.30}$$

Figure 8.4 can now be used (replace the zero vector at the right hand side with the ε vector) to analyze the direction of ε. Suppose the biomass conversion rate is the only rate that contains an error. If it is for example 4 instead of the true value 3, then ε becomes:

$$\varepsilon = \begin{pmatrix} 1.0 \\ 1.8 \\ 0.5 \\ 0.2 \end{pmatrix}$$

which is the elemental composition of biomass! If ε was 5 instead of 3, ε would have been twice as long. In the general situation when Equation (8.23) must be applied instead of Equation (8.30), ε will resemble the column of $\mathbf{R}'$ belonging to the erroneous rate (for example r_e, the conversion rate of compound e) in $\bar{\mathbf{r}}^m$. This column R'_e is related to the elemental composition of compound e and it will usually have a unique direction. Again, the length of ε, divided by the length of R'_e, is a measure for the absolute error in r_e. In conclusion:

> *By comparing the direction of ε with the directions of the columns of the reduced redundancy matrix $\mathbf{R}'$ (8.23), an erroneous conversion rate can be identified. The size of this error can be derived from the length of ε.*

In order to get an objective measure for the resemblance of two vectors, a statistical test was developed by Van der Heijden *et al.* [3]. This *vector comparison test* is illustrated in Figure 8.7. The circle around the arrow head of vector ε indicates the (for example) 95 % confidence region of the direction of ε. If (an extended) vector representing one of the columns of $\mathbf{R}'$ runs through this region, it can be concluded with 95 % certainty that ε resembles this column. If columns have about the same direction, it becomes difficult to distinguish between different possible error sources.

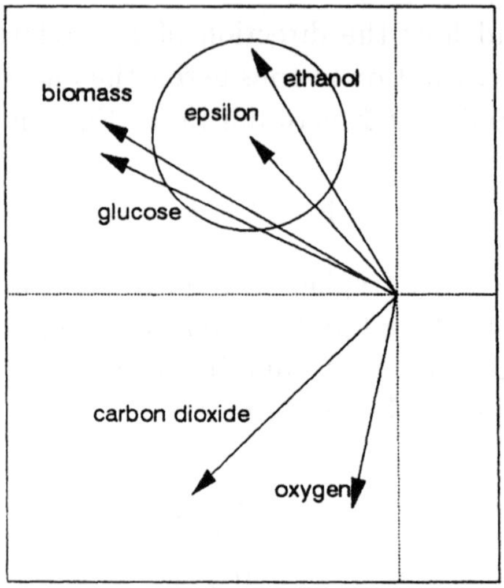

Figure 8.7: The vector comparison test used to locate errors in the measured and overdetermined conversion rates.

It is important to note that the vector comparison test can only be performed if the rank of $\mathbf{R}'$ is at least 2 (otherwise ε is a scalar and has no direction). This means simply that the system of Equations (8.23) must at least be two times overdetermined. It was mentioned before that this is a prerequisite for finding the error location.

Also, ε can be compared with linear combinations of the columns of $\mathbf{R}'$, in order to look for simultaneous errors. It should be noted, that when the number of columns in $\mathbf{R}'$ is larger than its number of rows, an arbitrary vector ε can be constructed with different linear combinations of columns of $\mathbf{R}'$. In other words, the higher the rank of $\mathbf{R}$ (the more the system is overdetermined), the more discriminative the error indication generally will be.

As an alternative, the *serial elimination method* can be used to look for a measurement error [9,11]. It is a very simple test. If an error is detected, one of the measured and overdetermined measurements is left out and the whole procedure is repeated (Equations (8.18) to (8.23)). If this results in a value for $h(\varepsilon)$ that is below the χ-square test value[2], then the rate that has been left out is concluded to be the erroneous one.

[2] The number of degrees of freedom for the χ-square test decreases by one for every measurement left out.

As this second test can only be performed on a balanceable set of equations, again, the rank of the original matrix $\mathbf{R}'$ (i.e., before one of the rates is left out) must be larger than 1. The serial elimination method has a few serious drawbacks compared to the vector comparison test:

- The serial elimination method may not always lead to the correct conclusion (this corresponds to the situation where two compare vectors have more or less the same direction).
- One can not see *a priori* which errors can be and which cannot be diagnosed.
- The magnitude of the gross error can not be estimated.

Moreover, errors in the elemental composition matrix can not explicitly be looked for with the serial elimination method. The next section will be used to show how and under which conditions such errors can be found with the vector comparison test.

8.6.2 System definition errors

The vector comparison test can also be employed to look for missing compounds in the system description, or a wrong specification of one or more of the elemental compositions.

To start with the omitted component, it is important to note that it can be of known composition (ethanol formation during yeast production) or not. If its composition is known, the system description can be extended by adding an extra column to $\mathbf{E}$ ($\mathbf{E}_{n+1}$) and an extra element to $\mathbf{r}$ (r_{n+1}, see Equation 8.17). This reduces the degree of overdetermination with one, and produces an estimate for r_{n+1}. If the new value of h is now acceptable, an indication is obtained (not a proof!) that the proposed component was lacking indeed. However, it should be noted that r_{n+1} will always get a non-zero value, due to the presence of noise on the measurements, ... It is therefore recommended not to extend the elemental composition matrix until an indication has been obtained for the presence of the missing compound. This can be achieved, by constructing a reference vector for an additional vector comparison test. The direction of this vector follows from [3]:

$$\varepsilon = \mathbf{K} \cdot (\mathbf{E}^c \cdot (\mathbf{E}^c)^{\#} - \mathbf{I}) \cdot \mathbf{E}_{n+1} \cdot k \tag{8.31}$$

k is some arbitrary scalar. If the test provides evidence for the existence of compound $n+1$, then k equals the estimated value for r_{n+1}. $\mathbf{K}$ is a matrix that can be used to obtain $\mathbf{R}'$ from $\mathbf{R}$:

$$\mathbf{R}' = \mathbf{K} \cdot \mathbf{R} \tag{8.32}$$

If Equation (8.31) is rewritten to:

$$\mathbf{E}_{n+1} \cdot k = [\mathbf{K} \cdot (\mathbf{E}^c \cdot (\mathbf{E}^c)^{\#} - \mathbf{I})]^{\#} \cdot \boldsymbol{\varepsilon} \tag{8.33}$$

it becomes clear that $\boldsymbol{\varepsilon}$ can also be used to estimate the composition of a missing compound, or rather the *holes* in the balances for a number of elements: if matrix $\mathbf{K} \cdot (\mathbf{E}^c \cdot (\mathbf{E}^c)^{\#} - \mathbf{I})$ does not have full rank, it will be impossible to estimate all elements from vector $(\mathbf{E}_{n+1} \cdot k)$. This matrix can only have full rank if rank($\mathbf{E}^c$) is zero, i.e., if all conversion rates are measured. Therefore, it is usually impossible to produce a complete estimate of the composition of the missing compound. To find out which elements can be estimated, matrix $\mathbf{K} \cdot (\mathbf{E}^c \cdot (\mathbf{E}^c)^{\#} - \mathbf{I})$ must be analyzed with singular value decomposition.

It is important to realize that changes in the elemental composition can occur during fermentation. A lumped product compound may have been specified that obtains gradually another composition, or –for example– the biomass composition can be subject to changes. The latter could be indicated in an industrial fermentation process with the methods described here [4].

8.7 Increasing the Test Sensitivity

The lower the standard deviations of the measurements, the higher the test sensitivity. If a time series of data is available for a (nearly) steady state situation, the mean value of the measurements, along with the standard deviation of that mean value can be used. This standard deviation is lower than the standard deviation of the individual measurements, as is shown in Appendix B.

Especially in batch or fed-batch processes, the *cumulatively produced or consumed amounts* of the compounds can be used instead of the conversion rates. The general expression for the converted amount p over the interval (t_{i-1}, t_i) is given by:

$$\begin{aligned} p_c(t_{i-1}, t_i) &= V(t_i) \cdot C(t_i) - V(t_{i-1}) \cdot C(t_{i-1}) \\ &+ \int_{t_{i-1}}^{t_i} [\Phi_{out}(t) \cdot C_{out}(t)]dt - \int_{t_i}^{t_{i+1}} [\Phi_{in}(t) \cdot C_{in}(t)]dt \end{aligned} \tag{8.34}$$

$p(t_{i-1}, t_i)$ should be relatively large with respect to the standard deviations of the relevant measurements. The stochastic variations on the multiple measurements for a compound in the interval (t_{i-1}, t_i) will *compensate each other* (the expected value of the variations is 0 if no systematic errors are present). Hence, the cumulative amount divided by the time interval can be considered as a filtered conversion rate. It was shown [4] that in an industrial fed-batch baker's yeast production process, systematic deviations in the order of 2 to 5 % can be detected using cumulative data.

8.8 Detecting Errors in the Primary Measurements

Sometimes, it is more appropriate to directly detect errors in the primary measurements of volumes, flow rates and concentrations (instead of in the conversion rates).

To investigate how errors in the primary measurements influence the conversion rate or accumulated conversion, Equation (8.34) is modified to yield:

$$\begin{aligned} p_c(t_i, t_j) &= p_c(t_i, t_j)^{\text{true}} \\ &+ e_V \cdot [C(t_j) - C(t_i)] \\ &+ e_C \cdot [V(t_j) - V(t_i) + \int_{t_i}^{t_j} \Phi_{out}(t)dt] \\ &+ e_{F_{out}} \cdot \int_{t_i}^{t_j} C_{out}(t)dt + e_{F_{in}} \cdot \int_{t_i}^{t_j} C_{in}(t)dt \\ &+ e_{C_{in}} \cdot \int_{t_i}^{t_j} \Phi_{in}(t)dt \end{aligned} \tag{8.35}$$

e_i indicates an error in primary measurement i (e_V refers to the volume, e_C to the concentration in the fermentor, $e_{F_{out}}$ to an error in the effluent concentration, ...).

The question is here, how a compare vector can be constructed to find an error in such a primary measurement. As an example, let us consider an error in the reactor volume (e_V has a non-zero value, while all others are zero). This error will affect the estimated production of all compounds. For each compound the deviation in the estimated production is proportional to the concentration difference $C(t_j) - C(t_i)$ in Equation (8.35). For each compound, this deviation leads to a contribution to the vector of residuals (ε) via the elemental composition of that compound. The comparison vector results from a linear combination of the elemental composition vectors and the concentration differences. For more details the reader should consult Van der Heijden [4].

8.9 Discussion and Conclusions

Linear constraint relations can elegantly be used for fault diagnosis and filtering by data reconciliation. In practical situations a pathological set of equations can be obtained. The balanceability and calculability of conversion rates depend on the division between measured and non-measured rates and on the location of these rates in the equations. Straightforward matrix calculations will often be hampered by singularities. The concept of the redundancy matrix provides a robust calculation scheme that will indicate under all conditions which measured rates are overdetermined (balanceable) and which non-measured rates are calculable.

For fault diagnosis the vector of residuals from the balance equations must be analyzed. In the well known serial elimination method, only the vector length is used in statistical tests. The vector comparison test described here analyzes the direction of this vector as well. For all possible errors in the measurements and/or system description *compare vectors* can be constructed. When the direction of a compare vector and the direction of the vector of residuals are equal (using some statistical criterion) the error represented by the compare vector can be present indeed. All compare vectors can be constructed so that an *a priori* analysis reveals which errors can be distinguished from one another, and which produce similar compare vectors. If in addition error combinations are considered (which can be represented by compare vectors as well) it may become clear that the vector of residuals can mostly be due to more than one error source. This is not a limitation of the vector comparison test, but of the method as such. The vector comparison test indicates all conclusions that can be drawn in a particular situation. The accuracy of the test can be increased by reducing the standard deviations relative to the measured values. Especially for (fed-)batch fermentations, the use of cumulative amounts can be attractive to achieve this.

In practical situations additional information will be available that can help to draw the right conclusions. In laboratory experiments, probably the measurements are very reliable, and the method can be used primarily to find errors in the system definition. For (on-line) fault detection in an industrial process, experience must grow for interpreting the test results.

Not in the last place, the right confidence level of the χ-square test must be established. The test must reveal errors as soon as possible, but, on the other hand, the operator should not be warned too often for false alarms. Knowledge with respect to the robustness of measurement devices, byproducts being formed in particular conditions, ... can help a lot to draw the right conclusions.

A drawback of the method is that the process must be relatively *clean*. Preferably only a few substrates and/or products must be converted that account for almost the complete turnover of the considered chemical elements, ... The more compounds, the more degrees of freedom, and the more measurements are needed to become overdetermined. Further, all relevant compounds must be known, which may be a problem for example with industrial substrates. Also, concentration variations in substrates must be carefully measured in order to be able to calculate conversion rates. Pseudo compounds may be included in the system description, to lump (minor) compounds, which is only convenient if the composition of such a pseudo compound is rather constant. Sometimes the biomass elemental composition (for example in terms of the C/N ratio) can vary during a process run [4]. In many cases, however, this effect will not be very important. Initial calculations (e.g., to determine which measurements should be carried out for a particular conversion) can often be performed with tabulated or estimated values for the biomass composition [10].

Here, the conversion process itself is considered as a black box. This means that the conversion rates are only related by their composition. If in addition metabolic knowledge is available, for example with respect to (lumped) energy consuming and energy producing reactions, extra requirements such as ATP or NADH conservation can be taken into account. Noorman [6] shows how this can be worked out mathematically. In doing so, the degrees of freedom of the system description are diminished, so that less measurements are necessary to become overdetermined, ...

8.10 Appendix A: Elementary Error Propagation Rules

In the following, a, b, and c are stochastic variables. Δx denotes the absolute error in x.

- Summation

$$\begin{aligned} c \pm \Delta c &= (a \pm \Delta a) + (b \pm \Delta b) \\ c &= a + b \\ \Delta c &= \Delta a + \Delta b \\ \Delta c/c &= (\Delta a + \Delta b)/(a + b) \end{aligned}$$

- Multiplication

$$\begin{aligned} c \pm \Delta c &= (a \pm \Delta a) \cdot (b \pm \Delta b) \\ c &= a \cdot b \\ \Delta c/c &= \Delta a/a + \Delta b/b + \Delta a \cdot \Delta b/(a \cdot b) \quad \Rightarrow \\ \Delta c/c &\approx \Delta a/a + \Delta b/b \end{aligned}$$

- Division

$$\begin{aligned} c \pm \Delta c &= (a \pm \Delta a)/(b \pm \Delta b) \\ c &= a/b \\ \Delta c/c &= (\Delta a/a + \Delta b/b)/(1 + \Delta b/b) \quad \Rightarrow \\ \Delta c/c &\approx \Delta a/a + \Delta b/b \end{aligned}$$

In words: the absolute error of a sum equals the sum of the absolute errors, and the relative error of a product or a quotient equals approximately the sum of the relative errors.

REMARK

Δx could be defined as the minimum and maximum value that were encountered for x, divided by 2. This value tends to become larger with the number of measurements, as chances increase to find an extremely low or high value, while the q % confidence interval decreases with an increasing amount of measurements (see Appendix B). ■

8.11 Appendix B: Effect of the Number of Samples on the Estimated Accuracy of a Stochastic Variable

A measured and thus stochastic variable x can be considered as the sum of the *true* and fixed value μ and a stochastic variable ε. The best estimate for μ is the arithmetic mean value of x: $\bar{x}$. In practical situations the standard deviation σ_x of x ($\sigma_\varepsilon = \sigma_x$) is unknown. Remember that σ is a function of the operating conditions. s provides us with an estimate of σ:

$$
\begin{aligned}
s &= \sqrt{\sum (x_i - \bar{x})^2/(n-1)} \\
\bar{x} &= (\sum x_i)/n
\end{aligned}
$$

n is the number of samples and x_i is the i-th sample of x. To calculate the boundaries of an interval around $\bar{x}$, where the chance for inclusion of μ is for example 97.5 %, the Student (or t-) distribution can be used:

$$-t_{97.5} \cdot s/\sqrt{n} < \mu < +t_{97.5} \cdot s/\sqrt{n}$$

t is tabulated as a function of the confidence interval and the number of samples n in various standard text books on statistics. If we call $\Delta x = t_{97.5} \cdot s/\sqrt{n}$ and assume that $\Delta x = 1$ in a particular situation for $n = 2$, then the values of Table 8.1 can be calculated for Δx as function of n. Here, it is assumed that (by pure coincidence) the calculated s remains constant (so, after two samples the same s is calculated as after 100 samples). Figure 8.8 shows how the confidence interval changes more realistically.

So, 5 samples can give a 8-fold increase in accuracy with respect to 2 samples. This effect is also illustrated in Figure 8.8.

The χ-square distribution, can be used as a base for evaluating the chance that the estimated variance s deviates from the real variance σ.

n	Δx
2	1.0000
3	0.2765
4	0.1771
5	0.1280
10	0.0784
100	0.0221

Table 8.1: Maximum possible improvement of the confidence for x as a function of n at the 97.5 % confidence level (Student distribution).

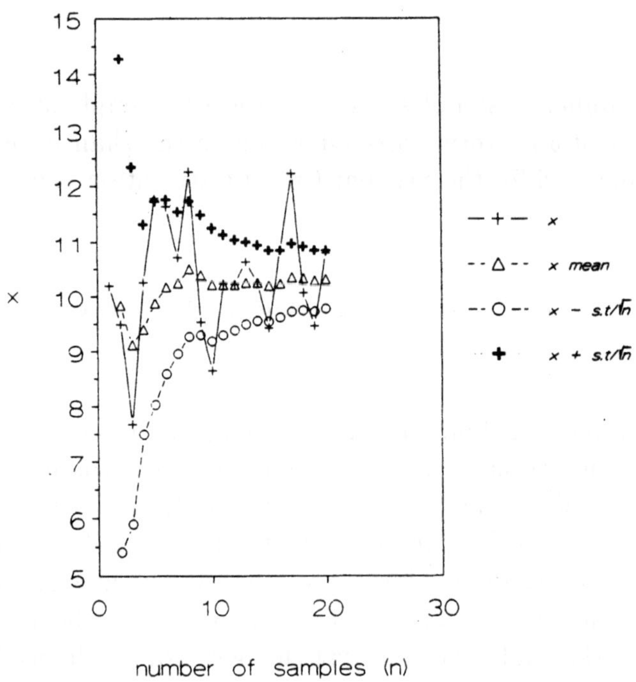

Figure 8.8: Actual decrease of the 97.5 % confidence interval for a normally distributed stochastic variable x as a function of the number of samples. Variance σ being estimated during sampling (Student distribution).

8.12 Appendix C: Macrobal - a Computer Program for Data Reconciliation and Gross Error Detection

MACROBAL has been developed at the Delft University of Technology, Department of Biochemical Engineering, for performing the system analysis as indicated in Figure 8.4 –using the concept of the redundancy matrix– and the subsequent data reconciliation. Fault detection is still limited to the serial elimination method. The program consists of 4 sections. Figure 8.9 shows the page where the elemental composition matrix can be defined (up to 16 conserved quantities and 16 compounds). In Figure 8.10 the *calculation sheet* is presented where the user defines which conversion rates have been measured, the values of these rates and their standard deviations. MACROBAL returns which rates are calculable and which are balanceable (a 'C' or a 'B' in the second column), and calculates the appropriate values along with their standard deviations. On the lower lines the results of the χ-square consistency test are summarized. The third and fourth section (not shown here) are for storing and retrieving data and for performing an automatic serial elimination procedure respectively.

MACROBAL runs on any IBM (compatible) PC. It can be obtained at the above mentioned department.

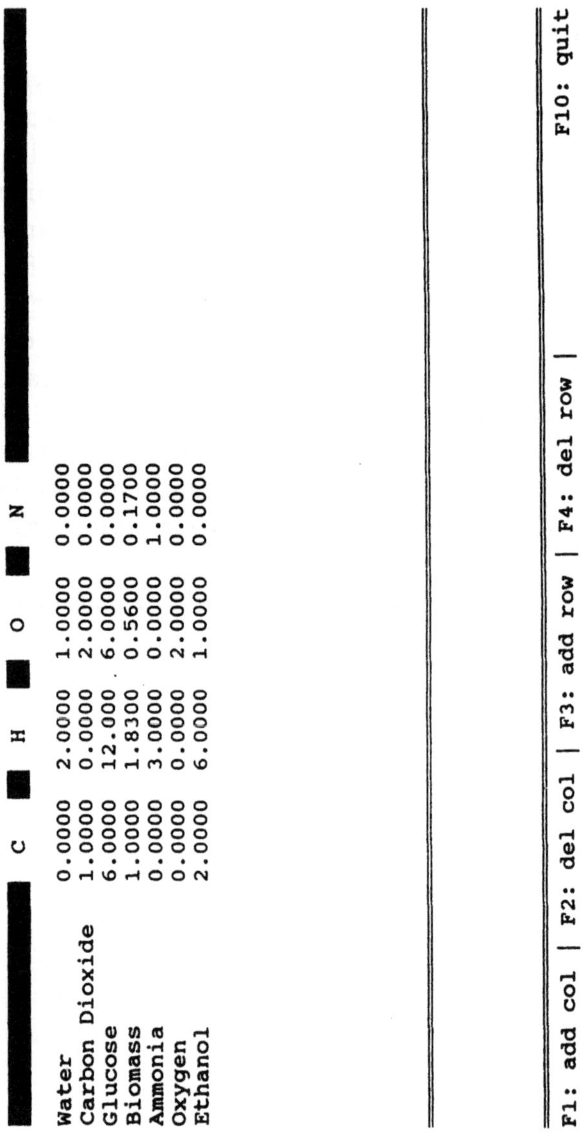

	C	H	O	N
Water	0.0000	2.0000	1.0000	0.0000
Carbon Dioxide	1.0000	0.0000	2.0000	0.0000
Glucose	6.0000	12.000	6.0000	0.0000
Biomass	1.0000	1.8300	0.5600	0.1700
Ammonia	0.0000	3.0000	0.0000	1.0000
Oxygen	0.0000	0.0000	2.0000	0.0000
Ethanol	2.0000	6.0000	1.0000	0.0000

F1: add col | F2: del col | F3: add row | F4: del row | F10: quit

Figure 8.9: MACROBAL screen for defining the set of constraint relations (elemental composition matrix).

```
net conv.         measured   st. dev      measured      estimated        st. dev
rate of         <spacebar!>              conv rate      conv rate

Water                    C   ********     *********     3.205E+01       2.54E+00
Carbon Dioxide         * B   3.99E+00     3.590E+01     3.384E+01       2.01E+00
Glucose                * B   7.20E-01    -1.200E+01    -1.213E+01       6.84E-01
Biomass                * B   1.06E+00     2.110E+01     2.106E+01       1.06E+00
Ammonia                  C   ********     *********    -3.580E+00       1.80E-01
Oxygen                 * B   2.71E+00    -2.320E+01    -2.383E+01       2.49E+00
Ethanol                  C   ********     *********     8.954E+00       1.74E+00

Degree of redundancy:  1.  Confidence level :  75     90     95     97.5    99%
Test h =    0.36   against  Chi square ( 1)  : 1.32   2.71   3.84   5.02    6.63
No proof for measurement errors or incorrect system definition

F1: calculate | F2: gross error | F3: show matrices                | F10: quit
```

Figure 8.10: MACROBAL calculation sheet.

8.13 Nomenclature

c	: number of rates to calculate	
C	: concentration	[mole/m^3]
h	: test quantity	
e	: number of independent equations in $\mathbf{E}$	
k	: arbitrary scalar	
m	: number of equations	
n	: number of samples / number of conversion rates	
O	: oxygen concentration in the gas phase	[vol%]
OTR	: oxygen transfer rate	
OUR	: oxygen uptake rate	[mole/s]
r	: conversion rate	[mole/s]
s	: estimation for the standard deviation	
Δx	: absolute error of x	
δ	: correction terms used in balancing	
μ	: actual mean value of a stochastic variable	
σ	: standard deviation	
Φ	: gas flow rate	[m^3/s]

vectors

$\mathbf{r}$	: conversion rates
$\boldsymbol{\varepsilon}$	: balance residuals

matrices

$\mathbf{E}$	: elemental composition matrix
$\mathbf{I}$	: identity matrix
$\mathbf{P}$	: covariance matrix
$\mathbf{R}$	: redundancy matrix

superscripts

$\hat{}$	: estimated value
$'$	: reduced matrix or vector
$\#$	: pseudo-inverse
^{c}	: (intended to) calculate
^{m}	: (intended to) measure
^{T}	: transposed
$\bar{}$	: measured value

subscripts

C : carbon dioxide
E : ethanol
G : glucose
H : water
in : influent
$NH3$: ammonia
O : oxygen
out : effluent
X : biomass

8.14 Bibliography

1. A.E. Bryson and Y.C. Ho 1975. *Applied Optimal Control*, Hemisphere, Washington D.C.

2. R.T.J.M. Van der Heijden, J.J. Heijnen, C. Hellinga, B. Romein and K.Ch.A.M. Luyben 1994. Linear Constraint Relations in Biochemical Reaction Systems: I. Classification of the Calculability and the Balanceability of Conversion Rates. *Biotechnol. Bioeng.*, 43, 3-10

3. R.T.J.M. Van der Heijden, B. Romein, J.J. Heijnen, C. Hellinga and K.Ch.A.M. Luyben 1994. Linear Constraint Relations in Biochemical Reaction Systems: II. Diagnosis and Estimation of Gross Errors. *Biotechnol. Bioeng.*, 43, 11-20

4. R.T.J.M. Van der Heijden, B. Romein, J.J. Heijnen, C. Hellinga and K.Ch.A.M. Luyben 1994. Linear Constraint Relations in Biochemical Reaction Systems: III. Sequential Application of Data Reconciliation for Sensitive Detection of Systematic Errors. *Biotechnol. Bioeng.*, 43, 781-791

5. F. Madron, V. Veverka and V. Vanicik 1977. Statistical analysis of material balances of a chemical reactor. *AIChE J.*, 23, 482-486

6. H.J. Noorman, J.J. Heijnen and K.Ch.A.M. Luyben 1991. Linear relations in microbial reaction systems: a general overview of their origin, form and use. *Biotechnol. Bioeng.*, 38, 603-618

7. W.H. Press, B.P. Flannery, S.A. Teukolsky and W.T. Vetterling 1989. *Numerical Recipes in Pascal*, Cambridge University Press, New York

8. P.M. Reilly and R.E. Carpani 1963. Application of statistical theory of adjustment to material balances. *13th Congress Chemical Engineering Conference Montreal*, Quebec, Canada

9. D.L. Ripps 1965. *Chem. Eng. Prog. Symp. Ser. 55*, 61, 8

10. R.A. Roels 1983. *Energetics and Kinetics in Biotechnology*, Elsevier Biomedical Press, Amsterdam

11. N.S. Wang and G. Stephanopoulos 1983. Application of macroscopic balances to the identification of gross measurement errors. *Biotechnol. Bioeng.*, 25, 2177-2208

Chapter 9

General Concepts of Bioprocess Modeling

G.C. VANSTEENKISTE
University of Gent
Coupure Links 653 B-9000 Gent (BELGIUM)

9.1 General Introduction

Every abstract model represents an attempt to logically rationalize the behavior and perhaps even the mechanism of one or other system under study. The successes in modeling are very much dependent on the nature of the system, probably not because some systems are too complex but rather because our mental constructs are too meager. Just as the equations relating to mass, force and velocity deal with the primitives in engineering, a mathematics that deals with the fundamental primitives of the system under study is needed. The concept of a model as a database has not yet been exploited to its full potential.

9.2 Simulation Scenario

In the first phase of modeling, the analist has to identify the structure, organization and/or the functional behavior of the system under consideration for example by means of objects, connections and transitions in between. In addition he has to declare the global objectives and to specify evaluation parameters. Based on a detailed abstracted (possibly formalized) prob-

J.F.M. Van Impe et al. (eds.), Advanced Instrumentation, Data Interpretation, and Control of Biotechnological Processes, 245-250.

lem specification, the analist has to select the most appropriate modelling method for calculation of the desired evaluation parameters. In this phase characteristic features and properties of modeling methods have to be compared and evaluated with respect to the actual modeling task. After this selection, a tool independent construction of a model (hierarchy) should be done. This description offers the advantage to adjust the structure of the model to requirements given by the problem to be analyzed and not by the availability of tools, machines, etc. In the next phase, solution techniques have to be taken into consideration, like numerical analysis versus simulation, event-oriented versus process-oriented simulation, deterministic versus stochastic simulation, etc., as well as tools for the realization of an executable model (hierarchy), like simulation languages versus domain specific modeling packages. These considerations have to take into account cost-benefit considerations, as well as flexibility, degree of detail of the model and deadlines for the analist. In the next phase the correctness of the executable model, mostly of a computer program, has to be verified; the computer model has to be validated. As a result the analist is informed about ranges of the input parameter values. In the experimentation finally, a detailed statistical analysis of modeling results has to be assumed as a basis for a correct interpretation and for further inferences.

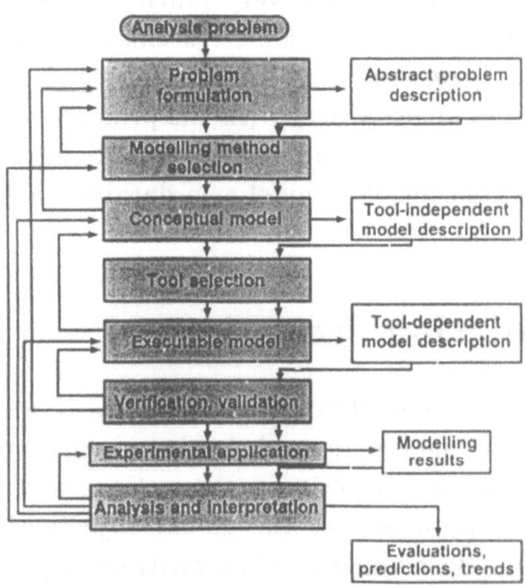

Figure 9.1: Simulation scenario.

The whole enterprise, initially undertaken by the human mind, becomes gradually a computer-aided modeling activity. The act of *performing experiments with computer models* -called simulation- upgrades the initial modeling activity; in other words, it promotes the knowledge representation of the process under consideration. Eventually the engineering goal is reached: system management and control, involving determination of the excitation, based on a system model and the predetermined response.

9.3 Modeling Methodologies

The paths followed to build models are basically detected by pondering the *sources* feeding the modeling activity: the *a priori knowledge* concerned with fundamental laws and previous modeling efforts, the *on-line and off-line* assembled measurements providing direct information on the specific system that is under investigation, and finally the *purpose* of the simulation activity providing clues with regard to the level of detail that has to be attained.

A necessary condition to reach the purpose eventually, with the help of a model, is the applicability of an experimental frame derived from the objective to the model. The more complex the model (higher descriptive detail), the more experimental frames there are that pertain to the model. How detailed has a model to be in order to be retained for a given experimental frame? (This question is related to validity.) The arguments tend to develop between: perceived complexity reflects reality, omissions might provide risks and perceived complexity obscures an underlying simplicity; it is the discovery of these simple underlying mechanisms that represents the art of scientific investigation.

Using information base management systems in simulation opens the possibility to either search for a model with prescribed features that optimally fit the user's wishes or to create new models from models previously stored. The activity *modeling* may imply the search of a model with certain specified characteristics. In simulation of large-scale systems a promising strategy is to build up model bases of *primitives*, i.e., submodels which can be considered to be standard building blocks for larger models. To choose the right primitive requires expert knowledge. Aspects concerning the relations between the various primitive submodels should also be taken into account. Some primitives might be expressed on the basis of non-numeric heuristic findings while others are modeled as dynamic systems (using basic laws).

The modeling context differs very much for technical systems in comparison with large-scale interdisplinary systems. The modeling methodology followed will very much depend on the respective amount of deductive, inductive and purpose-oriented information available. Structure information on the primitives is gathered for the major part from *a priori* knowledge. Structure identification is seen as model class selection on the basis of corresponding feature comparison and hence follows a pattern recognition frame.

9.4 Biotechnological Processes as Ill-defined Systems

Biosystems have a number of properties which make a quantitative approach cumbersome and difficult to carry out in a rigorous way: these properties are characterized by the term *ill-definition.* The processes are often of an intricate nature: lots of meaningful interconnections. Moreover the systems are as difficult to separate in space, in scale and in time.

The activities of an organism and the rates at which it carries on those activities are dependent on the organism itself as well as on the organism's environment. Hence, it is difficult to define clearly the boundaries of the system: separation for experimentation becomes almost unfeasible without substantially altering the phenomena.

Scale and time constant differences are small so that it is sometimes not clear which level is appropriate.

Finally biotechnological systems show low accessibility: experimentation is a nuisance not only because of the processes being difficult to separate, but also because of measurement techniques being inadequate and/or very expensive.

The model-building stages of the classic methodology do not explicitly take into account the specific problems that arise from grayness (leading to unknown system structure) and ill-definition (leading to missing details of the system framework, like system boundaries, input-output variables, disturbances). As a consequence, techniques and insight are required to infer unknowns in framework and structure from the available information sources.

9.5 Needs in Advanced Simulation of Ill-defined Systems

Advanced information processing techniques are needed in simulation of biosystems and environmental systems, as well as more advanced methodologies and new mathematical concepts. Such techniques will allow research workers to more and more concentrate on the disciplinary aspects of the implemented problem rather than on computer implementations and programming aspects. Computer models, which are growing in complexity by tapping discipline knowledge, should be capable of following the reasoning capability of the discipline expert. This requires parallel information processing and dedicated array processors based on pipelining techniques.

Computer models have to combine heuristic and discipline mathematical fundamental knowledge. Knowledge engineering has to substitute computational formalisms for human expertise knowledge. This is called *expertise modeling.* Mathematical formalisms implemented on computational engines provide the user a great amount of number crunching, but cannot provide insights into the problem solving process and interpretation of the results. Common rule based expert systems on the other hand employ empirical association and symbolic processing and are not well suited to deal with number crunching. Therefore there is a need to couple symbolic and numerical techniques.

Highly interactive tools are of utmost importance. The interactive environment changes fundamentally the way data are perceived and the way users behave during problem analysis and model design. A computer-aided modeling facility equipped with multicolor graphic displays gives the discipline expert/designer more or less finger touch control in the process. This gives him an opportunity to utilize his human abilities in pattern recognition and innovation. One must be able to construct models using entirely the language and thought patterns of the specialized field under study rather than of computers and mathematics. The future is to incorporate valid models into the structure of the reasoning. Model-building methodology could be incorporated for its major part in *intelligent*, *self-organizing* machines, so that human intervention could be brought to a bare small level. Hence, artificial intelligence could make the retrieval of an existant model from primitive submodels easier and more reliable, improve model networking, lump models together and finally assist in validation of tentative models by suggesting useful experiments.

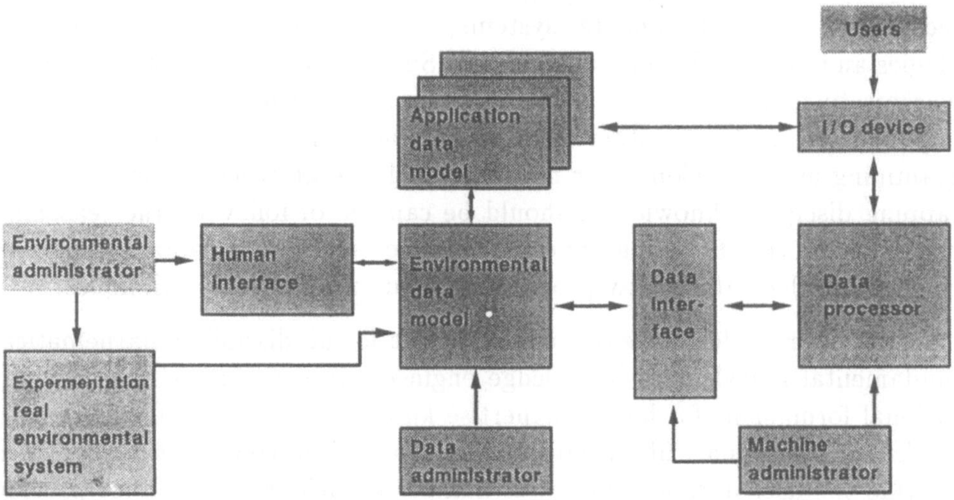

Figure 9.2: Simulator lay-out for ill-defined modeling.

9.6 Perspectives of the Simulation Tool

- Possibility of collecting and condensing information and controling the quality of additional data
- Gives access to quantities which are difficult to measure
- Permits predictions of responses to unusual events and conditions
- Allows a more extensive use of existing information and facilitates information transfer
- Can help or prepare decision-making
- Creates an intensification and improvement of the monitoring system

Chapter 10

Bioprocess Model Identification

P.A. VANROLLEGHEM
University of Gent
Department of Applied Mathematics, Biometrics and Process Control - BIOMATH
Coupure Links 653 B-9000 Gent (BELGIUM)
email: Peter.Vanrolleghem@rug.ac.be

D. DOCHAIN
Chercheur qualifié FNRS
Université Catholique de Louvain
CESAME Bâtiment Euler
Av. G. Lemaître 4-6 B-1348 Louvain-La-Neuve (BELGIUM)
email: dochain@auto.ucl.ac.be

ABSTRACT – *This text introduces different tasks that are to be performed in a full-fledged model identification exercise. Special emphasis is given to the typical problems inherent to the modeling of biotechnological processes: 1) insufficient a priori knowledge to deduce model structure so that recurrence is to be made to experimental data; 2) models that are non-linear both in the state variables and the parameters needing complex and computing intensive numerical algorithms to identify and apply the model (parameter estimation, simulation); moreover, the study of for instance the theoretical identifiability of the parameters is hampered due to a lack of theoretical development for the non-linear case; 3) insufficient measurement capability*

J.F.M. Van Impe et al. (eds.), Advanced Instrumentation, Data Interpretation, and Control of Biotechnological Processes, 251-318.

emphasizing the desire for optimal experimental design with the purpose of best use of the performed (expensive/labor intensive) experimental runs; 4) time-variance of the process which in the area of bioprocess control gives rise to the need for on-line identification of the models and, hence, imposing real time constraints on the procedures for experimental design and identification.

To illustrate the different methodologies taking part in the model identification, a particular case study taken from wastewater treatment processes is developed throughout the text.

KEYWORDS – *Mathematical Modeling, Identifiability, Structure Characterization, Parameter Estimation, Optimal Experimental Design*

10.1 Introduction

Mathematical models have become a major tool in biotechnology. They are currently used to facilitate an increase in the understanding of the underlying bioprocess mechanisms; to communicate using the model as a common ground; in the design of full-scale processes, control and operating strategies; for the training of operators and process engineers, and so forth. In these examples the mathematical model can only be applied successfully, if it is a proper description, in terms of both model structure and model parameters, of the underlying process. Hence, theoretical modeling is most often not enough. There is also a need for identification of the model from experimental data.

The goal of model identification is to find and calibrate a model for the system under investigation that is adequate for its intended purpose.

Although mathematical models can be classified according to many different criteria, an interesting classification is to distinguish between *models for understanding* and *models for prediction/forecasting* [9]. In this respect it is important to stress the point that prediction, e.g., in control applications, is possible with calibrated empirical, so-called *black box*, models without understanding the basic mechanisms of the system. Such models may serve the purpose they were meant for.

The models for understanding, on the other hand, do not reach the goal their name implies: full understanding is too ambitious. These models should therefore be considered no more than a (dense) description of hypotheses or conjectures that are acceptable explanations of process behavior. Such model will hold as long as there are no significant deviations between model predictions and measurements as it is put into jeopardy under different conditions [7]. In other words, a model for understanding can never be validated, it can only be disproved or confirmed [54]. The confidence in the model assumptions (and the mechanisms they encapsulate), however, increases as the model passes more and more severe tests. The bottom line is that accepting a hypothesis never means that the hypothesis is proved, but only that it was not possible to contradict the hypothesis with the aid of the available data.

10.1.1 Mathematical models

From a mathematical point of view, different approaches can be discerned for the quantitative description of biotechnological processes.

Black box (or input/output) models describe the dependency of the system output y at time t_k on past and present inputs $u(t_i)$. An example of such a model is:

$$y(t_k) = \frac{B(q)}{A(q)} u(t_k) \tag{10.1}$$

where $A(q)$ and $B(q)$ are polynomials in the backward shift operator q, i.e.,

$$q^{-j}\left(y(t_i)\right) = y(t_{i-j}) \tag{10.2}$$

$$A(q) = 1 + a_1 q^{-1} + a_2 q^{-2} + \ldots + a_n q^{-n} \tag{10.3}$$

$$B(q) = b_0 + b_1 q^{-1} + b_2 q^{-2} + \ldots + b_m q^{-m} \tag{10.4}$$

The a_i and b_i and the order of the polynomials n and m are to be determined from a set of input-output data.

Alternative black box models are artificial neural network models that have been shown to be capable to describe any non-linear function. This feature has attracted a lot of attention and intense research is going on for their use in biotechnological applications. More details are given in the chapter by te Braake *et al.* [63].

The essential feature of black box models is that they assume no knowledge of physical or internal relationships between the system's inputs and outputs other than that the inputs should produce observable responses in the output. Hence, the system is considered *black box* and no use is made of the available *a priori* knowledge.

Mechanistic or *white box* models have found wider acceptance due to the possibility to incorporate the increasing *a priori* knowledge of the bioprocesses into these mathematical descriptions. The dynamics of the variables

$\mathbf{x}$ considered important for the adequate description of the process can be described by the following state-space model:

$$\frac{d\mathbf{x}}{dt} = A\mathbf{x} + B\mathbf{u} \tag{10.5}$$

and the output observations $\mathbf{y}$ are given by

$$\mathbf{y} = C\mathbf{x} \tag{10.6}$$

In this model A, B and C are matrices containing the characteristic (possibly time varying) parameters of the system, $\mathbf{u}$ is the vector of system inputs or forcing functions and the state vector $\mathbf{x}$ of the system contains such variables as the biomass, substrate, oxygen, nitrate, etc. Often the basis of the model consists of mass balances of the reactor system.

The non-linearities of the bioprocesses involved however ask for a different representation than the linear one given above. A more general model is:

$$\frac{d\mathbf{x}}{dt} = \mathbf{f}(\mathbf{x}, \mathbf{u}, t, \theta) \tag{10.7}$$

$$\mathbf{y} = \mathbf{h}(\mathbf{x}, t, \theta) \tag{10.8}$$

One can observe the non-linear relations $\mathbf{f}$ and $\mathbf{h}$ between the state variables, inputs and outputs, and the model parameters θ.

The picture of mathematical models is only complete if one also considers *grey box* models. These are mainly considered as models in which part of the model structure is based on *a priori* knowledge of the process while another part consists of black box descriptions such as empirical rules. Such hybrid models combine the advantages of both approaches: identifiability and extrapolative power [8][64].

Fuzzy models may also be considered as grey box models though in a different sense. This is because these models incorporate insights in the internals of the processes under study in a qualitative way. Therefore these models are particularly useful for description of systems where the process mechanisms are not (yet) completely understood. The mathematics of fuzzy set theory and some applications are introduced in the chapter by te Braake *et al.* [63].

10.1.2 The model building exercise

Two primary types of reasoning are used to build a model [5]. The first involves the assembling of all prior hypotheses made on the mechanisms and phenomena that are believed to govern the system's behavior. This is the synthesis part of the exercise. The analysis part consists of refuting or confirming these hypotheses on the basis of a set of field data.

Typically during the initial phases of model building, linguistic models are built using qualitative reasoning. There are however no methods or advises available to guide the modeller in this exercise. The translation of these conceptual models into mathematical models is better supported. Textbooks provide plenty of basic kinetic expressions for the biological, physical and chemical interactions among the variables of interest. It is of course up to the analyst to extract the proper ones. While this is the creative task of the model builder, the model building exercise is at this stage only half way. It is equally important to demonstrate whether a particular candidate model (among many proposals that are being made) is a good or a poor approximation of reality. This second stage of evaluating the model against experimental data is best described as system identification (the term used in control theory). Simply stated, system identification may be considered as model calibration. However, as Beck stated [5]:

The word calibration suggests an instrument already well-designed and in need only of small, fine-tuning adjustments. Such a view weakens the role of experimental observation, relegating it to a minor test on what is the more or less inevitable path to apply the model with which the model-building procedure was first begun. It encourages the attitude that there is little to be learned from the field data that was not known before.

Our preferred view of system identification is quite the contrary: it is an integral part of the process of developing scientific theories about the behavior of a system; it, too, is about knowledge acquisition and hypothesis generation.

This statement brings us back to the fact that modeling is an iterative cycle in which experiments play the role of indicating areas of model deficiency that are subsequently tackled in a new hypothesis generating step. The diagram of Figure 10.1 summarizes the aspects of this second stage of model building. Three sources of information can be used to infer a model [78]:

a priori knowledge: general laws, principles and previous investigation

experimental data: information obtained from experiments performed to study the underlying phenomena

goal: information which is the result of requirements and specifications that have been set.

The task of system identification is to make optimal use of this information to come up with the most adequate model. System identification is an iterative task [7] consisting of the following steps that depend strongly on each other:

frame definition: choice of the system boundaries, input and output variables, type of models considered (e.g., linear/non-linear, input-output/ state-space, ...)

structure characterization: infer the level of model complexity (dimension of state vector, degrees of polynomials, ...) and determine the functional relationships between variables

parameter estimation: find numerical values for the constants in the functional relationships

validation: confront the resulting model performance with the purpose it was built for.

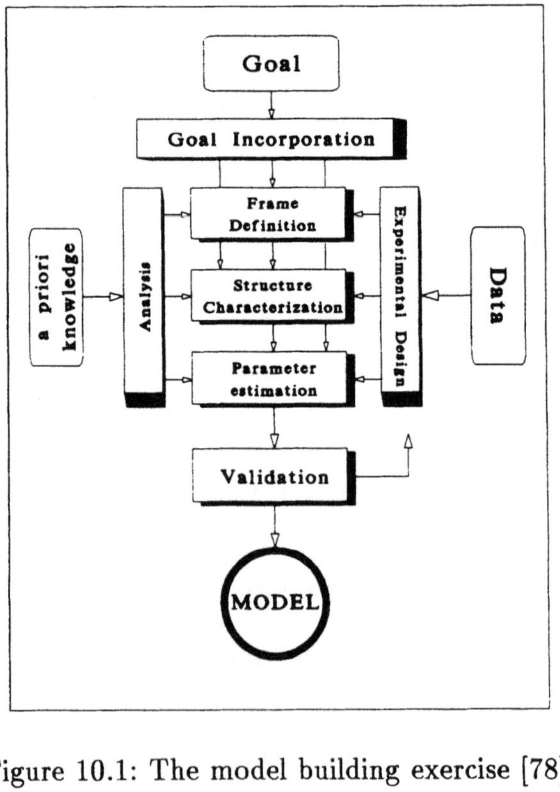

Figure 10.1: The model building exercise [78].

For most physical and chemical applications, the *a priori* knowledge is of such high quality that the system framework and most of the model structure can be deduced from it. The modeling methodology developed for these systems is adequate to estimate the parameters and solve the minor uncertainties in the model structure by using final validation experiments and eventually iterating a small number of times through the procedure.

In contrast with this, the inherent characteristics of bioprocesses, i.e., their non-linearity and non-stationarity, coupled with the lack of adequate measuring techniques, make that this mathematical modeling methodology cannot be applied without modification [78]: more emphasis must be given to inductive reasoning to infer a larger part of the model structure from the scarce (or harder to obtain) experimental data. Consequently, structure characterization methods become a more important tool, because the chance of obtaining an invalid model is much larger and, hence, the number of modeling iterations may increase substantially.

The data scarcity also induces an important problem in the parameter estimation step. Identifiability of model parameters, i.e., the possibility to give a unique value to each parameter of a mathematical model, is a general concern in bioprocess modeling efforts [3][23][26][42][45]. This problem is however more pronounced in on-line identification because one is relying much more on real time information to perform the parameter estimation whereas off-line model calibration also can take advantage of off-line acquired data.

10.1.3 Current research areas w.r.t. bioprocess models

Incorporation of latest insights. For wastewater treatment processes for instance important efforts are made to model hydrolysis of substrates, the fate of biopolymers and the sedimentation process with special emphasis on the interaction between the biological phenomena such as filament growth and the settling properties of the sludge [19]. In fermentation technology the increasing biochemical background is being incorporated in the models through the use of metabolic network analysis and flux distributions [14][67][68][69][79]. The dynamics of gene exchange between organisms [10][29][33][65] and the fate of plasmids in genetically modified production strains receive a lot of attention as well [31][32][48][49][57].

Identifiability. A discrepancy has grown between the amount of data needed to identify the increasingly complex models and the amount of information that can be obtained on behalf of the process. Especially if only on-line data can be used for model identification, serious problems may occur in finding unique parameter estimates and decisions on the most appropriate model structure [72]. Even combined on-line and off-line data may be insufficient for accurate modeling. Current research is therefore directed towards the development of new monitoring equipment and new off-line methodologies adapted to the information need of the new models.

Verifiability. The models that have been introduced recently are the result of considerable fundamental studies aimed at elucidating the mechanisms of certain microbial processes. In order to more precisely explain the detailed experimental findings, state variables and parameters have been introduced in the models which are not directly measurable, e.g., active heterotrophic organisms [3][26][46] or specific enzymes [82]. Hence, since verification of a model requires that all model predictions of the states can be compared with experimental data, current models have become intrinsically unverifiable. Here too, new experimental methods are being studied to cope with this problem.

Model reduction for process control. The identifiability and verifiability problems mentioned above ask for considerable efforts devoted to the development of new sensor technology and experimental methods so that the new process models can be used in adaptive model-based control systems. An alternative approach which attracts a lot of attention is directed at the reduction of the complexity of existing mechanistic models to such a level that on-line identification with existing technology is feasible, at the same time maintaining the necessary predictive capabilities of the major phenomena [26][39][46].

10.2 Case Study

As a red line through this chapter, a particular application is used to illustrate different stages of model identification. The case study consists of a modeling exercise that is to be performed on-line since the process model is to be updated (identified) on the basis of raw data that are produced at regular intervals in a sensor that is operated according to the principle of *In-Sensor-Experiments* [72]. The raw data are the result of short-term pulse experiments and reflect the impulse response of the system under study, allowing its mathematical description.

10.2.1 Process

The approach taken in this "bio"-sensor is the following [77]: within the apparatus a sample of activated sludge (taken from the wastewater treatment plant to control) is subjected to a pulse injection of wastewater. Using a dissolved oxygen electrode, the oxygen consumption for oxidation of the wastewater pollutants by the sludge is monitored. The resulting exogenous (i.e., substrate degradation induced) oxygen uptake rate (OUR_{ex}) data reflect the impulse response of the biological system and can be used for model identification. The typical impulse responses given in Figure 10.2 illustrate that models of different levels of complexity are required to describe the experimental data. Based on *a priori* knowledge, a set of candidate models was constructed as described below.

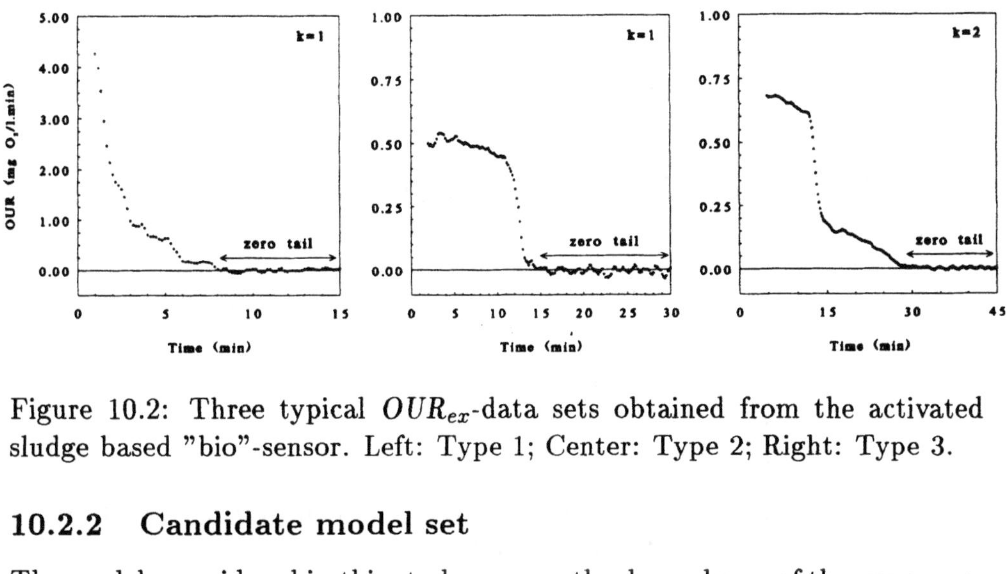

Figure 10.2: Three typical OUR_{ex}-data sets obtained from the activated sludge based "bio"-sensor. Left: Type 1; Center: Type 2; Right: Type 3.

10.2.2 Candidate model set

The models considered in this study express the dependence of the exogenous oxygen uptake rate OUR_{ex} on the biodegradation of k substrates S_i present in the mixed liquor:

$$OUR_{ex} = -\sum_{i=1}^{k}(1 - Y_i)\frac{dS_i}{dt} \tag{10.9}$$

In the above expression, Y_i (the yield coefficient) is the fraction of pollutant S_i which is not oxidized but converted into new biocatalyst X. As usual for wastewater treatment systems, all concentrations are expressed in chemical oxygen demand (COD) units. The different model complexities taken into account in this work express the different number k of pollutants to be considered and the degradation mechanism. The four types of wastewater/sludge interaction included are:

Type 1 (Exponential): One pollutant, first order kinetics (k=1)

$$\frac{dS_1}{dt} = -\frac{\mu_{max1}X}{Y_1}S_1 \tag{10.10}$$

Type 2 (Single Monod): One pollutant, Monod kinetics (k=1)

$$\frac{dS_1}{dt} = -\frac{\mu_{max1}X}{Y_1}\frac{S_1}{K_{s1}+S_1} \tag{10.11}$$

Type 3 (Double Monod): Two pollutants, Monod kinetics (k=2)

$$\frac{dS_1}{dt} = -\frac{\mu_{max1}X}{Y_1}\frac{S_1}{K_{s1}+S_1} \tag{10.12}$$

$$\frac{dS_2}{dt} = -\frac{\mu_{max2}X}{Y_2}\frac{S_2}{K_{s2}+S_2} \tag{10.13}$$

Type 4 (Modified IAWQ model $n°1$) [61]: 3 pollutants, 2 of them are hydrolyzed into the first substrate which is used for growth according to Monod kinetics (k=1)

$$\frac{dS_1}{dt} = -\frac{\mu_{max1}X}{Y_1}\frac{S_1}{K_{s1}+S_1} + k_rX_r + k_sX_s \tag{10.14}$$

$$\frac{dX_r}{dt} = -k_rX_r \tag{10.15}$$

$$\frac{dX_s}{dt} = -k_sX_s \tag{10.16}$$

Because of the hydrolysis terms in (10.15) and (10.16), OUR_{ex} (10.9) must be rewritten for the type 4 model as follows:

$$OUR_{ex} = -(1-Y_1)(\frac{dS_1}{dt} - k_rX_r - k_sX_s) \tag{10.17}$$

The μ_{maxi} (i = 1,2) and k_j (j = r,s) are rate constants, and the K_{si} (i = 1,2) are the so-called affinity constants expressing the dependency of the degradation rate on the concentration of pollutant S_i (i = 1,2). Note that the models as given above can be used to describe batch experiments only. Also note that the pulse of wastewater injected at the beginning of an experiment is included in the model via the initial conditions $S_i(0)$ and that the biocatalyst concentration X is assumed constant within the course of these short-term experiments (≈ 30 minutes, see Figure 10.2). Recall that the aim is to identify the above biodegradation models via measurement of the oxygen uptake rate in the "bio"-sensor, as these reflect the rates of substrate degradation.

10.3 Structure Characterization (SC)

Finding the *true model* $M_T(S_T, P_T)$ with model structure S_T and parameters P_T is utopian. Rather, one must aim at finding -from a finite set of N noisy data points- the partial descriptions that are purposeful within the application [35]. Settling for the best possible model $M(S_N, P_N)$, however, induces an error that has two components:

$$M_T(S_T, P_T) - M(S_N, P_N) = M_T(S_T, P_T) - M(S_N, P^*) + M(S_N, P^*) - M(S_N, P_N) \tag{10.18}$$

The first term is due to the error between the true model structure S_T and the model structure S_N chosen from the set of candidate models with restricted complexity. This so-called *bias error* reflects the unmodeled dynamics [18]. The second component, the *variance error*, is caused by the particular realization of the noise in the limited number of data used in the system identification. Each data set will result in different parameter estimates P_N that will only tend to the real P^* (for this structure) when there is no noise or when the number of data points tends to infinity. This variance error also includes the effect of the overparametrization: the more parameters included in the model, the more uncertain their values will be.

Typically the variance error decreases like $1/N$, but increases like p, with p the number of parameters in the model structure, a measure of its complexity. The bias error, on the other hand, will decrease as p increases, but is independent of N [34][35]. Hence, as the aim is to obtain the model structure giving the lowest total error, the goal of model structure characterization will be to find the compromise between bias error and variance error.

In this section a variety of objective decision tools that enable to find this trade-off will be evaluated. The picture of approaches, methods and results of structure characterization (SC) is very diverse and scattered over several scientific disciplines. Although a number of studies have evaluated different SC techniques, none of them has resulted in a clear recommendation on a definitely superior method [2][12][30][38][62]. In the case study, an analogous exercise was devoted to decide on a SC technique most adapted to the problem at hand (see below). Moreover, it was found necessary to develop new SC methods. A number of these are specific to the candidate models used in the study, while others potentially are more generally applicable.

The SC techniques have been classified according to their impact on the real time constraint of the model identification. Most existing methods for SC evaluate the quality of the different model structures after fitting each model to the data. Hence, these methods can be termed *a posteriori SC* [74]. Methods capable of selecting a model without the need of first estimating the parameters belong to the other class, the *a priori SC* techniques. In view of the observation that parameter estimation of the non-linear bioprocess models which are dealt with here takes considerable computing time, it is clearly advantageous to apply *a priori* methods, since they only require estimation of the parameters of the selected model.

10.3.1 A priori SC

The approach of *a priori* SC has also been termed model structure selection based on preliminary data analysis and has been reported to be an underdeveloped field [35].

Two groups of methods can be discerned: one type of methods is generally applicable, while the other SC methods take advantage of specific features of the model structures present in the set of candidate models.

10.3.1.1 General methods

The pattern recognition capabilities of neural networks [59] has incited a study on their applicability as SC technique. Details on the method developed for the case study can be found elsewhere [80], but the main principles are summarized here (Figure 10.3): a three-layer recurrent neural net architecture receiving preprocessed data was used. Different data preprocessing algorithms were tested, all aiming to perform data reduction (to decrease the number of input nodes) without loss of structure specific information. The learning stage was performed with 750 training-patterns obtained from Monte Carlo simulations of the different candidate models. Once the net was trained to recognize the most appropriate model, it could be used on new (preprocessed) data. The results for the case study are discussed below.

A next method that can be applied stems from recent developments in identification of linear state space models. Numerical algorithms for Subspace State Space System Identification (N4SID, read as *enforce it*) combine the estimation of the order of the state with the identification of the system matrices, but, in contrast to traditional identification schemes, the order is estimated first [70]. It is this aspect of this data-driven approach which may be useful for non-linear processes as well. Clearly, for the case study for instance, one cannot expect that N4SID will give a direct measure of the number of substrates to be included or the degradation mechanism involved.

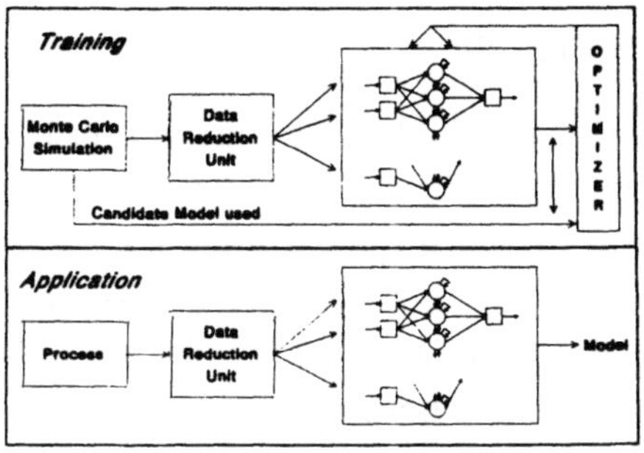

Figure 10.3: Training and application stages of a neural net used for model selection.

Rather it can be hoped that the order returned by the algorithm enables to differentiate among the models. The order is determined on the basis of the number of non-zero singular values SV in a matrix H composed of the following Block Hankel matrices containing *past* (Y_p) and *future* (Y_f) data:

$$Y_p = \begin{pmatrix} OUR_1 & \dots & OUR_j \\ OUR_2 & \dots & OUR_{j+1} \\ \vdots & & \vdots \\ OUR_i & \dots & OUR_{j+i} \end{pmatrix} \tag{10.19}$$

$$Y_f = \begin{pmatrix} OUR_{i+1} & \dots & OUR_{j+i} \\ OUR_{i+2} & \dots & OUR_{j+i+1} \\ \vdots & & \vdots \\ OUR_{2i} & \dots & OUR_{j+2i} \end{pmatrix} \tag{10.20}$$

Different H-matrices can be evaluated:

$$H = Y_f Y_p^T \tag{10.21}$$

$$H = Y_f Y_p^T (Y_p Y_p^T)^{-1} Y_p \tag{10.22}$$

$$H = L_f^{-T} Y_f Y_p^T L_p^{-1} \tag{10.23}$$

with L_f and L_p Cholesky factorizations of $Y_f Y_f^T$ and $Y_p Y_p^T$.

The choices of i and j are important: j must be as large as possible (for the case study j was set $N - 2i$); i must be much smaller than j, but larger than the largest model order expected (i was set 10). For the case study, the best results were obtained with the first form of H and singular values considered 0 when below 0.1; type 1 models were selected if $SV \ll 2$; type 2 if $SV = 2$ and type 3 if $SV \gg 2$.

10.3.1.2 Specific methods

Structure characterization on the basis of parameter invariant model features has been advocated, but has found little application due to the difficulty in finding such features from the models under consideration. For the candidate models in the case study, however, such feature can be found [76]: the number of inflection points is 0, 1 and 3 for models of type 1, 2 and 3 respectively (see also Figure 10.2). Though this is analytically correct, the determination of inflection points on noisy data is not straightforward. However, using a moving window regression in which linear and quadratic regressions are compared with an F-test, the significance of the second derivative can be assessed, leading to a more reliable estimate of the number of inflection points.

Another model-specific approach applied in the respirographic biosensor consists of fitting an exponential function, a Tanh and a Double Tanh to the data and comparing the resulting sum of squared residuals (SSR). In other words, this *a priori* method for selection among the biodegradation models is in fact an *a posteriori* method with respect to the Exp/Tanh/Double Tanh candidate model set. Empirical threshold values were set to the SSR-ratios so as to decide which model to select [76]. Ratios larger than 10 were considered necessary before a more complex model was accepted.

A third method developed for the models of the case study is based on the area below the OUR_{ex} versus time data. Some preprocessing of the (OUR_{ex}) data is performed prior to the determination of this criterion: first, the zero-tail (Figure 10.2) is cut from the data set using a t-test on the mean of a data window and, second, the reduced OUR_{ex}, t-dataset is scaled to the unit square. Type 1 data have the lowest area ($\ll 0.3$) and type 2 the largest area ($\gg 0.5$) under the curve (Figure 10.2). These threshold values for the areas enable the selection among the candidate models type 1, 2 and 3 [76].

10.3.2 A posteriori SC

Most methods described in this section have found widespread application. For the case study, however, they seem less appropriate because the real time constraint may be violated, especially as the number of candidate models may increase. However, it is not inconceivable that a SC strategy is devised in which *a priori* methods are used to make up a first ranking of the different structures, after which *a posteriori* methods are called in to make the final selection among the structures with the highest ranking. As such the real time constraint and the selection reliability can be sufficed.

10.3.2.1 Criteria with complexity terms

The criteria on which most of the *a posteriori* SC methods are based take one of the following two forms [60]:

$$\frac{SSR}{N}\left[1+\beta(N,p)\right] \tag{10.24}$$

$$N\log\left(\frac{SSR}{N}\right)+\gamma(N,p) \tag{10.25}$$

with SSR the sum of squared residuals, and N and p as defined before. For $N \gg p$ one can show that both representations are equivalent if [60]:

$$\gamma(N,p) = N\beta(N,p) \tag{10.26}$$

For both cases, the first term decreases with increasing p (increasing complexity) while the second term penalizes too complex (overparametrized) models. The model structure with the smallest criterion value is selected. Different authors have proposed several functional forms for the model selection criteria depending on the theoretical starting point. The two best-known are the Final Prediction Error, with [35]:

$$\beta(N,p) = \frac{2p}{N-p} \tag{10.27}$$

and Akaike's Information Criterion, with

$$\gamma(N,p) = 2p \tag{10.28}$$

FPE and AIC have been proved not to be consistent (i.e., do not guarantee that the probability of selecting the wrong model tends to zero as the number of data points tends to infinity) [27]. However, this disadvantage is compensated by the fact that AIC and FPE enjoy certain properties that allow to determine good prediction models in case the true model does not belong to the set of candidates [60].

An example of a consistent criterion is the Bayesian Information Criterion in which [2][58]

$$\gamma(N,p) = p\ \log(N) \tag{10.29}$$

If $p \geq 8$, it can be seen easily that BIC will tend to favor models of lower complexity than those chosen by AIC. Another consistent criterion is LILC where [20]

$$\gamma(N,p) = p\ \log\left(\log(N)\right) \tag{10.30}$$

10.3.2.2 Criteria that assess undermodeling

Recently, Goodwin *et al.* [18] decomposed the total error between the true model and a selected one in three components:

- the effect of the variance of the new noise realization
- the effect of the parameter errors due to the noise present in the identification data
- the effect of the undermodeling, corresponding with the bias term in (10.18).

This General Information Criterion, GIC, is defined as:

$$GIC(N,p) = \hat{\sigma}_v^2 + \frac{p}{N}\hat{\sigma}_v^2 + undermodeling \qquad (10.31)$$

In this equation it is essential to have an estimate of the residual error ${\sigma_v}^2$ that is independent of the undermodeling. Goodwin *et al.* [18] obtained this by fitting a high-dimensional model (assuming it to be the true model) and using the residual variance as an estimate. In the case study such model is not available, but, fortunately, the zero-tail end of the OUR_{ex}-data set can be used since this part of the data is characterized by the absence of biological dynamics [77]. As a result, the variance of this part of the data can be used for an accurate estimate of the residual error. Using the similarities shown by the authors between the expected value of FPE and GIC [18], it is possible to write an explicit formula for the undermodeling term [76]:

$$undermodeling = \frac{SSR}{N} - \left(\hat{\sigma}_v^2 + \frac{p}{N}\hat{\sigma}_v^2\right)\frac{N-p}{N+p} \qquad (10.32)$$

Structure characterization is performed by selecting the model with the lowest undermodeling value.

10.3.2.3 Statistical hypothesis tests

The F-test is probably the most frequently applied method to choose among model structures. The test-statistic

$$\frac{(SSR_j - SSR_i)\,/\,(p_j - p_i)}{SSR_j/\,(N - p_j)} \qquad (10.33)$$

is compared with the $F(p_j - p_i, N - p_j)$ distribution to decide whether the more complex model j is significantly (with a confidence level α) better than model i. The similarities that exist between F- and χ^2-tests and the equivalence between the AIC/FPE criteria and F-tests with a prespecified significance level have been shown [60].

10.3.2.4 Diagnostic checking (analysis of residuals)

When modeling, some assumptions are made concerning the properties of the noise. In most cases, the prediction errors $\epsilon(t)$ are assumed to be a realization of independent random variables with zero mean and a certain distribution. Then the quality of a model can be assessed by analysis of the properties of the calculated residuals. Two approaches can be used to check the independence of the residuals (white noise property): the autocorrelation and the run test. The autocorrelation test [60] is based on the fact that the covariance function for a white noise sequence $\epsilon(t)$:

$$\hat{r}_\epsilon(\tau) = \sum_{t=1}^{N-\tau} \epsilon(t-\tau)\epsilon(t) \tag{10.34}$$

is zero except for $\tau = 0$. Structure characterization with these tests is performed by selecting the model whose residuals are as white as possible. Two statistical tests allow to make objective decisions on the whiteness of the residuals.

(i) One test compares the covariance for each lag τ (10.34) with the limit value $N(0,1)/\sqrt{N}$, which for $\alpha = 0.05$ means that only 5 % of the autocorrelations may be larger than $1.96/\sqrt{N}$. In Figure 10.4, an example taken from the case study, the residuals for the three models fitted to a typical type 3 data set are given. In Figure 10.5 the corresponding autocorrelation function for lags 0 to 20 is depicted. Clearly, the residuals for models 1 and 2 are highly dependent, while for model 3 only the first 4 correlations (20 %) are significantly higher than the prescribed level (indicated by the horizontal lines). Hence, the residuals do not originate from a white noise sequence, indicating some unmodeled dynamics. Looking into some more detail to the residuals of model 3 (Figure 10.4), one distinguishes an oscillatory pattern, probably causing the significant autocorrelation.

(ii) The other autocorrelation test compares a combination of the first m covariances

$$\frac{N}{\hat{r}_\epsilon^2(0)} \sum_{i=1}^{m} \hat{r}_\epsilon^2(i) \tag{10.35}$$

with the $\chi^2(m)$ distribution giving a significance level for the independence of the residuals [35].

Another residuals test is a so-called non-parametric test in which the number of runs R, i.e., the number of sign changes in the sequence of residuals, is evaluated against the expected number of runs, $N/2$ [60]. To assess the significance of a deviation from this number the test-statistic

$$\frac{(R-N)/2}{\sqrt{N/2}} \tag{10.36}$$

can be compared with $N(0,1)$. *A posteriori* SC with this method selects the model with the test-statistic closest to zero or the simplest model with a non-significant test-statistic.

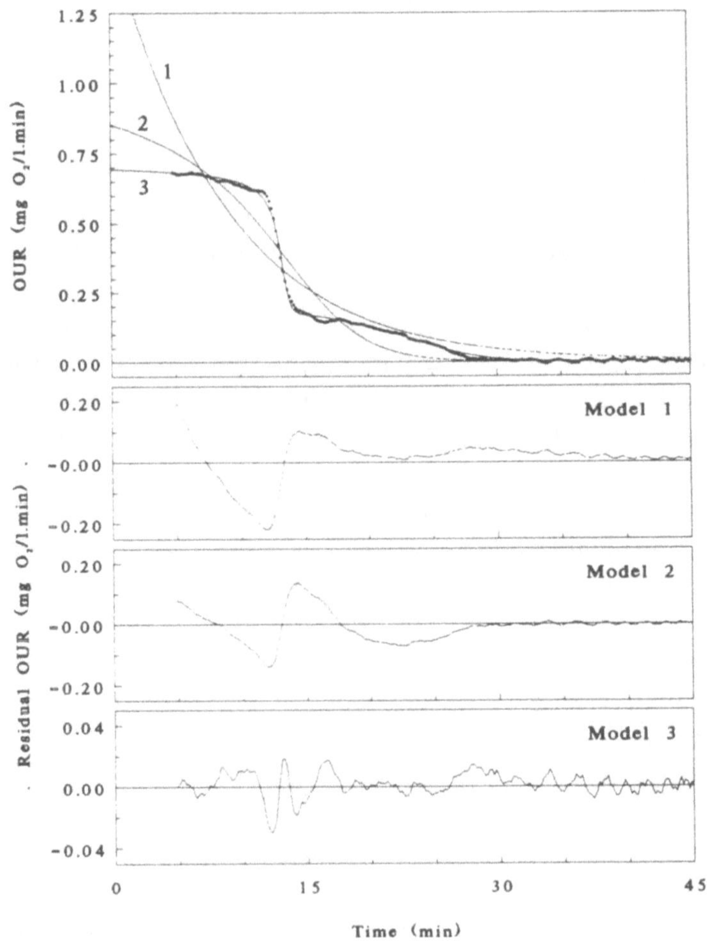

Figure 10.4: Typical model fitting results for three models to a *Double Monod* style data set.

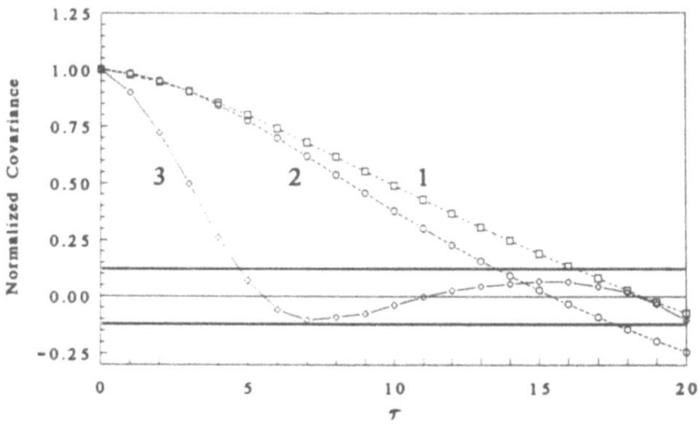

Figure 10.5: Normalized covariance function of the residuals for the models of Figure 10.4.

10.3.2.5 Results for the case study

The different SC methods introduced above were evaluated on the basis of 8 typical real-life OUR_{ex}-data sets [76]. Only the type 1, 2 and 3 models were included in the candidate model set. In Table 10.1 the selected models are compared with the advice of a human expert. Data set 4 is difficult to classify since it could be considered very close to both a type 1 and type 2 model. Hence, both were considered correct.

Except for the N4SID method, all *a priori* SC methods produced very good selection results. Among them, the Tanh and neural net approaches may be preferable considering the noisy data which may cause problems in estimating the number of inflection points. Neural nets have an additional advantage as a potentially general tool for a priori SC.

For the *a posteriori* methods, an important finding was that all *information* criteria, i.e., AIC, FPE, BIC and LILC, with the notable exception of GIC, result in overfitting of the model compared to the *expert advice*. This is probably due to the oscillations that can be observed in the OUR_{ex}-data (see Figure 10.2 and the residuals in Figure 10.4). Since the more complex models possess sufficient flexibility, some of these oscillations can be modeled. This reduces the residual error to such an extent that any penalty for model complexity is compensated for. As the observed oscillations cannot be explained by any biological process and are probably due to some hardware dependent process which is not of interest to the user, this *parasite process* should be eliminated from the data before model identification is initiated.

The F-test also suffers from the modeling of the parasite processes superimposed on the biological response. For the GIC-based method, these overfitting problems were non-existent because the effect of the oscillations is included in the estimate of the variance (from the zero-tail). In this way only the undermodeling of the biological phenomena is retained leading to model selections congruent with the observation of the human expert.

10.4 Parameter Estimation

The complex (mechanistic) models of bioprocesses that are currently available reflect the complexity of the systems in use today and the detailed process knowledge acquired during important research efforts in the last few decades. Besides the model structure, a large number of parameters in the model are to be inferred from *a priori* knowledge and experimental data. The quality of the estimation of parameters will depend on the amount and quality of (real time) data that is available to the identification algorithm.

SC Method	*Data set*								*Evaluation*		
	1	2	3	4	5	6	7	8	UF	OF	correct
N4SID	*1*	3	2	*1*	2	*2*	2	*2*	3	1	4
Neural Net	3	*2*	*3*	*1*	*3*	*2*	*3*	*2*	0	1	7
Inflection points	*1*	*2*	*3*	*2*	*3*	*2*	*3*	*2*	0	0	8
Tanh	*1*	*2*	*3*	*1*	*3*	*2*	*3*	*2*	0	0	8
OUR_{ex}-area	*1*	*2*	*3*	3	*3*	*2*	*3*	*2*	0	1	7
AIC	3	3	*3*	3	*3*	3	*3*	3	0	5	3
FPE	3	3	*3*	3	*3*	3	*3*	3	0	5	3
BIC	3	3	*3*	3	*3*	3	*3*	3	0	5	3
LILC	3	3	*3*	3	*3*	3	*3*	3	0	5	3
Undermodeling-GIC	*1*	*2*	*3*	*1*	*3*	*2*	*3*	*2*	0	0	8
F-test	3	3	*3*	3	*3*	3	*3*	3	0	5	3
Autocorrelation	*1*	*2*	*3*	*2*	*3*	*2*	*3*	3	0	1	7
Run-test	*1*	*2*	*3*	3	*3*	*2*	*3*	3	0	2	6
Human expert	*1*	*2*	*3*	*2(1)*	*3*	*2*	*3*	*2*	-	-	-

Table 10.1: Model selected on the basis of different *a priori* and *a posteriori* SC methods. The results are compared with the advice of a human expert (figures in italics indicate *right* choice, underfit (UF) and overfit (OF) refer to the complexity of the selected model compared to the choice of the expert).

Besides these limitations on the available information, another problem in the identification of the process model appears: the mathematical models required to describe the different bioprocesses are highly non-linear and the model parameters may exhibit a considerable correlation.

Because of the model complexity and the scarcity of measurements, the identifiability study of the dynamical models, prior to any parameter estimation, is certainly a key task. The central question of the identifiability analysis is the following:

Assume that a certain number of the state variables are available for measurement; on the basis of the model structure (theoretical identifiability) or on the type and quality of available data (practical identifiability), can we expect to give via parameter estimation a unique value to the model parameters?

Simply speaking, one would wonder what is the use of trying to calibrate the parameters of a model which is, structurally or practically, unidentifiable. This above formulation is quite crude, but the answer to the identifiability analysis is often more subtle: it is not just a *yes or no* answer, but when it results in some conclusions (which is not *a priori* obvious with non-linear models), these may tell that some subset or combinations of the model parameters are *a priori* identifiable.

10.4.1 Theoretical identifiability

The notion of theoretical identifiability is related to the possibility to give a unique value to each parameter of a mathematical model. In simple words, the theoretical (or structural) identifiability of a model can be formulated as follows (a rigorous definition can be found, e.g., in [16]): given a model structure and perfect (i.e., that fit perfectly to the model) data of model variables, are all the parameters of the model identifiable? From the structural identifiability analysis one may conclude that only combinations of the model parameters are identifiable. If the number of resulting combinations is lower than the number of original model parameters, or if there is not a one-to-one relationship between both parameter sets, then *a priori* knowledge about some parameters may be required to achieve identifiability. A simple example may illustrate this: in the model $y = ax_1 + bx_2 + c(x_1 + x_2)$, only the parameters $a + c$ and $b + c$ are structurally identifiable (and not the three parameters a, b, c; two parameters (e.g., a and b) will be identifiable if the value of a third one (here, c) is known *a priori*.

For linear systems, the structural identifiability is rather well understood, and besides classical identifiable models (like dynamical models in canonical form), there exists a number of tests for the identifiability (e.g., Laplace transform method, Taylor series expansion of the observations, Markov parameter matrix approach, modal matrix approach, ..., see, e.g., [16]). However, for models that are non-linear in the parameters (like the models used in this work), the problem is much more complex. There exist also several structural identifiability tests, but they are usually very complex to implement (they typically require the use of symbolic software packages [52], as will be illustrated below). In the following several similar approaches are used wherein the models are transformed into linear ones, after which the analysis is based on the linear model.

In this section attention will focus on the structural identifiability of the 4 models of the wastewater treatment case study introduced above. The study will allow to introduce possible approaches for the theoretical identifiability analysis.

The structural identifiability of the first model is rather straightforward. First note that the relation between OUR_{ex} and S_1 ((10.9) $OUR_{ex} = -(1-Y_1)dS_1/dt$) can also be written after multiplication of both sides by $d\tau$ and integration between time $\tau = 0$ and time $\tau = t$:

$$\int_o^t OUR_{ex}(\tau)d\tau = -(1-Y_1)(S_1(t) - S_1(0)) \tag{10.37}$$

which can be used to give an expression of $S_1(t)$:

$$S_1(t) = S_1(0) - \frac{1}{(1-Y_1)}\int_o^t OUR_{ex}(\tau)d\tau \tag{10.38}$$

Now define the variable $y(t)$:

$$y(t) = \int_o^t OUR_{ex}(\tau)d\tau \tag{10.39}$$

which also means that $OUR_{ex} = dy/dt$. Then by combining the model Equation (10.10) with the above expressions (10.38) and (10.39), one readily obtains the following dynamical linear equation for $y(t)$:

$$\frac{dy}{dt} = \theta_1 y(t) + \theta_2 \tag{10.40}$$

with:

$$\theta_1 = -\frac{\mu_{max1}X}{Y_1} \tag{10.41}$$

$$\theta_2 = \frac{\mu_{max1}X}{Y_1}(1-Y_1)S_1(0) \tag{10.42}$$

Because $dy(t)/dt$ is measured and $y(t)$ is readily calculated from these data, the parameters θ_1 and θ_2 of the linear dynamical model (10.40) are (structurally) identifiable, i.e., it is theoretically possible to generate signals for $y(t)$ and $dy(t)/dt$ (e.g., via an impulse response) such that the parameters θ_1 and θ_2 can be uniquely identified. This means, with respect to the original model parameters and by considering the above definitions (10.41) and (10.42), that the model parameter combinations $\mu_{max1}X/Y_1$ and $(1 - Y_1)S_1(0)$ are structurally identifiable.
The structural identifiability concept is illustrated via (real-life) data (Figure 10.6a) with an initial substrate concentration $S_1(0)$. Figure 10.6b shows the data pairs $(OUR_{ex}, \int_o^t OUR_{ex}(\tau)d\tau)$ corresponding to the $(OUR_{ex}(t), t)$-data presented in Figure 10.6a: θ_2 is given by the initial value of OUR_{ex}, and θ_1 is the slope.

Via the other three models, the two main approaches for theoretical identifiability analysis are introduced:

1. The Taylor series expansion [50],

2. The transformation of the models into models linear in the parameters.

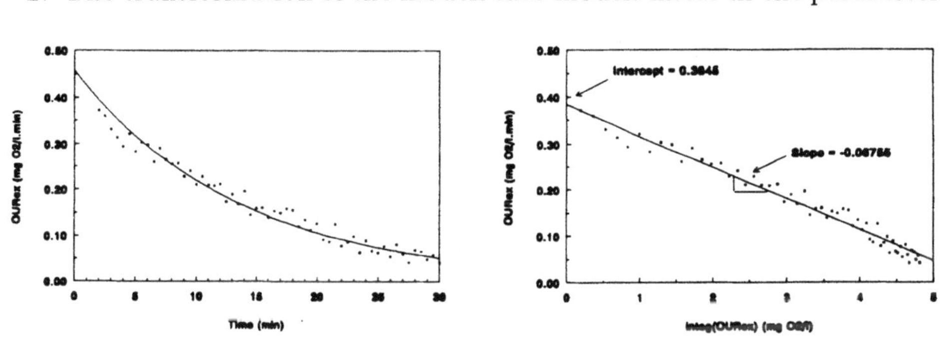

Figure 10.6: Transformation of OUR_{ex}-data corresponding to the exponential model (a) into a linear regression form (b).

10.4.1.1 Taylor series expansion

The method is based on Taylor series expansion of the observations (here, $OUR_{ex}(t)$) around time $t = 0$:

$$OUR_{ex}(t) = OUR_{ex}(0) + t\frac{dOUR_{ex}}{dt}(0) + \frac{t^2}{2!}\frac{d^2OUR_{ex}}{dt^2}(0) + \dots \qquad (10.43)$$

and consists of looking at the successive derivatives to check if they contain information about the parameters to be identified. The approach is illustrated with the second (Single Monod) model.

$$OUR_{ex}(0) = \frac{\mu_{max1}X(1-Y_1)}{Y_1}\frac{S_1(0)}{K_{s1}+S_1(0)} \quad (10.44)$$

$$\frac{dOUR_{ex}}{dt}(0) = -\frac{\mu_{max1}^2X^2(1-Y_1)}{Y_1^2}\frac{K_{s1}S_1(0)}{(K_{s1}+S_1(0))^3} \quad (10.45)$$

$$\frac{d^2OUR_{ex}}{dt^2}(0) = \frac{\mu_{max1}^3X^3(1-Y_1)}{Y_1^3}\frac{K_{s1}S_1(0)(K_{s1}-2S_1(0))}{(K_{s1}+S_1(0))^5} \quad (10.46)$$

There are five parameters to be identified: Y_1, μ_{max1}, X, K_{s1} and $S_1(0)$. Note:

$$z_i = \frac{d^iOUR_{ex}}{dt^i}(0),\ i = 0,\ 1,\ 2, \ldots \quad (10.47)$$

and consider the following combinations of the 5 above parameters:

$$\theta_1 = \frac{\mu_{max1}X(1-Y_1)}{Y_1} \quad (10.48)$$

$$\theta_2 = (1-Y_1)S_1(0) \quad (10.49)$$

$$\theta_3 = (1-Y_1)K_{s1} \quad (10.50)$$

Equations (10.44), (10.45) and (10.46) can then be rewritten under the following (equivalent) form:

$$z_0 = \frac{\theta_1\theta_2}{\theta_2+\theta_3} \quad (10.51)$$

$$z_1 = -\frac{\theta_1^2\theta_2\theta_3}{(\theta_2+\theta_3)^3} \quad (10.52)$$

$$z_2 = \frac{\theta_1^3\theta_2\theta_3(\theta_3-2\theta_2)}{(\theta_2+\theta_3)^5} \quad (10.53)$$

Therefore the *parameters* θ_1, θ_2 and θ_3 can be formally calculated from the values of z_i (which can be theoretically calculated from the $(OUR(t), t)$-data set via (10.47)) by inverting the above expressions, i.e.:

$$\theta_1 = \frac{z_0(z_0z_2-3z_1^2)}{z_0z_2-z_1^2} \quad (10.54)$$

$$\theta_2 = -\frac{2z_0^2z_1}{z_0z_2-3z_1^2} \quad (10.55)$$

$$\theta_3 = -\frac{4z_0^2z_1^3}{(z_0z_2-3z_1^2)(z_0z_2-z_1^2)} \quad (10.56)$$

It is important to note that this method could be applied successfully for the first two models by a symbolic algebra software package (Mathematica, Wolfram Research Ltd.) making it a fast and easy analysis of the problem. The complexity of the computer calculations for the identifiability analysis of the Double Monod model was such that the computer was not able to end up the computation procedure and to give any conclusive results.

10.4.1.2 Transformation of the non-linear models

Another way to analyze the structural identifiability is to transform the non-linear model into a model linear in the parameters, and then look at the identifiability of the linear model.

The line of reasoning is similar to the one considered for the exponential model hereabove (development (10.37) to (10.42)). Here, it is also applied to the Double Monod model (10.12) and (10.13). Recall that here the oxygen uptake rate OUR_{ex} is the sum of the contribution of two substrates S_1 and S_2:

$$OUR_{ex} = -(1 - Y_1)\frac{dS_1}{dt} - (1 - Y_2)\frac{dS_2}{dt} \tag{10.57}$$

A typical Double Monod type OUR_{ex}-profile is shown in Figure 10.7. In the following, it is assumed (as suggested in Figure 10.7) that one substrate (S_1) is absent in the mixed liquor after the first part of the experiment while the other is still present (note that there is only one pathological case when this assumption does not hold: when S_1 and S_2 are eliminated at exactly the same time). With this assumption the oxygen uptake rate can be subdivided in two parts corresponding with the degradation of each substrate. Hence, the identifiability analysis reduces to the analysis of the Single Monod model performed in two steps: for $t_1 \leq t \leq t_2$ for the second Monod model, and for $0 \leq t < t_1$ for the first one. First, the first step is tackled and the second term of the right hand side of (10.57) is denoted by OUR_2:

$$OUR_2 \quad = \quad -(1 - Y_2)\frac{dS_2}{dt} \tag{10.58}$$

The integration of the above equation gives:

$$S_2(t) \quad = \quad S_2(0) - \frac{1}{1 - Y_2}\int_0^t OUR_2(\tau)d\tau \tag{10.59}$$

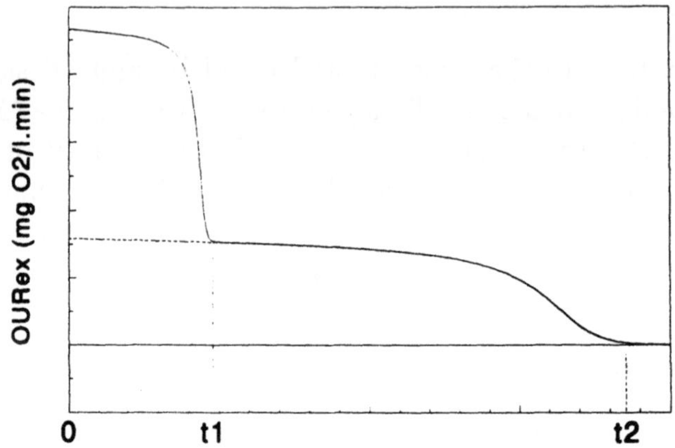

Figure 10.7: Typical OUR_{ex}-profile with Double Monod kinetics.

By introducing Equation (10.12), the oxygen uptake rate Equation (10.58) can be rewritten as follows:

$$OUR_2 = \frac{1 - Y_2}{Y_2} \frac{\mu_{max2} S_2}{K_{s2} + S_2} \tag{10.60}$$

By multiplying both sides of the above equation by $(K_{s2} + S_2)$, and by considering (in order to have more compact notations) the variable $y_2(t)$ defined as follows:

$$y_2(t) = \int_0^t OUR_2(\tau) d\tau \tag{10.61}$$

Equation (10.60) becomes:

$$y_2 \frac{dy_2}{dt} = \alpha_1 - \alpha_2 y_2 + \alpha_3 \frac{dy_2}{dt} \tag{10.62}$$

where the parameters α_i (i = 1, 2, 3) are indeed combinations of the parameters

$$\theta_4 = \frac{\mu_{max2} X (1 - Y_2)}{Y_2} \tag{10.63}$$

$$\theta_5 = (1 - Y_2) S_2(0) \tag{10.64}$$

$$\theta_6 = (1 - Y_2) K_{s2} \tag{10.65}$$

such that:

$$\alpha_1 = \theta_4 \theta_5 \tag{10.66}$$

$$\alpha_2 = \theta_5 \tag{10.67}$$

$$\alpha_3 = \theta_5 + \theta_6 \tag{10.68}$$

There is clearly a one-to-one relation between these two sets of parameters:

$$\theta_4 = \frac{\alpha_1}{\alpha_2} \tag{10.69}$$

$$\theta_5 = \alpha_2 \tag{10.70}$$

$$\theta_6 = \alpha_3 - \alpha_2 \tag{10.71}$$

Then it can be concluded that with independent data of $y_2\, dy_2/dt$, y_2 and dy_2/dt (generated via an appropriate experiment design), the parameters α_1, α_2, α_3, and therefore the parameters θ_4, θ_5, θ_6, are identifiable. This result corresponds to the one obtained in the preceding section for the Single Monod model (for which the Taylor series expansion approach was used).

One can proceed similarly for the second step $0 \leq t < t_1$, and using similar definitions for OUR_1 and $y_1(t)$, it is straightforward that the parameters θ_i (i $= 1, 2, 3$) as defined in (10.48), (10.49) and (10.50) are identifiable. This means that only six combinations of the nine original parameters (Y_1, $S_1(0)$, μ_{max1}, K_{s1}, Y_2, $S_2(0)$, μ_{max2}, K_{s2}, and X) are structurally identifiable.

As a second example of the theoretical identifiability analysis using the transformation approach, the analysis is carried out for the modified IAWQ model, based on the model Equations (10.14), (10.15), (10.16) and (10.17). As above, the first step in the analysis consists of considering that during a part of the experiment, the concentration of the rapidly hydrolyzable substrate X_r should be approximately zero.

However, the effects of the two substrates cannot be decoupled, unlike in the Double Monod model where this was possible due to the saturation in the kinetics of S_2. Therefore the analysis is started for time $t_1 \leq t \leq t_2$ when $X_r = 0$. This means that in the first step only Equations (10.14), (10.16) and (10.17) with $X_r = 0$ are considered. Note first that by introducing Equation (10.14), the OUR_{ex}-equation (10.17) can be rewritten as follows:

$$OUR_{ex} = \frac{\mu_{max1} X(1 - Y_1)}{Y_1} \frac{S_1}{K_{s1} + S_1} \tag{10.72}$$

Then the analysis is performed as follows:

1. Integration of Equations (10.17) and (10.16). This gives the following relation for $S_1(t)$:

$$\begin{aligned} S_1(t) &= S_1(0) - \frac{1}{1 - Y_1} \int_0^t OUR_{ex}(\tau) d\tau + \int_0^t k_s X_s(\tau) d\tau && (10.73) \\ &= S_1(0) - \frac{1}{1 - Y_1} \int_0^t OUR_{ex}(\tau) d\tau + k_s X_s(0)(1 - e^{-k_s t}) && (10.74) \end{aligned}$$

2. Linearization of the exponential term $e^{-k_s t}$ around $t = 0$ (in order to carry out the analysis with a model linear in the parameters):

$$e^{-k_s t} \cong 1 - k_s t + \frac{k_s^2 t^2}{2} \tag{10.75}$$

(the series expansion is stopped at the second order term since the additional terms do not add extra useful information for the analysis).

3. Introduction of these results in Equation (10.72) and rewriting by multiplying both sides by $(1 - Y_1)K_{s1} + (1 - Y_1)S_1(0) - \int_0^t OUR_{ex}(\tau)d\tau + (1 - Y_1)k_s X_s(0)(k_s t - \frac{k_s^2 t^2}{2})$ and considering $y(t)$ defined in (10.39):

$$y\frac{dy}{dt} = \beta_1 + \beta_2 y + \beta_3 \frac{dy}{dt} + \beta_4 t\frac{dy}{dt} - \beta_5 \frac{t^2}{2}\frac{dy}{dt} - \beta_6 t + \beta_7 \frac{t^2}{2} \tag{10.76}$$

$$\beta_1 = -\theta_1\theta_2 \tag{10.77}$$
$$\beta_2 = \theta_1 \tag{10.78}$$
$$\beta_3 = \theta_2 + \theta_3 \tag{10.79}$$
$$\beta_4 = \theta_7\theta_8^2 \tag{10.80}$$
$$\beta_5 = \theta_7\theta_8^3 \tag{10.81}$$
$$\beta_6 = \theta_1\beta_4 \tag{10.82}$$
$$\beta_7 = \theta_1\beta_5 \tag{10.83}$$

with θ_1, θ_2, θ_3 as defined in (10.48), (10.49) and (10.50), and θ_7 and θ_8 defined as follows:

$$\theta_7 = (1 - Y_1)X_s(0) \tag{10.84}$$
$$\theta_8 = k_s \tag{10.85}$$

Among the seven parameters β_i, only five are independent (β_6 and β_7 are related to β_4, β_5 and β_2). Therefore five parameter combinations are identifiable, i.e., β_1, β_2, β_3, β_4 and β_5, or equivalently θ_1, θ_2, θ_3, θ_7 and θ_8.

The second step for $0 \leq t \leq t_1$ considers that the dynamics are given by Equations (10.14), (10.15), (10.16) and (10.17) with $X_r \neq 0$, and that the values of the parameters β_i (i = 1 to 7) are given from the first step, i.e., data for times between t_1 and t_2. Then by following the same line of reasoning as in step 1, one obtains two more identifiable parameter combinations: $(1 - Y_1)X_r(0)$ and k_r.

10.4.1.3 Discussion of the results for the case study

The theoretical identifiability results for the four models of the case study are summarized in Table 10.2.

Exponential	Single Monod	Double Monod	Modified IAWQ
$(1-Y_1)S_1(0)$ $\frac{\mu_{max1}X(1-Y_1)}{Y_1}$	$(1-Y_1)S_1(0)$ $\frac{\mu_{max1}X(1-Y_1)}{Y_1}$ $(1-Y_1)K_{s1}$	$(1-Y_1)S_1(0)$ $\frac{\mu_{max1}X(1-Y_1)}{Y_1}$ $(1-Y_1)K_{s1}$ $(1-Y_2)S_2(0)$ $\frac{\mu_{max2}X(1-Y_2)}{Y_2}$ $(1-Y_2)K_{s2}$	$(1-Y_1)S_1(0)$ $\frac{\mu_{max1}X(1-Y_1)}{Y_1}$ $(1-Y_1)K_{s1}$ $(1-Y_1)X_r(0)$ k_r $(1-Y_1)X_s(0)$ k_s

Table 10.2: Identifiable parameter combinations of the 4 models.

The examples presented in the above sections are illustrative of the advantages and drawbacks of the considered methods. The implementation of the Taylor series expansion method has the advantage of being systematic in the sense that it follows a clearly identified route. The above example is illustrative of the potential difficulties with the Taylor series expansion:

- How many derivatives of OUR_{ex} are needed to obtain conclusive results? For certain models the question may rise whether one can achieve better identifiability properties by considering more terms in the expansion. Here it is not needed because the three terms yield sufficient information. Generally speaking the approach may imply more and more symbolic computations, and yet not lead to conclusive results (as experienced for the Double Monod model).

- How are the right combinations of identifiable parameters derived? There is indeed no general systematic rule for selecting these combinations, and therefore the procedure may look a little tricky. However, the structure of the different terms of the expansion is often a source for good initial guesses. For instance, the choice of θ_1 $(= -\frac{\mu_{max1}X(1-Y_1)}{Y_1})$ looks quite obvious from Equations (10.44), (10.45) and (10.46).

The non-linear transformation approach on the other hand may suffer from the difficulty to easily find out the transformation that will *a priori* suit to the problem (although in the proposed example, the choice of the transformation, i.e., multiplication by the denominator of the Monod model, is rather straightforward).

Some comments can be made about the obtained identifiability results.

First, note that the yield coefficient(s) Y_1 (Y_2) is (are) present in all the parameter combinations (except in k_r and k_s in the IAWQ model). This is not surprising since one can have no idea just on the basis of OUR_{ex}-data which quantity of substrate has been transformed into biomass.

Second, the parameter combinations $\frac{\mu_{max1} X(1-Y_1)}{Y_1}$ and $\frac{\mu_{max2} X(1-Y_2)}{Y_2}$ are an expression of the total activity of the sludge, and with that respect can be considered as giving an information different from that of the individual parameters.

Finally, it is important to point out that *a priori* information about some individual parameters (e.g., the yield coefficient(s) Y_1 (Y_2) value(s) obtained via separate experiments) can be incorporated in the parameter evaluation procedure. Then individual parameters (e.g., $S_1(0)$ or K_{s1}) can be estimated.

10.4.2 Practical identifiability

Practical identifiability, in contrast to the theoretical identifiability which is only dependent on the model, is related to the quality of the data and their *informative* content: are the available data informative enough for identifying the model parameters and for giving accurate values? In the model $y = ax_1 + bx_2$ the parameters are structurally identifiable but they will not be practically identifiable if the experimental conditions are such that x_1 and x_2 are proportional ($x_1 = \alpha x_2$; then only the combination $a\alpha + b$ is identifiable). A typical example is the Monod model in simple microbial growth processes, which has been shown to be structurally identifiable from (perfect) data of substrate and biomass [1], but which was found to be often not practically identifiable because of the usually poor quality of the data and the limited number of data points (e.g., [23]).

While the theoretical identifiability is studied under the assumption of perfect (i.e., noiseless) data, the problem with highly correlated parameters surfaces when a limited set of experimental, noise-corrupted data is used for parameter estimation. Under such conditions the uniqueness of parameter estimates predicted by the theoretical analysis may no longer be guaranteed, because a change in one parameter can be compensated almost completely by a proportional shift in another, still producing a satisfying fit between experimental data and model predictions. In addition, the numerical algorithms that perform the non-linear parameter estimation show poor convergence when faced with this type of ill-conditioned optimization problems, the estimates being very sensitive to the initial parameter values given to the algorithm [23][40]. Consequently, the estimated parameters may vary over a broad range and little physical interpretation can be given to the parameter values obtained.

In order to overcome the practical identifiability problem, different approaches have been suggested. Sometimes it is possible to transform the model into an equivalent form that is numerically more tractable leading to more reliable estimation [53]. It has been proposed to use additional *a priori* information -such as a known maximum growth rate- to impose parameter bounds [42].

A few studies were directed at the design of experiments by which more informative data can be collected. Vialas *et al.* [81] proposed to sample more frequently in defined periods of the experiment in order to increase the informative content of the collected data. By Holmberg [23], it was shown that the practical identifiability of Monod parameters from batch experiments depends on the initial substrate concentration. The author stated that the optimal initial substrate concentration depends on the noise level and the sampling instants. It is also obvious from her results that the experimental design is dependent on the parameter values, which, in view of the changing nature of the process studied in the wastewater treatment case study, implies that the experimental design is time-varying. Munack [42] proposed different modifications to batch experiments and it was shown that important improvements in parameter confidences can be achieved by optimal experimental design techniques. These techniques will be introduced in a subsequent section devoted to optimal experimental design for parameter estimation.

Finally, one may also circumvent the practical identifiability problem by changing the goal of the modeling exercise. Indeed, sometimes it may suffice only to give a reasonable description (*curve fit*) of the experimental data. Dedicated estimation algorithms (the so-called set membership methods) have been developed that yield sets of parameters that allow description of process *behavior*. The aim is then no longer to find unique estimates nor to give a physical interpretation to the obtained values [28].

10.4.2.1 Theoretical framework

The question addressed in this section is the following: with the available experimental data, can the parameters be given unique values? Or, in other words, if a small deviation in the parameter set occurs, does this have a considerable decrease of the fit as a consequence? Mathematically, this can be formalized as follows [43]. First, recall that parameter estimation can be formulated as the minimization of the following quadratic objective functional by optimal choice of the parameters θ:

$$J(\theta) = \sum_{i=1}^{N} (y_i(\hat{\theta}) - y_i)^T Q_i (y_i(\hat{\theta}) - y_i) \tag{10.86}$$

in which y_i and $y_i(\hat{\theta})$ are vectors of N measured values and model predictions at times t_i ($i = 1$ to N) respectively, and Q_i is a square matrix with user-supplied weighting coefficients. The expected value of the objective functional for a parameter set slightly different from the optimal one is given by [42]:

$$E[J(\theta + \delta\theta)] \cong \delta\theta^T \left[\sum_{i=1}^{N} (\frac{\partial y}{\partial \theta}(t_i))^T Q_i (\frac{\partial y}{\partial \theta}(t_i) \right] \delta\theta + \sum_{i=1}^{N} tr(C_i Q_i) \tag{10.87}$$

in which C_i represents the measurement error covariance matrix (Q_i is typically chosen as C_i^{-1}). An important consequence of (10.87) is that in order to *optimize the practical identifiability* [i.e., maximize the difference between $J(\theta + \delta\theta)$ and $J_{opt}(\theta)$] one has to *maximize the term between brackets* $[\cdot]$. This term is the so-called Fisher Information Matrix and expresses the information content of the experimental data [35]:

$$F = \sum_{i=1}^{N} (\frac{\partial y}{\partial \theta}(t_i))^T Q_i (\frac{\partial y}{\partial \theta}(t_i)) \tag{10.88}$$

This matrix is the inverse of the parameter estimation error covariance matrix of the best linear unbiased estimator [16]:

$$V = F^{-1} = \left[\sum_{i=1}^{N} (\frac{\partial y}{\partial \theta}(t_i))^T Q_i (\frac{\partial y}{\partial \theta}(t_i)) \right]^{-1} \tag{10.89}$$

The terms $\partial y / \partial \theta$ are the output sensitivity functions. These quantify the dependence of the model predictions on the parameter values. The evaluation of the sensitivity functions is a central task in the practical identifiability study and this will be dealt with in a subsequent section.

The approximation (10.87) of the objective function allows to draw lines of constant functional values in the parameter space. In case a two-parameter problem is addressed, e.g., estimation of μ_{max} and K_s in the Single Monod model, these lines form ellipses. As it was pointed out by Munack [42], the axes of the ellipses are given by the eigenvectors of the Fisher Matrix, and their lengths are proportional to the square root of the inverse of the corresponding eigenvalues. Hence, the ratio of the largest to the smallest (in absolute value) eigenvalue is a measure of the shape of the functional close to the optimal parameter estimates. Figure 10.8 shows a valley-like form corresponding with a large eigenvalue ratio.

It is important to note that many numerical optimization algorithms (needed to solve these non-linear parameter estimation problems) have difficulties in finding a global optimum in such valley-like functionals. The need to invert the Fisher Matrix in many of these algorithms is important in this respect [55]. Indeed, the above mentioned ratio of eigenvalues equals the Fisher Matrix's condition number which is a measure for the reliability by which the inversion can be made. Hence, if an appropriate experimental design could be found that alleviates this problem, increased estimation accuracy would result (see below).

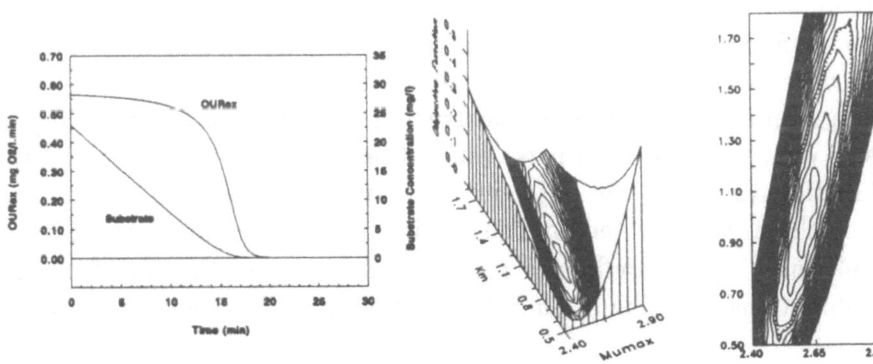

Figure 10.8: 3D (middle) and contour plot (right) of the objective function as function of the Monod biokinetic parameters for a typical respirogram (left).

10.4.2.2 Confidence region of the parameter estimates

A rather important remark considering practical identifiability concerns the parameter variance. Typically, to obtain an estimate of the parameter variance, one will make use of the covariance matrix V (10.89) and the residual mean square for which an estimate is available:

$$s^2 = \frac{J_{opt}(\theta)}{N-p} \tag{10.90}$$

with p the number of parameters in the model and $J_{opt}(\theta)$ as defined in (10.86). Approximate standard errors for the parameters can be calculated as:

$$\sigma(\theta_i) = s\sqrt{V_{ii}} \tag{10.91}$$

However, as Robinson comments [55], these standard errors are statistically optimistic due to the use of a linear approximation of the non-linear model in the neighborhood of the best parameter estimates. Alternative more robust techniques such as the jackknife and bootstrap methods produce parameter variances that are more realistic [11][15][41][47][55][56][66]. As a drawback one should mention that these methods are rather computing intensive. Another way to obtain the *true* confidence region of the parameters in non-linear models is by a systematic exploration of the objective functional for an extensive number of parameter combinations. This is a computing intensive task as well because the number of evaluations increases as a power function of the number of parameters [36].

The confidence region can be described as a hypervolume bound by a hypersurface in the parameter space. In case of two parameters to be estimated, a 2-dimensional region is constructed in a 3D parameter space. This allows to visualize some parameter estimation problems (see Figure 10.8 below). In this 3D space, the objective function is a surface. The confidence region is the area delimited by the confidence contour line. According to Beale [4] the $(1-\alpha)$ confidence region corresponds to the set of parameter combinations resulting in an objective functional less than the threshold value:

$$J_{opt}(1 + \frac{p}{N-p} F_{\alpha;p,N-p}) \tag{10.92}$$

where N and p are the number of measurements and model parameters respectively and F is the value of the F-distribution with p and $N-p$ degrees of freedom and a confidence level α. In Figure 10.8 contour lines are drawn for increasing α.

10.4.2.3 Sensitivity functions

The output sensitivity equations are central to the evaluation of practical identifiability as they are a major component of the Fisher Information Matrix, and hence, also of the parameter estimation covariance matrix. If the sensitivity equations are linearly dependent, the covariance matrix becomes singular and the model is not identifiable [55]. For many models used to describe biological phenomena, the sensitivity equations are nearly linearly dependent, resulting in parameter estimates that are highly correlated. This is also visualized in the error functional that looks like a valley, i.e., several combinations of parameters may describe the same data equally well.

Therefore, an easy way to study the practical identifiability of a model is to plot the sensitivity equations. In the literature numerous studies can be found in which this study is performed, especially for the Single Monod model considering measurements of both biomass and substrate concentrations [23][24][39][51][55][81].

For illustrative purposes, the sensitivity equations are deduced for the Single Monod and modified IAWQ model with OUR_{ex}-measurements as only source of information for the identification of the biokinetic parameters. The sensitivity of OUR_{ex} with respect to μ_{max1} is:

$$\begin{aligned}\frac{\partial OUR_{ex}}{\partial \mu_{max1}} &= \frac{\partial}{\partial \mu_{max1}}\left(-(1-Y_1)\frac{dS_1}{dt}\right)\\ &= -(1-Y_1)\frac{d}{dt}\left(\frac{\partial S_1}{\partial \mu_{max1}}\right)\end{aligned} \tag{10.93}$$

in which the state sensitivity $\partial S_1/\partial \mu_{max1}$ is obtained by integration of the differential equation (with zero initial value):

$$\begin{aligned}\frac{d}{dt}\left(\frac{\partial S_1}{\partial \mu_{max1}}\right) &= \frac{\partial}{\partial \mu_{max1}}\left(-\frac{\mu_{max1}X}{Y_1}\frac{S_1}{K_{s1}+S_1}\right)\\ &= -\frac{X}{Y_1}\left(\frac{S_1}{K_{s1}+S_1}+\frac{\mu_{max1}K_{s1}\dfrac{\partial S_1}{\partial \mu_{max1}}}{(K_{s1}+S_1)^2}\right)\end{aligned} \tag{10.94}$$

where the substrate concentration S_1 is calculated by integration of the substrate dynamic model:

$$\frac{dS_1}{dt} = -\frac{\mu_{max1}X}{Y_1}\frac{S_1}{K_{s1}+S_1} \tag{10.95}$$

Simultaneous solution of the differential Equations (10.94) and (10.95) allows to calculate the output sensitivities (10.93). One can proceed similarly for the sensitivity of OUR_{ex} with respect to K_{s1}. The following relations are obtained:

$$\frac{\partial OUR_{ex}}{\partial K_{s1}} = -(1-Y_1)\frac{d}{dt}\left(\frac{\partial S_1}{\partial K_{s1}}\right) \tag{10.96}$$

$$\frac{d}{dt}\left(\frac{\partial S_1}{\partial K_{s1}}\right) = -\frac{\mu_{max1}X}{Y_1}\left(\frac{K_{s1}\dfrac{\partial S_1}{\partial K_{s1}} - S_1}{(K_{s1}+S_1)^2}\right) \tag{10.97}$$

The equations show that the sensitivities of the Single Monod model are dependent on the parameter values. This is a general characteristic of non-linear models that has even been used to define non-linearity [13]. Consequently the Fisher Information Matrix depends on the parameter values and this feature has important implications for the optimal experimental design (see below).

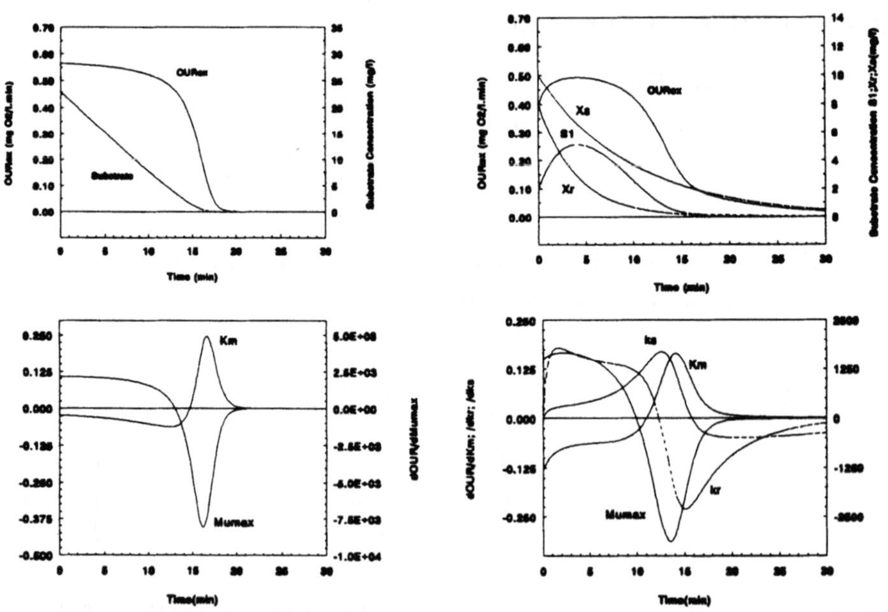

Figure 10.9: Left: output sensitivities (bottom) for a Single Monod type OUR_{ex}-profile (top). Right: output sensitivities (bottom) for an IAWQ type OUR_{ex}-profile (top).

An example of an OUR_{ex}-profile with the corresponding sensitivity function evolutions is given in Figure 10.9 (left). One observes that the sensitivity functions for K_{s1} and μ_{max1} are nearly linearly dependent, a well-known characteristic of the Monod model.

Intuitively, the sensitivity functions express the dependence of the output or state variable on a change in the parameters. Hence, the sensitivity functions indicate conditions where the dependence is the largest and therefore, under which conditions the most information can be gathered on the parameters. In the example of Figure 10.9 (left) these conditions prevail when the substrate concentration has dropped to a level close to the affinity constant K_{s1}.

From this, one can deduce a first approach to increase the information content of an experiment: choose the sampling times when the parameters are influent, i.e., in the high sensitivity zone [81].

The output sensitivities for the IAWQ model as modified by Sollfrank and Gujer [61] are deduced in a similar manner (in case OUR_{ex} is the only measured variable and the biokinetic parameters μ_{max1}, K_{s1}, k_r and k_s are to be inferred):

$$\begin{aligned}\frac{\partial OUR_{ex}}{\partial k_r} &= (1-Y_1)\frac{\partial}{\partial k_r}\left(-\frac{dS_1}{dt}+k_rX_r+k_sX_s\right)\\ &= (1-Y_1)\left(X_r+k_r\frac{\partial X_r}{\partial k_r}-\frac{d}{dt}\left(\frac{\partial S_1}{\partial k_r}\right)\right) \qquad (10.98)\end{aligned}$$

$$\frac{\partial OUR_{ex}}{\partial k_s} = (1-Y_1)\left(X_s+k_s\frac{\partial X_s}{\partial k_s}-\frac{d}{dt}\left(\frac{\partial S_1}{\partial k_s}\right)\right) \qquad (10.99)$$

State sensitivities needed for the calculation of the output sensitivities (10.98)(10.99) are:

$$\begin{aligned}\frac{d}{dt}\left(\frac{\partial X_r}{\partial k_r}\right) &= \frac{\partial}{\partial k_r}(-k_rX_r)\\ &= -\left(X_r+k_r\frac{\partial X_r}{\partial k_r}\right) \qquad (10.100)\end{aligned}$$

$$\frac{d}{dt}\left(\frac{\partial X_s}{\partial k_s}\right) = -\left(X_s+k_s\frac{\partial X_s}{\partial k_s}\right) \qquad (10.101)$$

The output and state sensitivities for μ_{max1} and K_{s1} are identical to (10.93) and (10.94), and (10.96) and (10.97) respectively.

In Figure 10.9 (right) the practical identifiability of the IAWQ model is studied by checking the output sensitivities for a short term batch experiment. No clear linear dependency between sensitivity functions is observed.

Stronger evidence can be obtained, however, by calculation of the rank of the Fisher Information Matrix. If no linear dependency exists, it should be full rank. This is indeed the case for this example. The condition number of the Fisher Matrix, or equivalently, the ratio of the largest to the smallest eigenvalue, indicates whether the sensitivities are nearly linearly dependent: the higher the condition number, the lower the practical identifiability.

So far, the initial conditions of the model variables were not included in the practical identifiability study, though they may be a highly desired outcome of parameter estimation. For the Single Monod model (10.11) for instance, one can write:

$$\begin{aligned}\frac{\partial OUR_{ex}}{\partial S_1(0)} &= -(1-Y_1)\frac{d}{dt}\left(\frac{\partial S_1}{\partial S_1(0)}\right)\\ &= (1-Y_1)\frac{\mu_{max1}X}{Y_1}\frac{\partial}{\partial S_1(0)}\left(\frac{S_1}{K_{s1}+S_1}\right)\end{aligned} \tag{10.102}$$

To solve this, one must introduce the initial condition, using the relationship:

$$S_1(t) = S_1(0) - \frac{\int_0^t OUR_{ex}(\tau)d\tau}{1-Y_1} \tag{10.103}$$

yielding:

$$\frac{\partial OUR_{ex}}{\partial S_1(0)} = \frac{(1-Y_1)\mu_{max1}X}{Y_1}\frac{\partial}{\partial S_1(0)}\left(\frac{(1-Y_1)S_1(0) - \int_0^t OUR_{ex}(\tau)d\tau}{(1-Y_1)(K_{s1}+S_1(0)) - \int_0^t OUR_{ex}(\tau)d\tau}\right) \tag{10.104}$$

and the final equation:

$$\frac{\partial OUR_{ex}}{\partial S_1(0)} = \frac{(1-Y_1)^2\mu_{max1}XK_{s1}}{Y_1}\left(\frac{(1-Y_1) - \int_0^t \frac{\partial OUR_{ex}(\tau)}{\partial S_1(0)}d\tau}{\left((1-Y_1)(K_{s1}+S_1(0)) - \int_0^t OUR_{ex}(\tau)d\tau\right)^2}\right) \tag{10.105}$$

10.5 Experimental Design

10.5.1 Introduction

So far, this chapter has devoted a lot of attention to the two main tasks of a model building exercise (Figure 10.1). However, these tasks can only be performed successfully if the sources of information are of a sufficiently high quality. Probably the definition of the objective is not the most difficult task albeit sometimes overlooked and not stated in a quantitative way. The analysis of the *a priori* knowledge leading to the construction of the set of candidate models may take some more time as knowledge must be acquired from literature, experts and so forth. The main efforts however are, certainly for bioprocesses, devoted to the collection of data. The design of the experiments that will provide the data is an increasingly important task for the model builder for the following reasons:

- the limited resources available, e.g., experimentation time and expenses
- the increased degrees of freedom that have to be exploited in an optimal way
- the trade-off that is to be made between the information content of the data and the pursued goal.

Goals pursued during an experimental design procedure can be categorized in the following areas that will be dealt with separately in the following subsections:

- experimental design to reliable selection of an adequate mathematical model structure of the process
- the design of experiments for precise estimation of model parameters
- the dual problem of structure characterization and parameter estimation.

Each of these goals is associated with one or another quantitative function which will determine the focus of the experimental efforts. Some authors [35] also suggest to consider separately:

- the optimal sampling frequency
- the best location of sensors and sampling points in a reactor system.

However, these items should be regarded as part of the degrees of freedom available and the constraints to which one has to adhere during the experimental design process. An experimental design typically consists of three steps:

1. Definition of an objective function that is the mathematical translation of the pursued goal

2. Enumeration of the available degrees of freedom and experimental constraints
3. Extremization of the objective function by appropriate variations of the degrees of freedom without violation of the applicable constraints.

Figure 10.10 schematizes the procedure used for (automated) optimization of the experimental conditions w.r.t. the defined objective function. An essential feature of the procedure is that a preliminary model is available that is identified on the basis of previously acquired data. Essentially, the algorithm performing the design performs *simulated experiments* to quantify the potential effect of proposed experimental conditions on the objective function. Using a (non-linear) optimization algorithm that is able to take into account the experimental constraints, an *optimal* experiment is proposed that is subsequently applied in real-life. Real-life may be a specific lab experiment, a full-scale process on which some degrees of freedom are available that can be exploited to gather informative data or, as in the case study developed in this chapter, real-life may be the respirographic sensor in which pulse experiments are conducted.

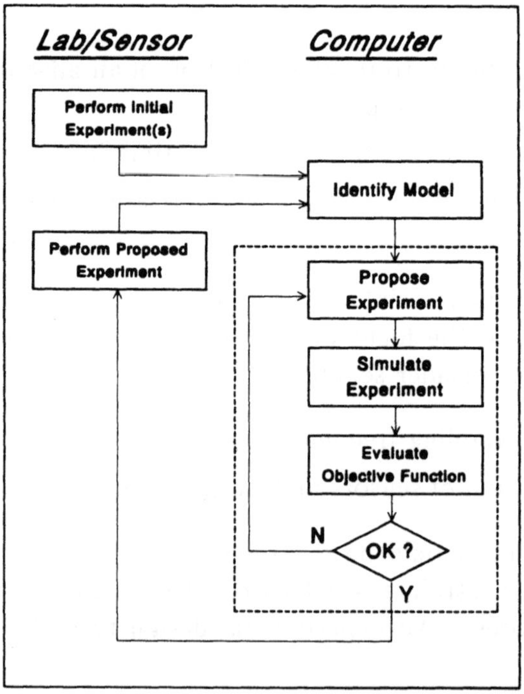

Figure 10.10: General procedure for experimental design.

Since the experimental design procedure depends to a large extent on a model of the process under study, it is evident that a prerequisite to good design is the availability of a good process model. Questions that may be of guidance when initiating an experimental design study are:

What to measure? This question stresses the importance of a good system definition (input, output and state variables, ...).

Where to measure? This addresses the problem of sensor location.

When to measure? Here the sampling strategy is paid attention to.

Which manipulations? In many cases this is the most important question and requires creative work related to the design of excitation signals that will perturb the process in such a way that highly qualitative information can be gathered. Examples are sine-waves, pulses, PRBS (pseudo random binary sequences [37]), ...

What data treatment? Though most of the time not considered as a part of the experimental design, pretreatment of the data is often applied before model identification starts. Typical data pretreatment includes noise rejection, outlier detection/removal and elimination of *uninteresting* dynamics.

Finally, some statements by Ljung [35] are given to rephrase the importance of a good experimental design:

While several different design variables associated with models and methods can be tried out on the computer, the experimental data can be changed only by a new experiment, which could be a costly and time-consuming procedure.

Therefore it is worthwhile to design the experiment thoughtfully so as to generate data that are sufficiently informative.

It is wise to let the experiment resemble the situation under which the model is to be used.

Careful experiment design yielding data with good information is the basis of a successful identification application.

10.5.2 Optimal experimental design for structure characterization

10.5.2.1 Theoretical background of OED/SC

To start with, the OED problem must be clearly stated and an objective function defined as its translation. The goal may be to

Design an experiment to discriminate between rival models y_1 *and* y_2.

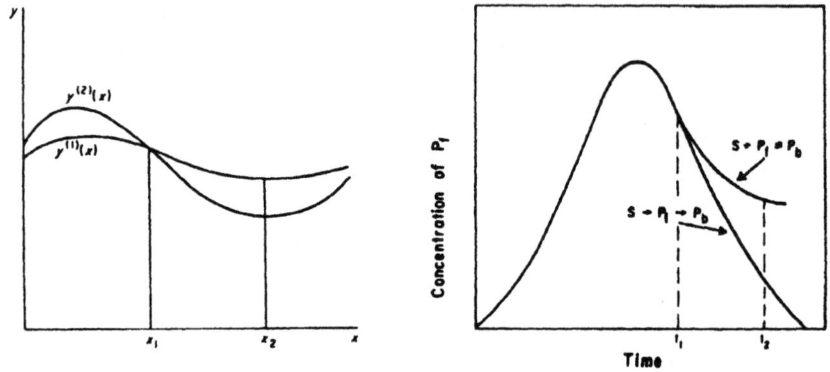

Figure 10.11: Hunter-Reiner approach for an optimal experimental design to discriminate between two rival models.

A very useful and intuitive translation into an objective function is given by the Hunter-Reiner approach [25] that chooses the experimental conditions χ_i such that the difference between the predictions by both models

$$(\hat{y}_1(\chi_i) - \hat{y}_2(\chi_i))^2 \tag{10.106}$$

is maximized.

The approach is illustrated in the examples of Figure 10.11. In both cases it is obvious that no selection can be made between the 2 models in competition at condition 1 while maximal discriminative power is obtained under experimental conditions 2.

While this approach is very appealing (and widely applied), a major drawback is that no consideration is given to the fact that uncertainty exists on the model predictions $\hat{y}_i$. Figure 10.12 shows the problem of uncertainty effects for two simple regression models that are in competition:

$$y_1 = bx + a \tag{10.107}$$

$$y_2 = bx \tag{10.108}$$

It is straightforward to compute the confidence regions for both models.

From the figure it is evident that the confidence regions are important for the assessment of the discriminative power of an experiment. For this particular case, it is clear that model selection will be most reliable under experimental conditions equal or lower than zero. Note that one would have preferred high positive values of the independent variable if the confidence regions were not included in the graph.

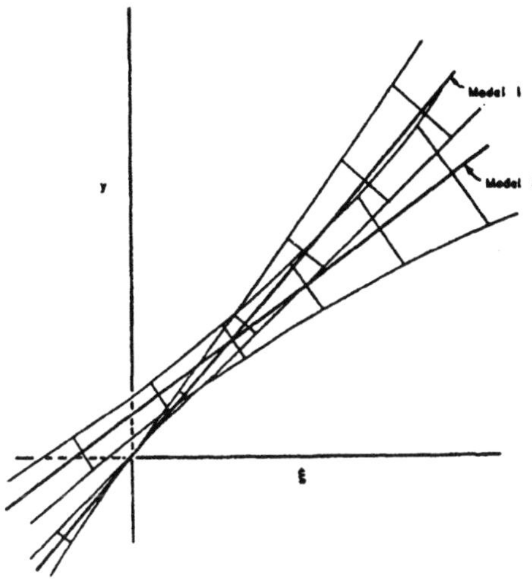

Figure 10.12: Uncertainty in model predictions of two rival regression models.

Box and Hill [6] developed a quantitative description that allows to compute the divergence between rival models under model prediction uncertainty. They extended the method to the case of discrimination among m rival models and the use of the combined data of n previous experiments so as to design the $(n+1)$th. It would lead us to far to include these complex objective functions for OED/SC here. They can be found in the mentioned reference together with some illustrative examples.

The OED/SC methods mentioned so far were mainly developed for static models (e.g., regression type empirical models). For dynamic models Munack [44] proposed two approaches, one based on the maximization of the difference in model predictions, i.e., similar to the Hunter-Reiner method, the other being based on the level of dependence of the parameter estimates on the experimental conditions. The latter in fact corresponds to a kind of adequacy evaluation of the rival models since adequate models are characterized by the fact that the parameters are independent of the experimental conditions.

For the first approach the optimization loop of OED/SC is the following [44]:

1. Propose experimental conditions and perform hypothetical experiments through simulation with the 2 models, giving rise to two sets of *raw* data

2. (a) Consider Model 1 correct, than fit Model 2 to the Model 1 generated data (evidently, Model 1 must not be fitted since the data originate from this model)
   (b) Consider Model 2 correct, than fit Model 1 to the data generated from Model 2 (here Model 2 must not be fitted)
3. For both cases calculate the difference between the (simulated) trajectories
4. Maximize the smallest of the two calculated differences.

It is important to note that the two parameter estimations are included in the procedure, because it is reasonable to accept that a data set generated by a particular model can be described best by the rival model after a new estimation of its parameters. However, these parameter estimations make that this OED/SC approach may ask for lengthy computations. These may not be desired for certain applications, e.g., for the design of the experiments in the respirographic biosensor (see below).

In the second method proposed by Munack [44], the change in the model parameter values as a result of the new fit necessary to accommodate for a proposed change in experimental conditions, is assessed and maximized. Since no *a priori* knowledge on the *best* model is available at the moment the experimental design is performed, the rival models are treated on an equal basis. The design calculations are identical to the above mentioned procedure but as an objective function to be maximized the parameter sets for the first (real) experiment and the second (simulated) experiment are compared for both models. For this, the Mahalanobis distance between the parameter vectors is calculated and the smallest of these distances is maximized. The Mahalanobis distance is a measure for the difference between two vectors which takes the estimation accuracy into account.

10.5.2.2 Application: real time OED/SC in a respirographic biosensor

In the respirographic biosensor of the case study developed throughout this chapter, the possibility exists to adjust the experimental conditions of the bioreactor. The aim is to maintain the quality of the identification of models describing the biodegradation processes. In this section attention will be given to a specific method that enables OED/SC even under the real time constraints imposed by the 30 minute interval between consecutive experiments. It was shown that the methods described above could not handle this constraint due to the excessive computational burden and therefore a dedicated method had to be developed [73]. It is important to note that, as for each OED/SC method, the objective function is closely related to the method of structure characterization applied during the identification stage. Here, the *a priori* structure characterization method using the number of inflection points is used as the basis of the OED/SC procedure.

It is good to recall that to determine the number of inflection points from the respirograms and their reliability, the second derivative (the curvature) must be calculated. To estimate its value and to determine whether it is significant, a moving window regression with window width n is applied and both a straight line and a parabola are fitted through points j till $j+n$.

Whether the parabolic fit is significantly better, and therefore, whether the estimated value of the curvature is significantly different from zero, is tested by [13]:

$$\frac{(SSR_1 - SSR_2)/1}{SSR_2/(n-2)} \simeq F(1; n-2) \tag{10.109}$$

where SSR_1 and SSR_2 are the not explained variances of the linear and parabolic regression respectively. $SSR_1 - SSR_2$ is the extra sum of squares due to the inclusion of the curvature in the regression. The null hypothesis that the parabolic fit is not significantly better than the linear one can be tested by referring this ratio of mean squares to the F-distribution with 1 and $n-2$ degrees of freedom. The 1 in the numerator is the difference in degrees of freedom between a straight line and a parabola. One should note that the approach used to discriminate between the parabola and the linear regression is based on the F-test that was presented as one of the traditional structure characterization methods.

If the parabolic fit is significantly better, the highest order coefficient (the curvature) is returned, otherwise its value is set to zero. An inflection point is defined by a point where the curvature crosses zero. The results for simulated (noiseless) example respirograms are displayed in Figure 10.13.

To define the reliability of an inflection point, it is to be noticed that, as illustrated in Figure 10.13, an inflection point is surrounded with two *pulses*, a positive one and a negative one. These two pulses can be used in several ways to define the reliability $r(f)$ of an inflection point f, e.g.:

- The surface of both pulses
- The total height of both pulses
- The height of the smallest pulse

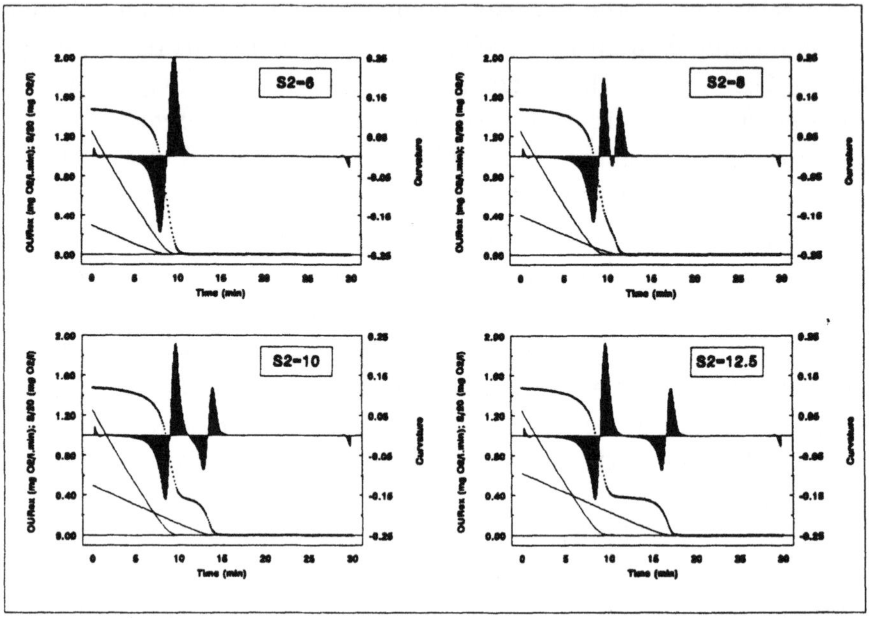

Figure 10.13: Curvature of respirograms obtained with different ratios of two substrates.

The first approach will be used in the sequel. In order to increase the discriminative power of the experiments, the aim is to determine the inflection points with the highest reliability, and hence the aim of the optimal experimental design is to maximize $r(f)$.

The examples given will be restricted to the cases where the wastewater influent contains two substrates S_1 and S_2 (hence, a Double Monod model can be used) and the aim is to design an experiment such that the Double Monod model will be selected reliably. Pulse injection of such wastewater to the batch reactor of the biosensor results in initial substrate concentrations noted as $S_1(0)$ and $S_2(0)$. Two related problems will be treated:

OED/SC for calibrations. Since both the initial concentrations $S_1(0)$ and $S_2(0)$ are at the disposal of the experimenter, two degrees of freedom for the optimization problem exist.

OED/SC for wastewater. in this case, since the wastewater composition can not be altered, the ratio $S_1(0)/S_2(0)$ is fixed. Only the amount of wastewater injected is variable and hence only one degree of freedom is left.

10.5.2.3 OED/SC for calibrations

In the respirographic biosensor used in this work, calibrations are regularly performed, mainly to verify the correct functioning of the apparatus. As shown by Vanrolleghem and Verstraete [75], this calibration can, however, also be used to independently characterize the two main groups of aerobic organisms constituting the activated sludge, i.e., heterotrophs and nitrifiers. To achieve this goal, a calibration mixture of ammonia and a readily biodegradable carbon source such as acetate is injected. The optimal experimental design is then aimed at obtaining the amount of each substrate such that the resulting OUR_{ex}-curve allows the extraction of the three inflection points with the highest reliability. This can be done by maximizing the reliability of the three inflection points, or by maximizing the least reliable inflection point. The first approach has been chosen and hence the following optimization problem can be formulated:

$$\max_{S_1(0),S_2(0)} [r(f_1) + r(f_2) + r(f_3)] \tag{10.110}$$

where $S_1(0)$ and $S_2(0)$ are the initial concentrations of the calibration substrates and $r(f_i)$ is the area of the positive and negative pulse determining the ith inflection point (see above).

This optimization problem can be solved approximately by computing the sum of the reliabilities for each substrate combination on the grid $S_1(0)$ = 5(5)50 (from 5 to 50 in steps of 5) and $S_2(0)$ = 5(5)50 mg/L. One obtains a response surface that points to the optimal substrate combination. In Table 10.3, the two sets of biokinetic parameters that were used in the simulations are summarized.

Parameter	Set 1	Set 2
X	4000 mg/L	4000 mg/L
μ_{max1}	5. e-4 /min	2.62 e-4 /min
K_{s1}	1.mg/L	0.226 mg/L
μ_{max2}	1. e-4 /min	2.85 e-4 /min
K_{s2}	0.2 mg/L	0.6 mg/L

Table 10.3: Parameter sets used in the simulations for OED/SC.

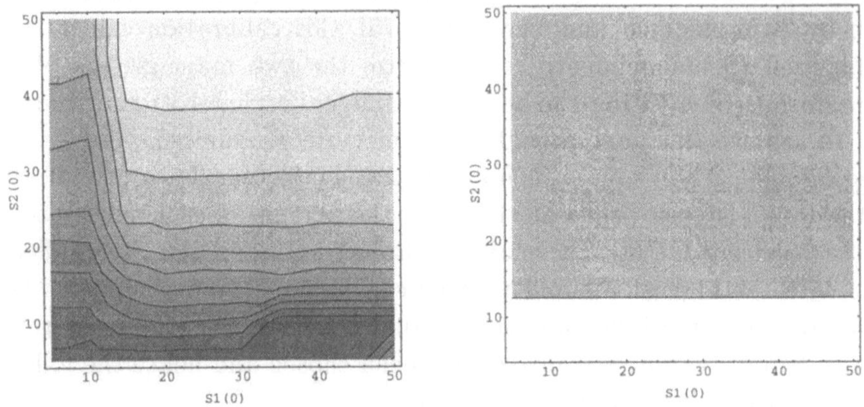

Figure 10.14: Contour plot indicating the reliability (left) and the length (right) of the respirogram for parameter Set 1. Black indicates lower reliability or a violation of the real time constraint.

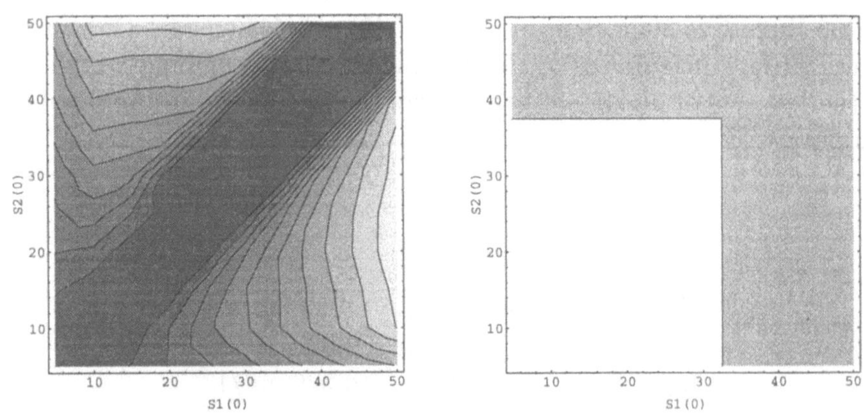

Figure 10.15: Contour plot indicating the reliability (left) and the length (right) of the respirogram for parameter Set 2. Black indicates lower reliability or a violation of the real time constraint.

For Set 1, the results are schematized in Figure 10.14. On the left, the sum of surfaces is given as a contour plot. Black indicates the experimental conditions to avoid and the lighter areas result in more reliable inflection points. The length of the respirogram is depicted on the right side. An experimental condition resulting in a respirogram longer than 30' is colored black, reflecting the importance given to the real time constraint in this optimization problem. From this it is clear that the line $S_2(0) = 5$ mg/L should be avoided and that $S_2(0)$ should be less than 12.5 mg/L.

In Figure 10.13 the OUR_{ex}-curve and corresponding substrate removal curves for four cases with increasing $S_2(0)$ are displayed ($S_1(0)$ is fixed at 25 mg COD/L). Note that the OUR_{ex}-profile in the top left case degenerates in a Single Monod model, implying that the parameters of the Double Monod model have become practically unidentifiable [75].
It is very important to note that this optimal experimental design needs to be performed on-line, because other biokinetic parameters result in a completely different advice. This is clearly illustrated by the OED/SC results (Figure 10.15) for the second set of parameter values of Table 10.3. For these sludge characteristics, the substrate concentrations to avoid, i.e., the line $S_1(0) = S_2(0)$, are clearly different compared to the ones obtained for the first parameter set (Figure 10.14).

This result emphasizes the need to perform OED/SC on-line and therefore stresses the requirement for *a priori* SC methods so as to meet the real time constraints.

10.5.2.4 OED/SC for wastewater

The OED/SC for wastewater can be derived from the OED/SC for calibrations. Suppose the wastewater composition is characterized by a ratio $r = S_1(0)/S_2(0) = \alpha$. This implies that only initial concentrations lying on the line $S_1(0) = \alpha S_2(0)$ need to be considered. This line can be drawn on the given contour plots (Figure 10.14 and Figure 10.15), reducing the two-dimensional contour plot to a one-dimensional plot. One of these is shown in Figure 10.16 (left) for the first set of parameters and for a fixed ratio $S_1(0)/S_2(0) = 1$. The reliability of the inflection point $r(f)$ is plotted versus $S_1(0)$, while $S_2(0)$ can be computed from the known ratio. Figure 10.16 (left) illustrates that the reliability increases monotonically with increasing $S_1(0)$. The marginal increase however, is variable, implying that the gain in reliability is not constant. No maximum is found, but $S_1(0)$ is limited by the same real time constraints as in Figure 10.14, limiting $S_1(0)$ to a maximum of about 12.5 mg/L.

The same wastewater composition with an activated sludge characterized by parameter Set 2, results in an uncharacterizable respirogram as illustrated in Figure 10.16 (right): no $S_1(0)$ can be found for which the resulting respirogram will yield significant inflection points and hence no Double Monod model will be selected, although the process may intrinsically be a Double Monod type process. Only increasing the degrees of freedom of the experimental design may help to solve the identification problem. This may consist of an artificial change of the wastewater's composition by deliberate change of one of the substrate's concentrations by a known amount.

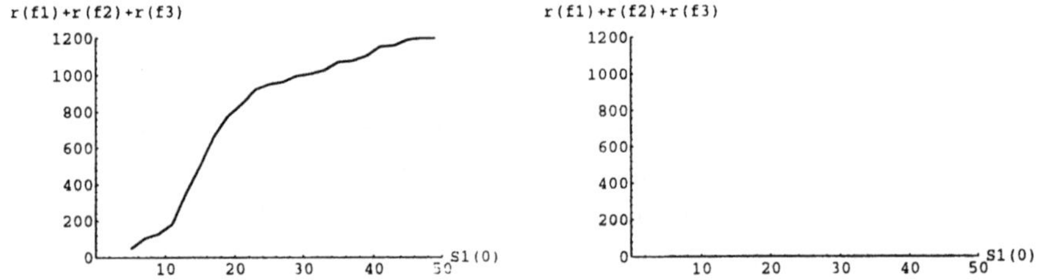

Figure 10.16: The reliability of inflection point determination for a given $S_1(0)$ and a ratio $S_1(0)/S_2(0) = 1$ for parameter Set 1 (left) and Set 2 (right).

10.5.3 Optimal experimental design for parameter estimation

If one aims to design experiments for optimal parameter estimation (OED/PE), it is good to recall that the variable describing the reliability of a parameter estimate, the approximate standard error of the parameter is given by (10.91):

$$\sigma(\theta_i) = s\sqrt{V_{ii}} \tag{10.111}$$

One observes two factors that can be manipulated to increase the parameter estimation accuracy. The first one, the residual mean square s^2 is readily calculated from the sum of squared errors between N model predictions and experimental data J_{opt} (10.90):

$$s^2 = \frac{J_{opt}(\theta)}{N - p} \tag{10.112}$$

It is evident that this term can be decreased by increasing the number of experimental data N, e.g., by repeating the experiment.

Alternatively one may aim to reduce the parameter estimation error covariance matrix V (10.89). The methods to decrease V by optimal experimental design are the focus of attention in this section.

10.5.3.1 Theoretical background of OED/PE

Different strategies have been developed to design experiments in such a way that the measurement data allow unique determination of the (combinations of) parameters that were shown to be theoretically identifiable, i.e., produce *informative* experiments. The Fisher Information Matrix F or, equivalently, the covariance matrix V are the corner-stones of the optimal experimental design procedures because these matrices summarize the information content of an experiment or the precision of the parameter estimates. Depending on the requirements imposed by the application different scalar measures of these matrices are optimized [43]:

$$\textit{A-optimal design criterion}: \quad min[tr(F^{-1})] \tag{10.113}$$
$$\textit{Modified A-optimal design criterion}: \quad max[tr(F)] \tag{10.114}$$
$$\textit{D-optimal design criterion}: \quad max[det(F)] \tag{10.115}$$
$$\textit{E-optimal design criterion}: \quad max[\lambda_{min}(F)] \tag{10.116}$$
$$\textit{Modified E-optimal design criterion}: \quad min[\frac{\lambda_{max}(F)}{\lambda_{min}(F)}] \tag{10.117}$$

in which $\lambda_{min}(F)$ and $\lambda_{max}(F)$ are the smallest and largest eigenvalue of the Fisher Information Matrix.

The following interpretation can be given to these optimal experimental design criteria [43]. The A- and D-optimal designs minimize the arithmetic and geometric mean of the identification errors respectively, while the E-criterion based experimental designs aim at minimizing the largest error. Because in both these criteria a maximization of eigenvalues of the Fisher Information Matrix is pursued, they guarantee the maximization of the distance from the singular (non-informative) case. The modified E-criterion should be interpreted in the frame of the objective functional shape. The ratio of the largest to the smallest eigenvalue is an indication of this shape. The objective is to have eigenvalues as close as possible to each other: the shape is then circular. When $\lambda_{min}(F)$ is zero, this ratio is infinite, i.e., an infinite number of parameter combinations can be used to describe the experimental data and hence the experiment is non-informative. The Fisher Matrix is then singular, and hence the D- and E-criteria are zero while the

A-criterion cannot be determined since inversion of F is impossible. This case points also to problems that can be encountered with the modified A-criterion: even if a non-informative and unidentifiable experiment is conducted, the modified A-criterion may still be maximized because one of the other eigenvalues has become large [17]. Finally, it should be mentioned that other design criteria can be proposed, e.g., reducing the estimation error of a particular parameter can be obtained by designing experiments with this particular variance component as design criterion.

10.5.3.2 Application: real time OED/PE in a respirographic biosensor

Designing identification experiments requires several choices, e.g., what outputs should be measured at what time instants and at what frequency, and what inputs to manipulate and in what way. In the case study the output and sampling frequency are no longer available to the experimenter since they are fixed by the hardware used in the study. The only degree of freedom left is the design of the input. Optimal experimental design therefore reduces to finding the input functions $u(t)$ that lead to the most informative experiments.

It must be stressed that optimal experimental designs for parameter estimation of non-linear models are influenced by the parameter values since the design criteria are based on the Fisher Information Matrix which is parameter dependent (via $\partial y/\partial\theta$). In the application studied here this feature has considerable implications. For instance, as the wastewater/sludge interaction is time-varying, the values of the biokinetic parameters of the biodegradation are changing, and therefore the optimal experimental design has to be changed as well.

10.5.3.3 OED/PE for the Single Monod model: introduction

In a previous section the theoretical identifiability of four kinetic models (Exponential, Single Monod, Double Monod and modified IAWQ $n°1$) was studied, based on OUR_{ex}-measurements. Here the practical identifiability is concentrated on and the optimal experimental design for parameter estimation (OED/PE) of one of these models, the Single Monod model is considered:

$$\frac{dS_1}{dt} = -\frac{\mu_{max1}X}{Y_1}\frac{S_1}{K_{s1}+S_1} \tag{10.118}$$

$$OUR_{ex} = -(1 - Y_1)\frac{dS_1}{dt} \tag{10.119}$$

The choice of the Monod model is, at least partially, motivated by its very large use in biotechnological applications. More specifically in the context of this study, this choice means that it is assumed that the experimental data are characterized by Single Monod kinetics, either because the real-life data are always characterized by this type of kinetics, or because a preliminary model structure characterization has been performed, leading to the selection of the Monod kinetics.

It was shown above that three combinations ($\mu_{max1}X(1 - Y_1)/Y_1$, $(1 - Y_1)S_1(0)$, $(1 - Y_1)K_{s1}$) of the five original parameters (μ_{max1}, X, Y_1, $S_1(0)$, K_{s1}) are structurally identifiable. In order to have a presentation as illustrative as possible (via, e.g., the use of 3D plots of the confidence regions), it is assumed here that the initial substrate concentration $S_1(0)$, the yield coefficient Y_1, and the biomass concentration X are known *a priori* (e.g., via some separate experiments). This leaves two parameters (μ_{max1}, K_{s1}) to be estimated. Two examples of OED/PE, based on real-life data, will now be developed here:

1. Optimal initial substrate
2. Optimal additional pulse with fixed initial substrate

In [71], other simulation examples with optimal additional pulse and initial pulse, and optimal design with multiple additional pulses are presented.

Start of a batch experiment by pulse injection of wastewater is included in the model via the initial conditions $S_1(0)$ (which will be the degree of freedom in the optimal experimental design). An additional term in the mass balance is required to describe the fed-batch experiments that are also treated in this chapter. In order to prevent numerical problems, a pulse injection of wastewater in the course of an experiment is described by a Gauss-like function:

$$S_{puls}e^{-\frac{(t - t_{puls})^2}{\sigma}} \tag{10.120}$$

in which t_{puls} is the time instant at which the pulse is given, σ is the width of the pulse and $S_{puls}\sqrt{\pi\sigma}$ is the total amount of substrate injected.

10.5.3.4 OED/PE for the Single Monod model: reference experiment

The reference experimental OUR_{ex}-profile consists of a batch experiment with an initial acetate concentration of 20 mg COD/L. Figure 10.17 presents the collected experimental data and the fit of the Single Monod model. Since the OED/PE study was mainly directed to the improvement of the numerical properties of the optimization problem via modified E-criterion based experimental designs, the objective functionals' shape was calculated for a grid of parameter combinations (μ_{max1}, K_{s1}) in the neighborhood of the optimum. It should be emphasized that the surface and corresponding contour plot depicted in Figure 10.17 is the result of systematic exploration of the error functional in parameter space and is not a mere representation of the linearized objective functional around the optimum as it is often found in the literature [36].

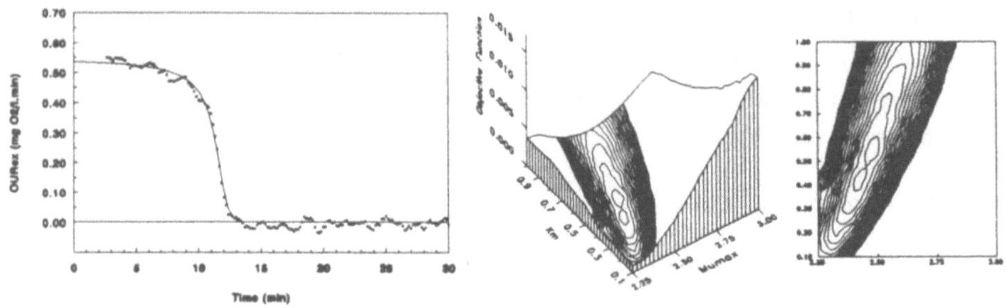

Figure 10.17: 3D (middle) and contour plot (right) of the objective function as a function of the Monod parameters for the validation reference respirogram (left).

This example of a flat valley in the parameter space may be the source of considerable problems to certain optimization algorithms. Experience with the cases studied so far tells that there exist adequate optimization algorithms, such as the direction set method of Brent (1973), which converge to the global minimum ($\mu_{max1} = 2.457\ 10^{-4}$ /min; $K_{s1} = 0.456$ mg COD/L). Still, the valley is undesirable and the aim of the study was to see whether the proposed OED/PE methods would result in improved properties.

The Fisher Information Matrix corresponding with this experiment and the deduced values of the different OED/PE criteria are summarized herebelow:

$$F = \begin{pmatrix} 3.475\ 10^8 & -12250.1 \\ -12250.1 & 0.57715 \end{pmatrix}$$

$$V = \begin{pmatrix} 1.148\ 10^{-8} & 2.443\ 10^{-4} \\ 2.443\ 10^{-4} & 6.926 \end{pmatrix}$$

$$tr(F) = 2.005\ 10^8,\ det(F) = 5.03\ 10^7,\ tr(V) = 7.95\ 10^{-8} \quad (10.121)$$

$$\lambda_{min}(F) = 0.145\ 10^{-2},\ \frac{\lambda_{max}}{\lambda_{min}}(F) = 2.39\ 10^9 \quad (10.122)$$

10.5.3.5 OED/PE for the Single Monod model: example 1 - initial substrate

As a first validation test, the effect of a change in initial concentration on the error functional shape and the estimation accuracy is assessed. For this purpose a batch experiment was conducted with half the initial concentration of the reference experiment (see Figure 10.18). The modified E-criterion value calculated from the experimental results was 2.42 times lower than the reference value, i.e., lower substrate concentrations give rise to batch experiments in which the error functional is more cone-like. However, it can also be pointed out that this numerical improvement is at the expense of estimation accuracy. Hence, if the parameter variances are calculated it is found that the variances increase, especially for the parameter μ_{max1} (increased with a factor 3.82) and to a lesser extent also for the affinity constant K_{s1} (a factor 1.52). Let us note that OED based on the modified E-criterion may sacrifice parameter estimation accuracy for improved numerical properties.

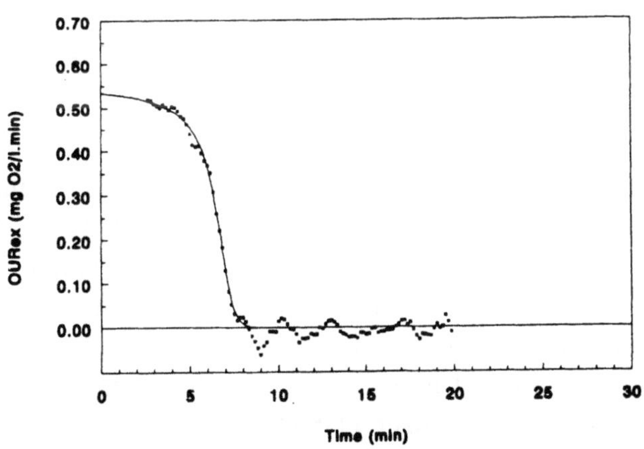

Figure 10.18: Experimental respirogram in the initial substrate case.

10.5.3.6 OED/PE for the Single Monod model: example 2 - additional pulse

The effect of an additional pulse of substrate was validated with three experiments in which the substrate concentration in the bioreactor was increased with 2 mg COD/L. Different injection times t_{puls} were tested in order to illustrate the effect of an optimal t_{puls}.

Suppose first that the data of the reference example are available and that an additional experiment has to be designed with the possibility of adding one more substrate pulse. The calculations result in curves of criterion values versus injection time, summarized in Figure 10.19. These graphs show the differences in optimum injection time for the different design criteria. The A-, D- and E-criteria propose to inject a substrate pulse after 14.6 minutes (after complete degradation of the initially present substrate). The modified A-criterion based experiment consists of a prolonged batch-phase. And the modified E-criterion OED results in a respirogram in which the oxygen uptake reaccelerates just before complete disappearance of the initial amount of substrate.

The three experiments that were performed had injection times of 13, 14.1 and 14.6 minutes respectively. The resulting OUR_{ex}-profiles are given in Figures 10.20 and 10.21. A first important observation is that the model extension for fed-batch operation is capable of simulating the behavior remarkably well. The pulse is described very well and microbial metabolism does not seem affected by the important transients imposed.

The following conclusions can be drawn by focusing on the effect of these fed-batch experiments on the error functional shape and parameter variances. Although the expected values for the modified E-criterion and the variances may change to a certain extent from the actually observed values due to changes in noise level, experimental error and biological changes, the trends set by the theoretical analysis are confirmed with these results. Predicted modified E-criterion values for instance were approximately 20 % underestimated compared to the actual values. However, the data given in Table 10.4 clearly illustrate that still a significant improvement in shape of the error functional is obtained with fed-batch experiments. Moreover, as Figure 10.19 illustrates, the times of pulse addition that were evaluated were in the secondary minimum and more important effects could be achieved if the substrate had been injected after 11.8 minutes.

A second conclusion concerns the variances. The experimental results confirm that significant improvements in parameter estimation accuracy can be obtained by this small extension of the experiment. The variances have decreased with more than 50 % (Table 10.4). One should note that a similar effect on the parameter variance can be obtained by repeating the experiment twice, recall (10.90) and (10.91). However, this would double the experimentation time while the approach taken here increases the experiment duration by only 3 minutes, i.e., 10 % of normal operation.

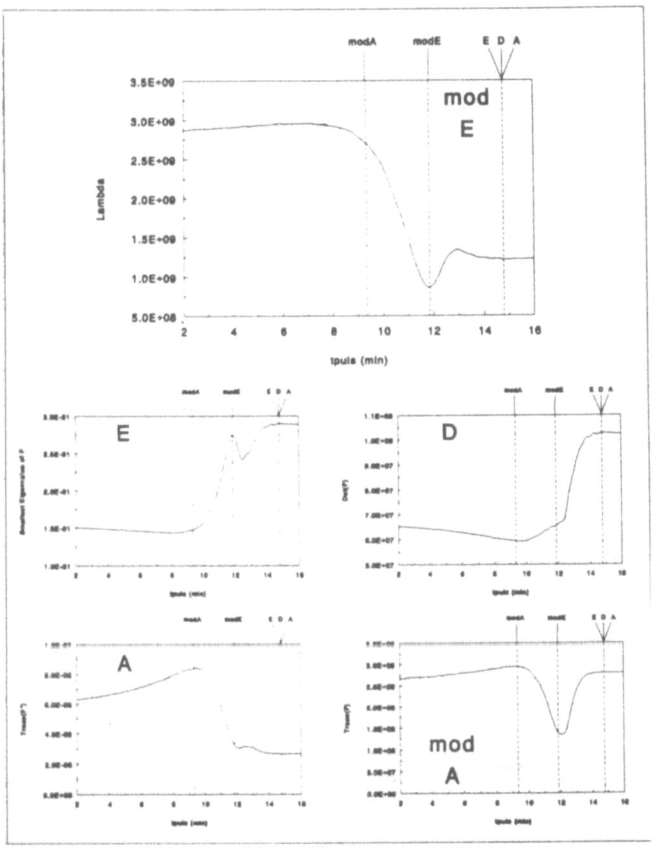

Figure 10.19: Evolution of the different OED/PE criteria as a function of the pulse addition time (vertical lines = optimal time of addition for the different criteria).

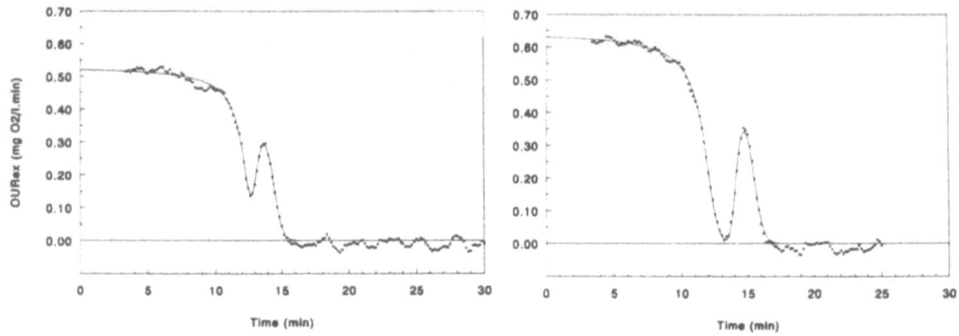

Figure 10.20: Experimental respirogram obtained with a fed-batch experiment with additional pulse. Left: after 13 minutes. Right: after 14.6 minutes.

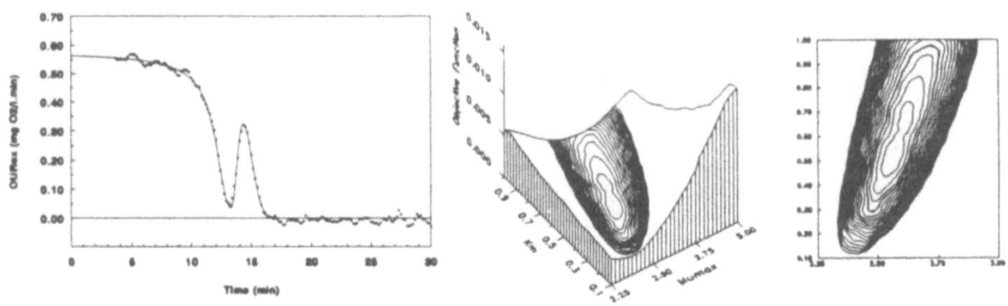

Figure 10.21: 3D (middle) and contour plot (right) of the objective function as a function of the Monod parameters for the validation respirogram with additional pulse after 14.1 minutes (left).

t_{puls}	Modified E	Var(μ_{max1})	Var(K_{s1})	Covariance
No pulse	1	1	1	1
13	0.676	0.4111	0.422	0.381
14.1	0.624	0.535	0.465	0.468
14.6	0.619	0.480	0.409	0.417

Table 10.4: Dependence of the modified E-criterion and of the parameter variances with respect to the pulse addition time.

10.6 Optimal Experimental Design for the Dual Problem of Structure Characterization and Parameter Estimation

As it was presented so far, the goal of experimental design was geared to *either* parameter estimation *or* structure characterization. Frequently, however, investigators want to perform experiments that will shed light on both questions simultaneously.

The simplest approach, which is often intuitively taken, is to apply first the discrimination criterion to select the best model and then use an OED/PE procedure for design of experiments for reliable parameter estimation [21]. However, one can imagine that during the structure characterization geared experiments important information is already collected on parameter values, and, vice versa, during the experiments for parameter estimation additional data are gathered on the most appropriate model structure. Most importantly, it is likely that the same model accuracy can be reached with fewer experiments if a joint design criterion were used.

The more integrated approach presented by Hill *et al.* [22] is based on a design criterion of the form

$$C = w_D D + w_E E \tag{10.123}$$

where D is a measure for the discriminative power of an experiment and E reflects the parameter estimation accuracy. Such joint design criterion should emphasize discrimination among rival models when there is substantial uncertainty as to which model is best and should emphasize parameter estimation accuracy when one particular model of the model set is definitely superior over the others. Hence, the values of the relative weights w_D and w_E should gradually change as experimentation proceeds to reflect the relative emphasis that is given to each of the objectives. A sequential design procedure where an $(n+1)$th experiment is designed should therefore find the experimental conditions χ_{n+1} such that C is maximized. Hill *et al.* [22] proposed a particular form for this joint criterion that is fit for the general case of m rival models:

$$C = w_D \frac{D}{D_{max}} + (1 - w_D) \sum_{j=1}^{m} \Pi_{j,n} \frac{E_j}{E_{j,max}} \tag{10.124}$$

where D_{max} and E_{max} are the maximum attainable values of D and E_j over the allowed χ-region. $\Pi_{j,n}$ is the probability of model j after n experiments, for which an iterative formula is given in [21]. The weighting w_D given to discrimination is suggested to be [22]:

$$w_D = \left(\frac{m}{m-1} (1 - \Pi_{max,n}) \right)^{\lambda} \tag{10.125}$$

in which $\Pi_{max,n}$ is the largest probability of the $\Pi_{j,n}$ after n experiments. The tuning parameter λ gives a degree of freedom to the experimenter on the rate at which interest shifts from OED/SC to OED/PE.

Very few results have been reported on the performance of such criteria. The example given by Hill *et al.* [22] indicates that no conflict arises as to the experimental conditions optimal for model selection or parameter estimation. In [21] on the other hand it is mentioned that design conditions that are best suited for discrimination may not be optimal for accurate parameter estimation. Clearly, some more experience with such joint criterion is necessary.

10.7 Conclusions

The different tasks one has to perform for model identification have been reviewed and illustrated with a real time application from wastewater treatment systems. Particular interest was given to the problems of model structure characterization, which in case of bioprocess modeling is a more important task than in many other areas due to the lack of *a priori* knowledge and the scarce experimental data. During the review of the existing methods, a clear distinction was made between *a posteriori* and *a priori* structure characterization methods. The former require preliminary parameter estimations for all rival models and therefore may take considerable computing time due to the non-linear nature of the bioprocess models.

For the parameter estimation stage it was stressed that one should be very cautious w.r.t. the theoretical identifiability of the models. It was illustrated that in many cases not all parameters can be given unique values but only a set of parameter combinations. While the evaluation of the theoretical identifiability is a difficult task it is important to be aware of the problems lack of identifiability may cause. To improve the accuracy of (combinations of) parameters that can be estimated (from a theoretical point of view at least), it was indicated that two possibilities exist. One is to repeat experiments, while the other consists of improving the quality of the performed experiment(s). While the former is a straightforward approach, the latter was paid a lot of attention to and it was illustrated using the case study developed throughout the text that important improvements in parameter estimation accuracy can be expected.

A similar optimal experimental design methodology was introduced for the goal of model structure discrimination. It was indicated that the main problem with the mentioned techniques concerned the lengthy computations necessary in case *a posteriori* structure characterization methods are used for the model identification. With the case study, for which specific *a priori* SC methods had been developed, it was shown that optimal experimental design does not necessarily have to violate the real time constraints.

In the last section attention was paid to the application of a joint experimental design criterion for model discrimination and parameter estimation. Because its use may substantially reduce the necessary experiments in case of an overall model identification exercise, this approach warrants more attention than given so far.

10.8 Nomenclature

AIC	: Akaike's information criterion	
BIC	: Bayesian information criterion	
C	: Joint experimental design criterion for structure characterization and parameter estimation	
C_i	: Measurement error covariance matrix	
D	: Discriminative power of an experiment	
$det(F)$	: Determinant of F	
E	: Parameter estimation accuracy of an experiment	
F	: Fisher Information Matrix	
FPE	: Final prediction error	
GIC	: General information criterion	
J	: Objective functional	
J_{opt}	: Minimal value of the objective functional	
k	: Number of substrates in the OUR_{ex}-model	
k_r	: Rapid hydrolysis rate constant	[1/min]
k_s	: Slow hydrolysis rate constant	[1/min]
K_{S_i}	: Affinity constant for substrate i	[mg COD/L]
$M(S_N, P_N)$	: Identified model from N data points	
$M(S_N, P^*)$	: Identified model from N data points with converged parameter estimates	

$M_T(S_T, P_T)$	: *True* model	
n	: Number of data in moving window regression	
N	: Number of measured data	
$N(0,1)$	: Normal distribution with zero mean and unit variance	
$N4SID$	: Numerical algorithms for subspace state-space system identification	
OED/PE	: Optimal experimental design for parameter estimation	
OED/SC	: Optimal experimental design for structure characterization	
OUR_{ex}	: Exogenous oxygen uptake rate	[mg O_2/L.min]
p_i	: Number of parameters of model i	
PE	: Parameter estimation	
q	: Backward shift operator	
Q_i	: Weighting matrix	
R	: Number of runs in the sequence of residuals	
$r(f)$	: Reliability of an inflection point f	
$r_\epsilon(\tau)$	: Covariance with lag τ	
s^2	: Residual mean square	
SC	: Structure characterization	
S_i	: Concentration of substrate i	[mg COD/L]
SSR_i	: Sum of squared residuals of model i	
t	: Time	[min]
t_{puls}	: Time of pulse addition	[min]
$tr(F)$	: Trace of F	
u	: Input vector	
V	: Parameter estimation covariance matrix	
w_D	: Weight attributed to model discrimination	
w_E	: Weight attributed to parameter estimation	
$\mathbf{x}$	: State vector	
X	: Concentration of biomass	[mg COD/L]
X_r	: Rapidly hydrolyzable substrate concentration	[mg COD/L]
X_s	: Slowly hydrolyzable substrate concentration	[mg COD/L]
Y_i	: Yield coefficient on substrate i	[mg COD_X/mg COD_{S_i}]
y	: Measurement vector	
$\hat{y}_i(\theta)$	: Model prediction vector for model i	
$\beta(N,p)$	: Complexity term in information criteria	
$\epsilon(t)$	: Residual	
$\gamma(N,p)$	: Complexity term in information criteria	
$\lambda_{max}, \lambda_{min}$	: Largest, resp. smallest eigenvalue of F (in absolute values)	

μ_{maxi}	: Maximum specific growth rate on substrate i	[1/min]
$\sigma(\theta_i)$	: Standard error of parameter i	
$\sigma_v{}^2$	: Residual error	
θ_i	: Parameter i	
χ_i	: Experimental conditions for experiment i	

Acknowledgments

This chapter presents research results of NFWO Project G.0286.96, and of the Belgian Programme on Interuniversity Poles of Attraction, initiated by the Belgian State, Prime Minister's Office for Science, Technology and Culture. The scientific responsibility rests with its authors.

10.9 Bibliography

1. S. Aborhey and D. Williamson 1978. State and parameter estimation of microbial growth processes. *Automatica*, 14, 493-498
2. H. Akaike 1978. A Bayesian analysis of the minimum AIC procedure. *Ann. Inst. Statist. Math.*, 30, 9-14
3. E. Ayesa, J. Florez, L. Larrea and J.L. Garcia-Heras 1993. Evaluation of sensitivity and observability of the state vector for system identification and experimental design. *Wat. Sci. Tech.*, 28(11-12), 209-218
4. E.M.L. Beale 1960. Confidence regions in non-linear estimation. *J. Roy. Stat. Soc.*, B22, 41-88
5. M.B. Beck 1989. System identification and control. In: G.G. Patry and D. Chapman (Eds.), *Dynamic Modelling and Expert Systems in Wastewater Engineering*, 261-323, Lewis Publishers, Chelsea, Michigan
6. G.E.P. Box and W.J. Hill 1967. Discrimination among mechanistic models. *Technometrics*, 9, 57-71
7. W.C. Boyle and P.M. Berthouex 1974. Biological wastewater treatment model building: Fits and misfits. *Biotechnol. Bioeng.*, 16, 1139-1159
8. J. Carstensen, H. Madsen, N.K. Poulsen and M.K. Nielsen 1994. Identification of wastewater treatment processes for nutrient removal on a full-scale WWTP by statistical methods. *Wat. Res.*, 28, 2055-2066
9. H. Caswell 1976. The validation problem. In: B.C. Patten (Ed.), *Systems Analysis and Simulation in Ecology, Vol. IV*, 313-325, Academic Press, New York
10. L.J. Clewlow, N. Cresswell and E.M.H. Wellington 1990. Mathematical model of plasmid transfer between strains of streptomycetes in soil microcosms. *Appl. Environ. Microbiol.*, 56, 3139-3145

11. A. Cornish-Bowden and J.T. Wong 1978. Evaluation of rate constants for enzyme-catalysed reactions by the Jackknife technique. Application to liver alcohol dehydrogenase. *Biochem. J.*, 175, 969-976
12. J.G. de Gooijer, B. Abraham, A. Gould and L. Robinson 1985. Methods for determining the order of an autoregressive-moving average process: A survey. *Int. Statist. Rev.*, 53, 301-329
13. N.R. Draper and H. Smith 1981. *Applied Regression Analysis.* Wiley, New York, 709 p.
14. T. Egli, U. Lendenmann and M. Snozzi 1993. Kinetics of microbial growth with mixtures of carbon sources. *Antonie van Leeuwenhoek*, 63, 289-298
15. R. Gehr, M.D. Cailas and K.S. Narasiah 1986. Statistical analysis of oxygen uptake data. In: Z. Dubinsky and Y. Steinberger (Eds.), *Environmental Quality and Ecosystem Stability*, 923-931, Bar-Ilan University Press, Ramat-Gan, Israel
16. K.R. Godfrey and J.J. DiStefano III 1985. Identifiability of model parameters. In: *Identification and System Parameter Estimation*, 89-114, Pergamon Press, Oxford
17. G.C. Goodwin 1987. Identification: Experiment design. In: M. Singh (Ed.), *Systems and Control Encyclopedia, Vol. 4*, 2257-2264, Pergamon Press, Oxford
18. G.C. Goodwin, M. Gevers and B. Ninness 1992. Quantifying the error in estimated transfer functions with application to model order selection. *IEEE Trans. Autom. Control*, 37, 913-928
19. K. Grijspeerdt, P. Vanrolleghem and W. Verstraete 1994. Measuring settleability of activated sludge using image analysis. In: *Proceedings Workshop on Modelling, Monitoring and Control of Wastewater Treatment Plants. Med. Fac. Landbouww. Univ. Gent*, 59, 2037-2045
20. E.J. Hannan 1980. The estimation of the order of an ARMA process. *Ann. Statist.*, 8, 1071-1081
21. P.D.H. Hill 1978. A review of experimental design procedures for regression model discrimination. *Technometrics*, 20, 15-21
22. W.J. Hill, W.G. Hunter and D.W. Wichern 1968. A joint design criterion for the dual problem of model discrimination and parameter estimation. *Technometrics*, 10, 145-160
23. A. Holmberg 1982. On the practical identifiability of microbial growth models incorporating Michaelis-Menten type nonlinearities. *Math. Biosci.*, 62, 23-43
24. A. Holmberg and J. Ranta 1982. Procedures for parameter and state estimation of microbial growth process models. *Automatica*, 18, 181-193
25. W.G. Hunter and A.M. Reiner 1965. Designs for discriminating between two rival models. *Technometrics*, 7, 307-323
26. U. Jeppsson and G. Olsson 1993. Reduced order models for on-line parameter identification of the activated sludge process. *Wat. Sci. Tech.*, 28(11-12), 173-183

27. R.L. Kashyap 1980. Inconsistency of the AIC rule for estimating the order of autoregressive models. *IEEE Trans. Autom. Control*, 25, 996-998

28. K.J. Keesman and G. van Straten 1987. Modified set theoretic identification of an ill-defined water quality system from poor data. In: M.B. Beck (Ed.), *Systems Analysis in Water Quality Management*, 297-308, Pergamon Press, Oxford

29. G.R. Knudsen, M.V. Walter, L.A. Porteous, V.J. Prince, J.L. Armstrong and R.J. Seidler 1988. Predictive model of conjugative plasmid transfer in the rhizosphere and phyllosphere. *Appl. Environ. Microbiol.*, 54, 343-347

30. A.B. Koehler and E.S. Murphree 1988. A comparison of the Akaike and Schwarz criteria for selecting model order. *Appl. Statist.*, 37, 187-195

31. G. Korte, U. Rinas, H.-A. Kracke-Helm and K. Schügerl 1991. Structured model for cell growth and enzyme production by recombinant Escherichia coli. *Appl. Microbiol. Biotechnol.*, 35, 185-188

32. G. Korte, U. Rinas, H.-A. Kracke-Helm and K. Schügerl 1991. Simulation of cell growth, substrate consumption and product formation with recombinant Escherichia coli. *Appl. Microbiol. Biotechnol.*, 35, 189-199

33. B.R. Levin, F.M. Stewart and V.A. Rice 1979. The kinetics of conjugative plasmid transmission: fit of a simple mass action model. *Plasmid*, 2, 247-260

34. H. Linhart and W. Zucchini 1986. *Model Selection.* Wiley, New York, 301 p.

35. L. Ljung 1987. *System Identification - Theory for the User.* Prentice-Hall, Englewood Cliffs, New Jersey, 519 p.

36. J.R. Lobry and J.P. Flandrois 1991. Comparison of estimates of Monod's growth model parameters from the same data set. *Binary*, 3, 20-23

37. A. Lübbert 1991. Characterization of bioreactors. In: K. Schügerl (Ed.), *Biotechnology, A Multi-volume Comprehensive Treatise, Vol. 4. Measuring, Modelling and Control*, 107-148, VCH, Weinheim

38. A.T. Marino, J.J. DiStefano III and E.M. Landaw 1992. DIMSUM: An expert system for multiexponential model discrimination. *Am. J. Physiol.*, 262, E546-E556

39. S. Marsili-Libelli 1989. Modelling, identification and control of the activated sludge process. *Adv. Biochem. Eng. Biotechnol.*, 38, 90-148

40. S. Marsili-Libelli 1992. Parameter estimation of ecological models. *Ecol. Modelling*, 62, 233-258

41. R.G. Miller 1974. The jackknife - a review. *Biometrika*, 61, 1-15

42. A. Munack 1989. Optimal feeding strategy for identification of Monod-type models by fed-batch experiments. In: N. Fish, R. Fox and N. Thornhill (Eds.), *Computer Applications in Fermentation Technology: Modelling and Control of Biotechnological Processes*, 195-204, Elsevier, Amsterdam

43. A. Munack 1991. Optimization of sampling. In: K. Schügerl (Ed.), *Biotechnology, a Multi-volume Comprehensive Treatise, Vol. 4. Measuring, Modelling and Control*, 251-264, VCH, Weinheim

44. A. Munack 1992. Some improvements in the identification of bioprocesses. In: M.N. Karim and G. Stephanopoulos (Eds.), *Modelling and Control of Biotechnical Processes*, 89-94, Pergamon Press, Oxford
45. M. Nihtila and J. Virkkunen 1977. Practical identifiability of growth and substrate consumption models. *Biotechnol. Bioeng.*, 19, 1831-1850
46. G. Olsson 1992. Model verification in activated sludge systems. In: *Proceedings Workshop Modelling, Monitoring and Control of the Activated Sludge Process. Med. Fac. Landbouww. Rijksuniv. Gent*, 57, 2249-2256
47. L. Oppenheimer, T.P. Capizzi and G.T. Miwa 1981. Application of jackknife procedures to inter-experiment comparisons of parameter estimates for the Michaelis-Menten equation. *Biochem. J.*, 197, 721-729
48. S.H. Park, D.D.Y. Ryu and S.B. Lee 1991. Determination of kinetic parameters related to the plasmid instability: for the recombinant fermentation under repressed condition. *Biotechnol. Bioeng.*, 37, 404-414
49. S.W. Peretti and J.E. Bailey 1987. Simulations of host-plasmid interactions in Escherichia coli: copy number, promoter strength, and ribosome binding site strength effects on metabolic activity and plasmid gene expression. *Biotechnol. Bioeng.*, 29, 316-328
50. H. Pohjanpalo 1978. System identifiability based on the power series expansion of the solution. *Math. Biosci.*, 41, 21-33
51. C. Posten and A. Munack 1990. On-line application of parameter estimation accuracy to biotechnical processes. In: *Proceedings American Control Conference 1990*, 2181-2186
52. A. Raksanyi, Y. Lecourtier, E. Walter and A. Venot 1985. Identifiability and distinguishability testing via computer algebra. *Math. Biosci.*, 77, 245-266
53. D.A. Ratkowsky 1986. A suitable parameterization of the Michaelis-Menten enzyme reaction. *Biochem. J.*, 240, 357-360
54. P. Reichert 1994. *Concepts underlying a computer program for the identification and simulation of aquatic systems.* Schriftenreihe der EAWAG Nr 7, EAWAG, Dübendorf, Switzerland, 386 p.
55. J.A. Robinson 1985. Determining microbial parameters using nonlinear regression analysis: Advantages and limitations in microbial ecology. *Adv. Microb. Ecol.*, 8, 61-114
56. T. Roy 1994. Bootstrap accuracy for non-linear regression models. *J. Chemometrics*, 8, 37-44
57. W. Ryan and S.J. Parulekar 1991. Recombinant protein synthesis and plasmid instability in continuous cultures of Escerichia coli JM103 harboring a high copy number plasmid. *Biotechnol. Bioeng.*, 37, 415-429
58. G. Schwarz 1978. Estimating the dimension of a model. *Ann. Statist.*, 6, 461-464
59. Z. Shi and K. Shimizu 1992. Neuro-fuzzy control of bioreactor systems with pattern recognition. *J. Ferment. Bioeng.*, 74, 39-45

60. T. Söderström and P. Stoica 1989. Model validation and model structure determination. In: *System Identification*, 422-467, Prentice-Hall, Englewood Cliffs, New Jersey

61. U. Sollfrank and W. Gujer 1991. Characterisation of domestic wastewater for mathematical modelling of the activated sludge process. *Wat. Sci. Tech.*, 23, 1057-1066

62. J.A. Spriet and P. Herman 1983. Simulation study of structure characterisation methods. *IMACS 1983*, 452-459

63. H. te Braake, R. Babuska, E. van Can and C. Hellinga 1994. Predictive control in biotechnology using fuzzy and neural models. In: J. Van Impe, P. Vanrolleghem and D. Iserentant (Eds.), *Advanced Instrumentation, Data Interpretation and Control of Biotechnological Processes*, Kluwer Academic Publishers, London

64. M.L. Thompson and M.A. Kramer 1994. Modelling chemical processes using prior knowledge and neural networks. *AIChE J.*, 40, 1328-1340

65. E. Top, P. Vanrolleghem, M. Mergeay and W. Verstraete 1992. Determination of the mechanism of retrotransfer by mechanistic mathematical modelling. *J. Bacteriol.*, 174, 5953-5960

66. A.G. Valdecasas and A. Baltanas 1990. Jacknife and bootstrap estimation of biological index of water quality. *Wat. Res.*, 24, 1279-1283

67. J.J. Vallino and G. Stephanopoulos 1993. Metabolic flux distributions in Corynebacterium glutamicum during growth and lysine overproduction. *Biotechnol. Bioeng.*, 41, 633-646

68. K. van Dam, N. Jansen, P. Postma, P. Richard, G. Ruijter, M. Rutgers, H.P. Smits, B. Teusink, J. van der Vlag, M. Walsh and H.V. Westerhoff 1993. Control and regulation of metabolic fluxes in microbes by substrates and enzymes. *Antonie van Leeuwenhoek*, 63, 315-321

69. J.P. van Dijken, R.A. Weusthuis and J.T. Pronk 1993. Kinetics of growth and sugar consumption in yeasts. *Antonie van Leeuwenhoek*, 63, 343-352

70. P. Van Overschee and B. De Moor 1994. N4SID: Two subspace algorithms for the identification of combined deterministic-stochastic systems. *Automatica*, 30, 75-93

71. P.A. Vanrolleghem 1994. *On-line Modelling and Control of Activated Sludge Processes: Development of an Adaptive Sensor*, PhD thesis. Faculty of Agricultural and Applied Biological Sciences, University Gent, Belgium, 201 p.

72. P.A. Vanrolleghem and F. Coen 1995. Optimal design of in-sensor-experiments for on-line modelling of nitrogen removal processes. *Wat. Sci. Tech.*, 31(2), 149-160

73. P.A. Vanrolleghem and M. Van Daele 1994. Optimal experimental design for structure characterization of biodegradation models: on-line implementation in a respirographic biosensor. *Wat. Sci. Tech.*, 30(4), 243-253

74. P.A. Vanrolleghem and J.F. Van Impe 1992. On the use of structured compartment models for the activated sludge biodegradation process. In: *Proceedings Workshop Modelling, Monitoring and Control of the Activated Sludge Process. Med. Fac. Landbouww. Rijksuniv. Gent*, 57, 2215-2228
75. P.A. Vanrolleghem and W. Verstraete 1993. Simultaneous biokinetic characterization of heterotrophic and nitrifying populations of activated sludge with an on-line respirographic biosensor. *Wat. Sci. Tech.*, 28(11-12), 377-387
76. P.A. Vanrolleghem, M. Van Daele, P. Van Overschee and G.C. Vansteenkiste 1994. Model structure characterization of nonlinear wastewater treatment systems. In: *Preprints 10th IFAC Symposium on System Identification*, Copenhagen, Denmark, July 4-6 1994, 1, 279-284
77. P.A. Vanrolleghem, J.F. Van Impe, J. Vandewalle and W. Verstraete 1992. Advanced monitoring and control of the activated sludge process: on-line estimation of crucial biological variables in a structured model with the RODTOX biosensor. In: M.N. Karim and G. Stephanopoulos (Eds.), *Modelling and Control of Biotechnical Processes*, 355-358, Pergamon Press, Oxford
78. G.C. Vansteenkiste and J.A. Spriet 1982. Modelling ill-defined systems. In: F.E. Cellier (Ed.), *Progress in Modelling and Simulation*, 11-38, Academic Press, London
79. A. Varma, B.W. Boesch and B.O. Palsson 1993. Biochemical production capabilities of Escherichia coli. *Biotechnol. Bioeng.*, 42, 59-73
80. L.K. Vermeersch, B. Kroes and P.A. Vanrolleghem 1994. Feature based model identification of nonlinear biotechnological processes. *Ecol. modelling*, 75, 629-640
81. C. Vialas, A. Cheruy and S. Gentil 1985. An experimental approach to improve the Monod model identification. In: A. Johnson (Ed.), *Modelling and Control of Biotechnological Processes*, 155-160, Pergamon, Oxford
82. D. Wild, R. von Schulthess and W. Gujer 1994. Synthesis of denitrification enzymes in activated sludge: modelling with structured biomass. *Wat. Sci. Tech.*, 30(6), 113-122

Chapter 11

Optimal Control of Fed-Batch Fermentation Processes

J.F. VAN IMPE
Katholieke Universiteit Leuven
Department of Food and Microbial Technology - BioTeC
Kardinaal Mercierlaan 92 B-3001 Leuven (BELGIUM)
email: jan.vanimpe@agr.kuleuven.ac.be

ABSTRACT – *In this chapter the design of a substrate feeding rate controller for a class of biotechnological processes in stirred tank bioreactors characterized by a decoupling between biomass growth and product formation is under study. As a typical example, we consider the most common case of monotonic specific growth rate and non-monotonic specific production rate. In general this yields a singular optimal control problem. Due to state inequality constraints –which reflect physical limitations– complete solutions are especially hard to obtain and almost non-existent in the literature.*

However, the techniques proposed in the literature to convert the problem into a non-singular one all suffer from problems with respect to practical implementation. Therefore, we adopt the traditional point of view, i.e., singular optimal control with the volumetric substrate feed rate as the manipulated variable. We consider performance indices of both the yield type and the productivity type. The optimization of the initial substrate concentration is included also.

J.F.M. Van Impe et al. (eds.), Advanced Instrumentation, Data Interpretation, and Control of Biotechnological Processes, 319-346.

We illustrate that the determination of the switching times in the feed rate profile can be simplified a lot by incorporating as much as possible microbiological knowledge concerning the process under consideration. The results we present here are in this way general that they are written in terms of the specific reaction rates. In other words, an exact analytical expression for these rates is not required a priori. In addition, no assumptions have been made concerning the metabolism involved for biomass survival and possibly product synthesis.

The computational algorithms presented have some marked advantages over algorithms reported in the literature. In addition, we are able to deduce a complete general solution in the realistic case of both substrate and biomass concentration constraints. As an example, simulation results are given for the penicillin G fed-batch fermentation.

It is shown in Chapter 13 how a detailed analysis of the profiles presented here leads in a natural way to the design of model-independent control strategies which combine nearly optimal performance with excellent robustness properties: optimal adaptive control.

KEYWORDS – *Fermentation Processes, Fed-batch, Optimal Control, Minimum Principle, Inequality Constraints*

11.1 Motivation

In this chapter the optimization of fed-batch fermentation processes is under study. These are of particular importance when the products are secondary metabolites. Fed-batch operation enables superior control as it provides different avenues in the substrate plane for control through distinct phases of biomass growth and metabolite production. By programming substrate feeding, one can control important phenomena such as substrate inhibition, glucose effect and catabolite repression.

Control opportunities in fed-batch operated fermentations have been reviewed in detail in a number of articles [1][2] ... Yet the determination of an optimal operating strategy is not a straightforward matter and is often further complicated by the presence of constraints imposed on the state variables.

Optimization of fed-batch fermentations has been traditionally sought with respect to the volumetric substrate feed rate while keeping the substrate concentration in the feed fixed. Formulating the problem with the feed rate as the control variable yields a *singular* optimal control problem.

Fishman and Biryukov considered the problem of optimal glucose feed rate strategy for penicillin production [3]. However, they did not give any general characteristics nor a numerical scheme and ignored volume changes arising from the feed. Ohno *et al.* used a method based on Greens' theorem for the production of lysine [4]. This method is applicable only to problems with two mass balance equations and fixed boundary conditions.

When the fermentation process is described by no more than three mass balance equations, one can take advantage of the structure of a particular problem. Weigand *et al.* reduce the problem of optimal feed rate determination for single cell protein production (three-dimensional problem) to one in which the time to switch from the maximum rate to a singular feed rate is iterated [5]. This approach provides the optimal solution with relative ease [6][7].

Due to limited success in singular control theory, most of the fed-batch fermentation processes considered so far have been limited to low-order processes, described by no more than four mass balance equations [8][9]. General computational algorithms have been developed by Modak *et al.* and Lim *et al.* for the optimal volumetric feed rate profiles for such fermentation processes [10][11].

A number of techniques have been proposed to convert the singular control problem into a non-singular one. To avoid a singular control problem, variables other than the feed flow rate may be taken as the control variable. Yamane *et al.* considered the specific growth rate as the control variable [12]. However, this can lead to a problem in physically realizing the flow rate from the optimum specific growth rate unless proper physical constraints are placed on the specific growth rate. If the specific growth rate is a non-monotonic function of substrate concentration or depends on other component concentrations as well, this approach cannot be used. Furthermore, constraints on the flow rate cannot be taken into account. Guthke and Knorre considered the substrate concentration as the manipulated variable and omitted the substrate and the overall mass balance equations [13]. This method runs into a problem of realizing the optimal substrate concentration profile due to the physical constraints on feeding rate and volume that were ignored. Modak and Lim proposed an alternative set of state variables and used the culture volume as the control variable [14]. This approach may be useful to optimize fed-batch cultures described by any number of mass balance equations involving different kinds of constraints. However, the success of the method depends greatly on the ability to numerically solve highly non-linear non-singular control problems with constraints on the manipulated variable and its time derivative. San and Stephanopoulos proposed to solve the problem in two steps [15]. First the optimal control problem with the fermentor substrate concentration as the control variable must be solved to yield an optimal substrate trajectory. State inequalities on both substrate and biomass concentration can be included with relative ease. In a second step they propose to realize the optimal reactor substrate profile by manipulating the substrate concentration in the feed while keeping the volumetric feed rate fixed. However, they do not solve this second step. Kurtanjek also proposed a two step procedure for a multidimensional optimization problem with constraints imposed on control and state variables based on the use of the orthogonal collocation technique [16]. The first step is the non-singular optimization of temperature and substrate profiles. The second part is the determination of the optimal substrate concentration in the feed rate and the optimal volumetric substrate feeding rate. However, this method critically relies on an exact knowledge of the specific growth and production rates.

As a matter of fact, these methods all suffer from problems with respect to practical implementation. Therefore, in this chapter we adopt the traditional point of view, i.e., singular optimal control with the volumetric substrate feed rate as the manipulated variable. As a case study, we consider processes with a decoupling between growth and production, characterized by a monotonic specific growth rate and a non-monotonic specific production rate. The main contributions are the following. We show that the determination of the switching times in the feed rate profile can be simplified a lot by incorporating as much as possible microbiological/biochemical knowledge concerning the process under consideration. The optimization of the initial substrate concentration has been included. The resulting computational algorithm has some advantages over the algorithm proposed by Lim *et al.* [11]. Performance indices of both yield and productivity type are considered. We also present for the first time a complete general solution in the case of substrate and biomass inequality constraints. As an example, numerical results are given for the penicillin G fed-batch fermentation process. It is shown elsewhere how the profiles deduced in this chapter lead in a natural way to the design of model-independent feedback control strategies which combine nearly optimal performance with excellent robustness properties: *optimal adaptive control* (see Chapter 13 and [17][18][19]).

11.2 Problem Statement

Consider the class of fed-batch fermentations described by a model of the form:

$$\begin{aligned}\frac{dS}{dt} &= -\sigma X + C_{S,in}u \\ \frac{dX}{dt} &= \mu X \\ \frac{dP}{dt} &= \pi X - kP \\ \frac{dV}{dt} &= u\end{aligned} \tag{11.1}$$

For an explanation of all symbols used, we refer to the nomenclature at the end of this chapter. S is considered the only limiting substrate for cell mass growth and production. All other nutrients are present in excess but not in toxic quantities throughout the fermentation period. Note that decay of product is explicitly taken into account using a first order model, where k denotes the specific product decay rate. At this point we allow the specific rates σ, μ and π to be arbitrary functions which may depend on substrate, cell and product concentrations, interrelated by:

$$\sigma = \frac{\mu}{Y_{X/S}} + m + \frac{\pi}{Y_{P/S}}$$

This can represent *any* metabolism (maintenance, endogenous, or mixed) for biomass survival and product synthesis.

The *shape* of the specific rates $\mu(C_S)$ and $\pi(C_S)$ is as depicted in Figure 11.1: the enzyme catalyzed production is *not associated* to the microbial growth.

By defining the following 4×1 vectors ($\mathbf{T}$ denotes the transpose):

$$\begin{aligned}\mathbf{x} &\triangleq [S \;\; X \;\; P \;\; V]^{\mathbf{T}} \\ \mathbf{f} &\triangleq [-\sigma X \;\; \mu X \;\; (\pi X - kP) \;\; 0]^{\mathbf{T}} \\ \mathbf{b} &\triangleq [C_{S,in} \;\; 0 \;\; 0 \;\; 1]^{\mathbf{T}}\end{aligned}$$

we obtain the following state space model linear in the control input u (i.e., the volumetric substrate feed rate):

$$\frac{d\mathbf{x}}{dt} = \mathbf{f}(\mathbf{x}) + \mathbf{b}\,u \tag{11.2}$$

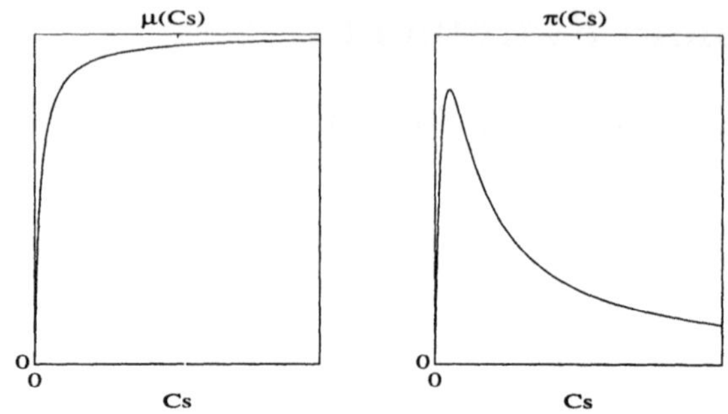

Figure 11.1: Specific rates for growth (μ) and production (π).

Observe that we can obtain a completely equivalent state space model using concentrations as state variables. However, it is easily seen that the corresponding column vector **b** then becomes an explicit function of the state variables which complicates analytical computations.

The initial values for biomass X and product P are *fixed*, i.e., (the initial time $t_0 = 0$):

$$\begin{aligned} x_2(0) &= X_0 \\ x_3(0) &= P_0 \end{aligned}$$

However, in contrast to classical optimal control problems, we consider some components of the initial state as *additional control parameters* that can be manipulated to minimize some performance index. More specifically, we assume that the initial value for substrate S, $x_1(0)$, is *free*; the initial values for S and V are only constrained by:

$$x_4(0) = V_* + \frac{x_1(0)}{C_{S,in}} \tag{11.3}$$

with V_* the initial volume *without* substrate. Note that substrate is added as a solution with concentration $C_{S,in}$.

Two general forms for the objective function are:

$$J_1[u, x_1(0)] = g_1[\mathbf{x}(t_f), t_f] = -\gamma_X X(t_f) - \gamma_P P(t_f) + \gamma_t t_f \tag{11.4}$$

and

$$J_2[u, x_1(0)] = g_2[\mathbf{x}(t_f), t_f] = -\frac{\gamma_X X(t_f) + \gamma_P P(t_f)}{t_f} \tag{11.5}$$

with γ_X, γ_P the unit *prices* associated with the cell mass and metabolite product respectively, and γ_t the unit *operating cost*. The optimization problem is then to determine for the set of differential equations (11.2) the optimal initial value S_0^* –or, equivalently, $C_S^*(0)$– and volumetric substrate feed rate profile $u^*(t)$ that minimize the performance index (11.4) or (11.5). Note that minimizing performance index (11.5) is equivalent to maximizing the *average productivity* $\mathcal{P}$, weighted by the unit prices of cell mass and product.

The cost index (11.4) or (11.5) must be minimized subject to an additional set of constraints.

1. The final time t_f is *free*.

2. As all variables represent non-negative physical quantities, we must have:

$$\forall t \in [0, t_f]: \quad x_i(t) \geq 0, \text{ for } i = 1, \ldots, 4 \quad \wedge \quad u(t) \geq 0$$

3. The total amount of substrate available for the fermentation (denoted with α) is *fixed*:

$$x_1(0) + \int_0^{t_f} C_{S,in} u(t)\, dt = \alpha \tag{11.6}$$

   The last *isoperimetric* constraint is equivalent to a *physical* constraint of the form (see the differential equation for V):

$$x_4(t_f) \equiv V(t_f) = V_f, \; V_f \text{ fixed} \tag{11.7}$$

Observe that, in order to obtain a well-posed optimal control problem, we must fix either t_f or α. We always fix α, while letting t_f free. This choice can be motivated as follows:

1. By fixing α, we can easily take into account the physical constraint on V (11.7).

2. By letting t_f free, we can derive more general properties of the optimal control profile in terms of microbial/biochemical knowledge of the process (see, e.g., Van Impe [19]).

3. From an industrial point of view the final time t_f is very often the most important factor. This can be taken into account in two ways. First, we can increase the unit operating cost γ_t in cost index (11.4). Second, we can use the additional degree of freedom $x_1(0)$ to calculate a *trade-off* between the value of the cost J and the final time t_f.

The *average yield* $\mathcal{Y}$ is defined as the ratio of the final amount of biomass and product weighted by their unit prices to the total amount of substrate consumed during fermentation α. If we neglect the unit operating cost γ_t and fix the total substrate amount α, then minimizing performance index (11.4) is equivalent to maximizing the average yield $\mathcal{Y}$.

11.3 Optimal Control

11.3.1 Statement of the two point boundary value problem

Pontryagin's minimum principle [20] states that the following Hamiltonian $\mathcal{H}$ must be minimized over all admissible control inputs $u(t)$:

$$\begin{aligned} \mathcal{H} &= \boldsymbol{\lambda}^{\mathbf{T}}[\mathbf{f}(\mathbf{x}) + \mathbf{b}u] \stackrel{\Delta}{=} \phi + \psi u \\ \phi &= \lambda_1 f_1 + \lambda_2 f_2 + \lambda_3 f_3 \\ \psi &= C_{S,in}\lambda_1 + \lambda_4 \end{aligned} \tag{11.8}$$

The adjoint vector $\boldsymbol{\lambda}$ satisfies the following system of differential equations:

$$\frac{d\boldsymbol{\lambda}}{dt} = -\frac{\partial \mathcal{H}}{\partial \mathbf{x}} = -\frac{\partial \mathbf{f}}{\partial \mathbf{x}}^{\mathbf{T}} \boldsymbol{\lambda} \tag{11.9}$$

Together with the state equations (11.2), we obtain a system of $2 \times n$ first order differential equations –where n denotes the dimension of the state vector $\mathbf{x}$, here $n = 4$– which needs of course $2 \times n$ boundary conditions. These can be specified as follows:

1. $x_2(0)$, $x_3(0)$, and $x_4(t_f)$ (see Equation (11.7)) are given; $x_1(0)$ and $x_4(0)$ are interrelated by (11.3)

2. $\lambda_i(t_f)$, $i = 1, \ldots, 3$ are given by (g is the terminal cost):

$$\lambda_i(t_f) = \frac{\partial g}{\partial x_i}[\mathbf{x}(t_f), t_f] \tag{11.10}$$

The boundary condition complementary to (11.3) is (Van Impe [19]):

$$C_{S,in}\lambda_1(0) + \lambda_4(0) \equiv \psi(0) = 0 \tag{11.11}$$

Since the boundary conditions are specified either at $t = 0$ or at $t = t_f$ this is a *two point boundary value problem (TPBVP)*. The Hamiltonian $\mathcal{H}$ does not depend explicitly on time t. Then on an extremal trajectory we have [20]:

$$\forall t \in [0, t_f] \;:\; \mathcal{H} \equiv \boldsymbol{\lambda}^{*\mathbf{T}}(t)[\mathbf{f}[\mathbf{x}^*(t)] + \mathbf{b}\; u^*(t)] = \mathcal{H}^*$$

with $\mathcal{H}^*$ a constant. If in addition the cost index is independent of the final time t_f (e.g., a cost index of the form (11.4) with $\gamma_t = 0$) we have [20]:

$$\forall t \in [0, t_f] \;:\; \mathcal{H} \equiv \boldsymbol{\lambda}^{*\mathbf{T}}(t)[\mathbf{f}[\mathbf{x}^*(t)] + \mathbf{b}\; u^*(t)] = 0$$

11.3.2 Extremal controls

An *extremal* control $u^*(t)$ follows from:

$$\min_{\text{all admissible } u} \mathcal{H}(\mathbf{x}^*, \boldsymbol{\lambda}^*, u) = \mathcal{H}(\mathbf{x}^*, \boldsymbol{\lambda}^*, u^*)$$

while $\mathbf{x}$ and $\boldsymbol{\lambda}$ satisfy the above two point boundary value problem. Minimization of the Hamiltonian $\mathcal{H}$ is considerably simplified since the control $u(t)$ appears *linearly* in it: an extremal control $u^*(t)$ can be determined by examining the coefficient of u in (11.8), $\psi \equiv \boldsymbol{\lambda}^{\mathbf{T}}\mathbf{b}$. In a real life application, there is a constraint on the feeding pump capacity $u(t)$:

$$0 \equiv U_{MIN} \leq u(t) \leq U_{MAX}, \quad U_{MAX} \text{ given}$$

In that case, an extremal control $u^*(t)$ can be characterized as follows. If ψ is positive we set $u = 0$, while if ψ is negative we have $u = U_{MAX}$. However, if over a finite time interval $[t_i, t_{i+1}]$ the coefficient ψ remains identically zero, the minimum principle fails to provide $u(t)$ during this interval. Such intervals are called *singular* intervals. In summary, we have:

$$u^*(t) = \begin{cases} U_{MAX} & \text{if} \quad \psi < 0 \\ u_{sing} & \text{if} \quad \psi = 0 \quad t_i \leq t \leq t_{i+1} \\ 0 & \text{if} \quad \psi > 0 \end{cases} \tag{11.12}$$

On any singular interval $[t_i, t_{i+1}]$ the *singular* control $u_{sing}(t)$ is obtained by repeatedly differentiating the function ψ until u appears explicitly. We obtain successively:

$$\frac{d\psi}{dt} = \dot{\boldsymbol{\lambda}}^{\mathbf{T}}\mathbf{b} = -\boldsymbol{\lambda}^{\mathbf{T}}\frac{\partial \mathbf{f}}{\partial \mathbf{x}}\mathbf{b} \triangleq \lambda^{\mathbf{T}}\mathbf{d} = 0$$

$$\frac{d^2\psi}{dt^2} = \dot{\boldsymbol{\lambda}}^{\mathbf{T}}\mathbf{d} + \boldsymbol{\lambda}^{\mathbf{T}}\frac{\partial \mathbf{d}}{\partial \mathbf{x}}\dot{\mathbf{x}} = -\boldsymbol{\lambda}^{\mathbf{T}}\frac{\partial \mathbf{f}}{\partial \mathbf{x}}\mathbf{d} + \boldsymbol{\lambda}^{\mathbf{T}}\frac{\partial \mathbf{d}}{\partial \mathbf{x}}(\mathbf{f} + \mathbf{b}u) = 0$$

$$u_{sing}(t) = \frac{\boldsymbol{\lambda}^{\mathbf{T}}[(\partial \mathbf{f}/\partial \mathbf{x})\mathbf{d} - (\partial \mathbf{d}/\partial \mathbf{x})\mathbf{f}]}{\boldsymbol{\lambda}^{\mathbf{T}}[\partial \mathbf{d}/\partial \mathbf{x}]\mathbf{b}} \tag{11.13}$$

A necessary condition for existence of a singular arc is [20]:

$$\frac{\partial}{\partial u}[\frac{d^2\psi}{dt^2}] \geq 0$$

We conclude that the TPBVP has been reduced to the determination of the *optimal control sequence* $[U_{MAX}, U_{MIN}, u_{sing}]$ and the corresponding *switching times.*

REMARKS

1. As a generic result, the denominator of the singular control (11.13) is indeed different from zero. Obviously, this problem is a *singular problem of order 2.*

2. In expression (11.13) both the numerator and the denominator are *linear* in the costate vector $\boldsymbol{\lambda}$. On any singular interval, there exist three linear algebraic equations between the costate variables. These equations are:

$$\begin{aligned} \psi \equiv \boldsymbol{\lambda}^T \mathbf{b} &= 0 \\ \frac{d\psi}{dt} \equiv \boldsymbol{\lambda}^T \mathbf{d} &= 0 \\ \phi \equiv \boldsymbol{\lambda}^T \mathbf{f} &= \mathcal{H}^* \end{aligned} \tag{11.14}$$

with $\mathcal{H}^*$ a constant. If the cost index is independent of final time t_f (e.g., a cost index of the form (11.4) with $\gamma_t = 0$) then $\mathcal{H}^* = 0$. Then the homogeneous system (11.14) specifies three costates as functions of the remaining fourth one. As a corollary, the singular control (11.13) is a *non-linear feedback law of the state variables* $\mathbf{x}$ *only.* ■

11.3.3 Optimal control sequence for monotonic μ and non-monotonic π

In determining the optimal control sequence $[U_{MAX}, U_{MIN}, u_{sing}]$ the decoupling between growth and production is a quite useful feature. Initial work along the same lines has been reported by Modak *et al.* [10] and Lim *et al.* [11]. The computational algorithms presented here are at the basis of the development of nearly optimal, but easier to implement and robust *adaptive* feedback control strategies (see Chapter 13 and [17][18][19]). Obviously, derivation of extremal control sequences can always be done without using *a priori* microbiological/biochemical knowledge concerning the process, i.e., by only using the necessary conditions for optimality. More specifically, Equations (11.12) and (11.13) in principle fully specify the optimal control profile. However, a complete numerical analysis of an extremal solution is in general a very tedious job, which provides little additional insight useful in designing nearly optimal feedback controllers. An example of verifying all necessary conditions for optimality is given in Van Impe [19]. In this text, we exploit the available microbiological/biochemical knowledge –more specifically, the experimentally observed biphasic behavior– in order to deduce extremal solutions as fast as possible. It is very helpful to interpret the differential equation for product formation (see Equations (11.1)) within the concept of a biphasic fermentation process:

$$\frac{dP}{dt} = \pi X - kP \tag{11.15}$$

Clearly, product synthesis is a result of two effects: (i) formation of biomass X (during a first *growth* phase), and (ii) product formation (during a second *production* phase) with specific rate π. A performance index of type (11.4) or (11.5) can then be optimized as follows.

1. During a first *growth* phase, we want to produce as much as possible biomass X. From the differential equation for biomass X (model equations (11.1)) we know that growth is governed by the specific growth rate μ. As a result, during the growth phase the feed rate must be determined so as to realize the values for μ as a function of C_S –observe that in general μ may depend on other component concentrations as well– which result in an optimal biomass accumulation.

2. During a second *production* phase, we want the cells to produce the desired product as much as possible. As a result, the feed rate must be used to optimize the specific production rate π.

REMARKS

1. The optimal *switching time* between these phases –given the total amount of substrate available for fermentation α (11.6)– is obviously

dictated by a *trade-off* between the gain in product P by further increasing the biomass X with little product degradation (as P remains small during growth), and the gain in P by optimizing the specific production rate π while product degradation takes place. This can be readily seen from Equation (11.15).

2. Up to this point, the above reasoning is independent of the exact analytical expression of the specific rates μ and π. ■

11.3.3.1 Bounded input u and fixed initial state $\mathbf{x}_0$

First, we determine the optimal control sequence in the (realistic) case of a constraint on the input

$$0 \equiv U_{MIN} \leq u(t) \leq U_{MAX}, \quad U_{MAX} \text{ given}$$

while we consider the complete initial state $\mathbf{x}_0$ as given: $x_1(0)$ is fixed, while $x_4(0)$ follows from relation (11.3). As a result, condition (11.11) is redundant. Obviously, the optimal sequence depends on the initial substrate concentration $C_S(0) = x_1(0)/x_4(0)$. Further, we tacitly assume that the initial biomass amount $x_2(0)$ is low.

Low initial substrate concentration $C_S(0)$

As the specific growth rate μ is monotonically increasing with respect to substrate concentration C_S, it is clear that at time $t = 0$ the control must take on its maximum possible value $u(t) = U_{MAX}$ to realize the maximum possible biomass accumulation. At some time instant $t = t_1$, the control is set equal to its minimum value $u(t) = 0$, until the production phase starts at time $t = t_2$. During this batch phase $[t_1, t_2]$, the substrate concentration decreases, until at time $t = t_2$ singular substrate feeding starts using a control law of the form (11.13). This can be motivated as follows. We already mentioned that during the production phase the specific production rate π is optimized. It is shown that under certain limiting conditions, the substrate concentration during production remains constant at the value C_S^* which maximizes π/σ [19]. If some of these conditions are not satisfied, we can expect that the optimal substrate concentration –although time-varying– will be at least in a neighborhood of the value C_S^*. Obviously, the substrate concentration can not be kept in this neighborhood using a simple bang-bang type control, but instead a singular feed rate profile will be required. This singular control is stopped at time $t = t_3$ when all substrate available α (11.6) is used, or equivalently, when the fermentor volume reaches its final value $x_4(t_3) = V_f$ (11.7). As there may be some substrate left in the reactor, the fermentation continues in batch mode ($u(t) = 0$) until the stopping condition –depending on the performance index under consideration– is met at $t = t_f$. This substrate feeding scheme and the corresponding state trajectories are summarized in Figure 11.2.

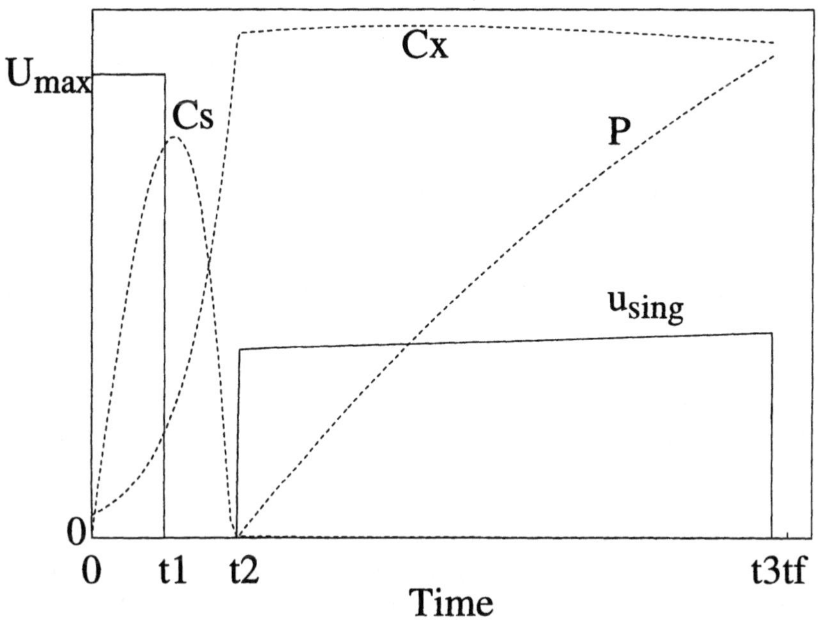

Figure 11.2: General feed rate profile and corresponding state trajectories for low initial substrate concentration.

High initial substrate concentration $C_S(0)$

If the initial substrate concentration is high enough, the first interval $[0, t_1]$ of feeding at the maximum rate should be omitted. The growth phase is then a complete batch phase, until at time $t = t_2$ the substrate concentration is small enough and the biomass concentration is high enough so that singular feeding can start.

Initial conditions on the singular hyperplane

In this case the growth phase can be omitted completely, while singular feeding starts from $t = 0$ on. Clearly, this situation will result in the lowest value of the performance index. However, the comparison with both previous cases is not completely fair, as we start with a biomass concentration high enough to omit the growth phase so that all substrate available can be used for product synthesis. Therefore, in the following this case will not be considered.

REMARKS

1. The first switching time t_1 can be interpreted as follows. It is a measure for the fraction of the total amount of substrate available α (11.6) which is consumed during the growth phase: $\alpha_{growth} \triangleq C_{S,in} \times U_{MAX} \times t_1$. Note that in general the switching times t_1 and t_2 are *not* known as functions of the state $\mathbf{x}$.

2. In the control sequence as shown in Figure 11.2 we tacitly assumed that the upper bound on the feed rate U_{MAX} is never violated on the singular arc $[t_2, t_3]$. The singular control (11.13) must be set equal to U_{MAX} if this upper bound is reached before all substrate available is added at $t = t_3$. Of course, the optimal switching times t_1 and t_2 will be changed.

3. The last batch phase $[t_3, t_f]$ $(u(t) = 0)$ is terminated if some *stopping condition* –depending on the performance index under consideration– is satisfied. Using the transversality condition $\mathcal{H}[\mathbf{x}(t_f), \boldsymbol{\lambda}(t_f), u(t_f)] + \partial g/\partial t[\mathbf{x}(t_f), t_f] = 0$ [20], we obtain, since the Hamiltonian $\mathcal{H}$ (11.8) remains constant along an extremal trajectory:

$$\forall t \in [0, t_f] \;:\; \mathcal{H}[\mathbf{x}(t), \boldsymbol{\lambda}(t), u(t)] \equiv \phi + \psi u = -\frac{\partial g}{\partial t}[\mathbf{x}(t_f), t_f] \equiv \mathcal{H}^*$$

with $\mathcal{H}^*$ a constant and $\phi = (-\lambda_1\sigma + \lambda_2\mu + \lambda_3\pi)x_2 - \lambda_3 k x_3$. Using terminal conditions (11.10), we obtain at $t = t_f$ for a performance index of the form (11.4):

$$\gamma_X \frac{dX}{dt}(t_f) + \gamma_P \frac{dP}{dt}(t_f) = \gamma_t$$

In other words: the fermentation is stopped when *the average gain of producing an additional unit of biomass and product (i.e., weighted by their unit prices) has become equal to the unit operational cost.* As a special case, if we want to optimize the final product amount $P(t_f)$ (so $\gamma_X = 0$ and $\gamma_t = 0$), then the fermentation is stopped if $dP/dt(t_f) = 0$. If the performance index is of the form (11.5), the process is stopped if:

$$\gamma_X \frac{dX}{dt}(t_f) + \gamma_P \frac{dP}{dt}(t_f) = \frac{\gamma_X X(t_f) + \gamma_P P(t_f)}{t_f}$$

which means that the fermentation is stopped when *the average gain of producing an additional unit of biomass and product has become equal to the average productivity $\mathcal{P}$.*

■

11.3.3.2 Computational algorithm

Once the optimal control sequence has been determined, the corresponding optimal switching times can be obtained using the following straightforward computational algorithm. First, we consider the generic case shown in Figure 11.2.

Algorithm

Step 1 Guess t_1, or equivalently, determine the amount of substrate α_{growth} consumed during growth. Integrate the state equations (11.2) from $t = 0$ to $t = t_1$ with $u(t) = U_{MAX}$.

Step 2 Make a guess of t_2. Integrate the state equations (11.2) from $t = t_1$ to $t = t_2$ with $u(t) = 0$. This completes the *growth* phase.

Step 3 Integrate the state equations (11.2) using singular control (11.13) until all substrate available α (11.6) is added, or equivalently, until the bioreactor is completely filled (11.7) at $t = t_3$.

Step 4 Complete the integration with $u(t) = 0$ until the stopping condition –depending on the cost index– is satisfied at time $t = t_f$. This completes the *production* phase. Store the value of the cost index $J[u]$ (11.4) or (11.5).

Step 5 Repeat Steps 2 to 4, considering $t_{2,new} = t_{2,old} \pm \delta t$, with δt as small as required. Save the time t_2 at which the cost index $J[u]$ (11.4) or (11.5) reaches its minimum.

Step 6 Repeat Steps 1 to 5 with a new guess of t_1 in order to minimize $J[u]$.

REMARKS

1. Obviously, the optimal solution of the two point boundary value problem is obtained using a *straightforward* computational algorithm. Due to the incorporation of all *a priori* knowledge available, there is no need for an integration backwards in time, as required in, e.g., gradient type methods for solving optimal control problems.

2. An initial guess of the switching times t_1 and t_2 can be obtained as follows. For t_1, an upper boundary is of course $t_1 \leq (V_f - V_0)/U_{MAX}$. This condition is very conservative, but in fact the starting value for t_1 is much less crucial than the corresponding starting value for t_2. If, e.g., t_2 is chosen too large, the substrate concentration may become so small that the singular control is unable to bring it back within the region which is optimal for production. Under certain limiting conditions the switching time t_2 can be obtained as a function of state variables [19]. In a general case, a good initial guess is therefore the time t_2 at which the substrate concentration $C_S(t_2)$ maximizes either the specific production rate π or the ratio π/σ. Note that in general the specific rates may depend on other component concentrations as well.

3. Calculation of the singular control (11.13) as required in Step 3 in general needs a complete knowledge of both the state $\mathbf{x}$ and the costate $\boldsymbol{\lambda}$. However, λ_4 is not required since $f_4 \equiv 0$. As a result, Equations (11.14) define λ_1 and λ_2 in function of λ_3 and $\mathcal{H}^*$. We can distinguish the following two cases.

   (i) The performance index is independent of final time t_f.

   Then we know that $\mathcal{H}^* = 0$ along an extremal trajectory. Since both the numerator and the denominator of the singular control law (11.13) are linear in the costate $\boldsymbol{\lambda}$, the costate cancels out and the singular control is a *non-linear feedback law of state variables only.* An example is a performance index of the form (11.4) with $\gamma_t = 0$.

   (ii) The performance index depends on final time t_f.

   If the performance index has the form (11.4), then $\mathcal{H}^* = -\gamma_t$ which is known. The third costate equation in (11.9) can be solved using

$\lambda_3(t_f) = -\gamma_P$:

$$\lambda_3(t) = -\gamma_P \exp[k(t - t_f)]$$

This can be used in Equations (11.14). As a result, all costates are known up to the unknown final time t_f. In other words, the singular feed rate is a *non-linear feedback law of the state and the final time.* The algorithm is then modified as follows. For each couple of switching times (t_1, t_2), the final time t_f must be determined iteratively in order to obtain agreement between the estimated and calculated value.

If the performance index has the form (11.5), we have:

$$\mathcal{H}^* = -[\gamma_X X(t_f) + \gamma_P P(t_f)]/t_f^2$$

Solving the third costate equation using boundary condition $\lambda_3(t_f) = -\gamma_P/t_f$ leads to:

$$\lambda_3(t) = -(\gamma_P/t_f) \exp[k(t - t_f)]$$

Both expressions are used in Equations (11.14) to obtain all costates in function of the unknowns t_f and $\mathcal{H}^*$, or equivalently t_f, $X(t_f)$, and $P(t_f)$. As a result, the singular feed rate is a *non-linear feedback law of the state and the unknowns t_f, $X(t_f)$, and $P(t_f)$.* Again, for each couple of switching times (t_1, t_2) these unknowns must be determined iteratively.

In summary, in all cases the complete optimal control can be calculated *without an explicit need for any costate variable.*

4. The algorithm proposed by Lim *et al.* [11] makes explicitly use of the costates. During singular control $[t_2, t_3]$ some costate equations must be integrated together with the state equations, while during the terminating batch phase $[t_3, t_f]$ all costate equations must be integrated. Furthermore, at the switching time t_2 an initial guess for some costate variables is required. Our algorithm represents a significant improvement for the following reasons. First, it is well-known that integrating the costate equations (11.9) in the forward direction of time t can give rise to (numerical) instabilities. From this point of view, our algorithm is numerically very reliable since it does not use any costate variable at all. Second, in the algorithm of Lim *et al.* some costates must be guessed at the beginning of the singular arc. Obviously, since we cannot assign a clear physical interpretation to the costate variables of

this problem, it may be very difficult to obtain a good initial guess. In our algorithm, we sometimes need an initial guess for quantities with a clear physical interpretation such as final time t_f and final product amount $P(t_f)$. As a result, good starting values can be obtained, e.g., from knowledge about the process before optimization. In summary, this algorithm reduces to a *two-dimensional optimization problem of switching times* t_1 *and* t_2. ■

11.3.3.3 Unbounded input u and free initial substrate concentration $C_S(0)$

This case can be seen as a limiting case of the previous bounded input problem with $U_{MAX} \to \infty$. As a matter of fact, the upper limit U_{MAX} on the input –representing an upper bound on the feeding pumps– can be removed by installing more or larger pumps to deliver the requested influent flow rate. The only place where the upper limit U_{MAX} is active is during the time interval $[0, t_1]$ (see Figure 11.2). Suppose now that the value of U_{MAX} becomes larger by installing more or larger feeding pumps. As a consequence, the value of t_1 will decrease. In the limit, when $U_{MAX} \to \infty$, t_1 goes to 0, and the control during the interval $[0, t_1]$ becomes a *Dirac* impulse at time $t = 0^+$ satisfying $t_1 \times U_{MAX} = \alpha_{growth}/C_{S,in}$. Obviously, this impulse can be replaced by an equivalent set of *initial conditions*:

$$\begin{aligned} x_1(0) &\equiv \alpha_{growth} \text{ (which will be optimized)} \\ x_4(0) &= V_* + x_1(0)/C_{S,in} \end{aligned}$$

where the last equation is constraint (11.3). So, omitting the upper bound on the input leads to the injection of all substrate consumed during growth at the beginning of the fermentation, the growth phase becoming a *batch* phase. Of course, if the initial substrate concentration is free, this is the way to achieve the maximum amount of biomass X with a given amount of substrate α_{growth}, as C_S and thus μ take on their highest possible value for all t during growth. In the above computational algorithm, the only modification is to replace the optimization of the first switching time t_1 by the optimization of the *equivalent* initial substrate concentration $C_S(0) \equiv x_1(0)/x_4(0)$ under constraint (11.3). In other words, the TPBVP is reduced to the *two-dimensional optimization of initial substrate concentration* $C_S(0)$ *and switching time* t_2. Obviously, this represents also the optimal control sequence in case of a *bounded input with free initial substrate concentration.*

11.4 Optimal Control Sequence with State Inequality Constraints

11.4.1 Substrate concentration constraint $C_S(t) \leq C_{S,MAX}$

In many practical applications, too high values for the substrate concentration C_S can lead to a lot of undesired effects, such as a *solubility* problem, an *inhibition* problem, *unproductive side-reactions*, ... This suggests the introduction of an upper limit on the substrate concentration during process optimization:

$$\forall t \in [0, t_f] \ : \ C_S(t) \leq C_{S,MAX}$$

Needless to say that for the type of fermentations under consideration, this constraint can only be violated during the growth phase. As a general result, if C_S reaches its upper limit $C_{S,MAX}$, we use a control of the form

$$u_{C_S}(t) = \frac{\sigma X}{C_{S,in} - C_S} \tag{11.16}$$

which keeps C_S constant at its upper bound $C_S = C_{S,MAX}$ (see model equations (11.1)). As an example, consider the case of a free initial substrate concentration $C_S(0)$. If the optimization as described above results in an optimal initial concentration higher than the upper boundary $C_{S,MAX}$, the control sequence must be modified as follows. The best we can do in order to ensure the highest possible growth rate without violating the constraint is to start with $C_S(0) = C_{S,MAX}$. The initial conditions are then, using (11.3):

$$\begin{aligned} x_1(0) &= \frac{C_{S,in} C_{S,MAX} V_*}{C_{S,in} - C_{S,MAX}} \\ x_4(0) &= \frac{x_1(0)}{C_{S,MAX}} \end{aligned}$$

We then apply a control of the form (11.16) which keeps the substrate concentration level constant. This control is shut off when all substrate for growth α_{growth} has been added at time $t = t_1$, whereafter the fermentation continues as described higher. If the input is bounded and the control resulting from expression (11.16) should exceed the upper limit U_{MAX} before $t = t_1$, obviously $u(t)$ must be set equal to U_{MAX} until $t = t_1$, resulting in a substrate concentration decrease.

11.4.2 Biomass concentration constraint $C_X(t) \leq C_{X,MAX}$

Under certain conditions the optimal control sequence as described here may result in high biomass concentrations in the bioreactor. However, it is well known that the volumetric mass transfer coefficient in, e.g., penicillin fermentations drops by as much as a factor of 10 with biomass concentrations in the region of 20 g dry weight/L. As a result, for a given mounted power, the oxygen transfer capacity of the fermentor will drop substantially as the biomass builds up. Hence, very often these fermentations can be continued only to a limit dictated by the oxygen transfer capacity. The correct way to include the oxygen limitation would be to incorporate additional terms into the model equations (11.1), and to take an oxygen balance around the fermentor. An example for the penicillin fermentation process is described by Bajpai and Reuß [21].

However, since at present accurate oxygen limitation kinetics models are not available, we impose an upper limit $C_{X,MAX}$ on the biomass concentration C_X in order to get an idea about the productivities in fermentors with limited oxygen transfer capacity:

$$\forall t \in [0, t_f] \ : \ C_X(t) \leq C_{X,MAX}$$

To cope with this inequality constraint, we could at first sight adopt a similar line of reasoning as presented for bounded substrate concentration. From model equations (11.1) we can easily deduce the control law which keeps biomass concentration C_X constant:

$$u_{C_X}(t) = \mu V \tag{11.17}$$

However, we now show that *the optimal control sequence under biomass concentration limitation is characterized by a biomass concentration profile C_X which is tangent to the upper bound $C_{X,MAX}$ at at most only one time instant.* In other words, $C_X = C_{X,MAX}$ for at most one value of time t. As an example, consider again the case of a free initial substrate concentration $C_S(0)$. For another case a similar argumentation can be given. The optimal control sequence can be summarized as: (i) high initial substrate concentration, (ii) batch growth phase, (iii) singular production phase, and (iv) batch terminating phase. Suppose that, without taking the biomass inequality constraint into account, this sequence is characterized by the optimal couple $(C_S^*(0), t_2^*)$. We can distinguish the following four cases.

The constraint is not violated

Clearly, in this case the optimal control sequence characterized by $(C_S^*(0), t_2^*)$ remains optimal under biomass limitation.

The constraint is violated during the batch growth phase

As a result, for some value of t within the interval $[0, t_2]$ the control must switch from zero to a positive value given by (11.17). The differential equation for C_S then reduces to:

$$\frac{dC_S}{dt} = -\sigma C_{X,MAX} + (C_{S,in} - C_S)\mu$$

Depending on the specific rates, the substrate concentration C_S will decrease at a lower rate or may even start increasing. As a result, the singular conditions will either be reached only at some time $t_2 > t_2^*$, or will never be reached at all. More substrate will be added during the growth phase, in the first place to keep C_X constant due to dilution, which is then lost for the production phase, resulting in a higher value of the performance index to be minimized.

The constraint is violated during the singular production phase

During the production phase, a singular control of the form (11.13) keeps the substrate concentration within the region of optimal productivity. If the biomass concentration is decreasing, there is no problem. If however the biomass increases, then the upper bound may be violated. At this time the feed rate should switch from the singular control to the control (11.17). In general, the input (11.17) required to keep biomass concentration constant is larger than the singular control (11.13). As an example, if the product decay rate k equals zero and the specific rates are function of the substrate concentration C_S only, then during singular control C_S remains constant at the value C_S^* which maximizes π/σ [19]. The singular control is then of the form

$$u_{sing}(t) = \frac{\sigma X}{C_{S,in} - C_S} \equiv \frac{\sigma C_X}{C_{S,in} - C_S} V$$

In general, we have

$$\frac{\sigma(C_S^*) C_{X,MAX}}{C_{S,in} - C_S^*} < \mu(C_S^*)$$

As a result, the substrate concentration will move away from the singular hyperplane, resulting in a decreasing performance.

The constraint is violated during the terminating batch phase

Obviously, this situation must be avoided also since there is no substrate available anymore to dilute the liquid phase.

From this more or less intuitive discussion, we can deduce the following procedure to obtain the optimal sequence under biomass limitation.

First, we determine the optimal couple $(C_S^*(0), t_2^*)$ without taking the biomass constraint into account.

Second, we verify if the constraint is violated for some value of t. If so, the optimal substrate concentration $C_S^*(0)$ must be decreased to such a level that the resulting biomass concentration profile and the constraint are tangent to each other for at most one value of t, or:

$$\exists!\ t^* \in [0, t_f] \ : \ C_X(t^*) = C_{X,MAX}$$

Then the control (11.17) will never become active. Of course, the corresponding optimal switching time t_2^* will decrease accordingly.

11.5 Example

Consider the penicillin G fermentation process as modeled by Bajpai and Reuß [21][22]:

$$\mu = \mu_{max} \frac{C_S}{K_X C_X + C_S} \text{ (Contois)}, \quad \pi = \pi_m \frac{C_S}{K_p + C_S(1 + C_S/K_i)} \text{ (Haldane)}$$

Parameters and initial conditions used during simulations are summarized in Table 11.1.

parameters						*initial conditions*		
μ_{max}	0.11	[1/h]	K_X	0.006	[g/g DW]	X_0	10.5	[g DW]
π_m	0.004	[g/g DW h]	k	0.01	[1/h]	P_0	0	[g]
K_p	0.0001	[g/L]	K_i	0.1	[g/L]	S_0	*free*	[g]
$Y_{X/S}$	0.47	[g DW/g]	$Y_{P/S}$	1.2	[g/g]	V_*	7	[L]
m	0.029	[g/g DW h]	$C_{S,in}$	500	[g/L]	α	1500	[g]

Table 11.1: Parameters and initial conditions used during simulations.

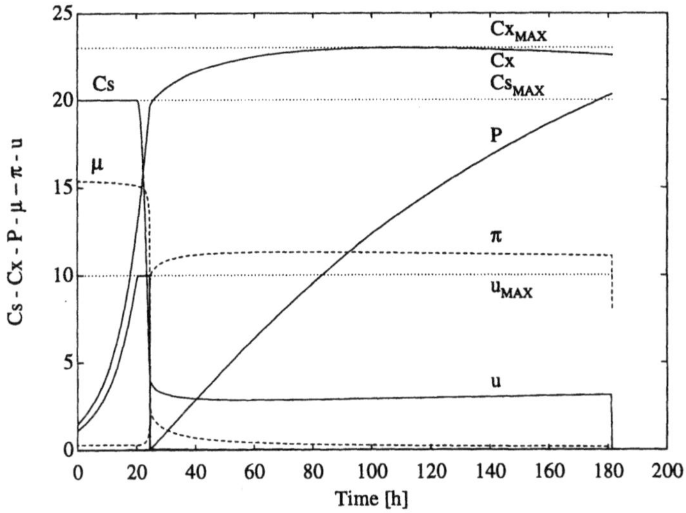

Figure 11.3: Optimal glucose feed rate and corresponding cell, glucose, product, μ and π profiles, with $u_{MAX} = 0.05$ L/h, $C_{S,MAX} = 4$ g/L and $C_{X,MAX} = 23$ g/L.

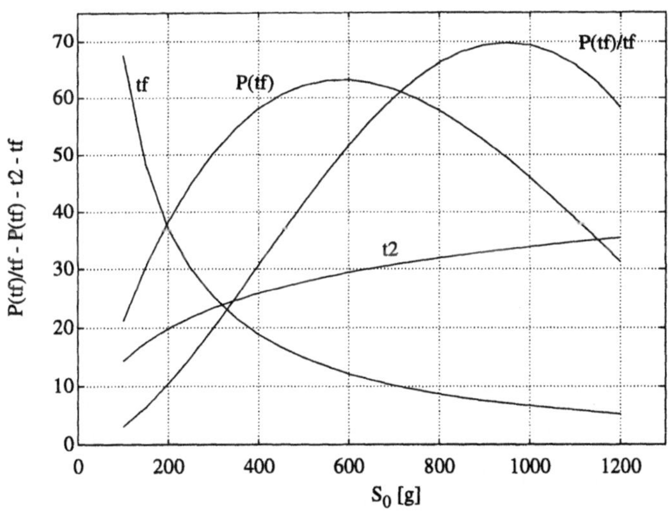

Figure 11.4: Extremal values for $P(t_f)/t_f$, $P(t_f)$, t_2 and t_f as functions of S_0. Scaling: $t_f/10$, $(P(t_f)/t_f) * 100$.

First, we consider the case of an unbounded input u and state $\mathbf{x}$ with free initial substrate concentration $C_S(0)$. If we maximize the final product amount $P(t_f)$ (which is a performance index of type (11.4)), we obtain $S_0 = 528$ g, $t_2 = 28.271$ h, $t_f = 132.033$ h, $\mathcal{Y} \equiv P(t_f) = 63.846$ g.

Using exactly the same initial and operational conditions, Lim *et al.* [11] obtain the following optimal control results with their computational algorithm: $S_0 = 561.2$ g, $t_f = 124.9$ h and $P(t_f) = 86.99$ g. This final product amount differs a lot from our result, and contradicts the conclusion of Bajpai and Reuß [21][22] and Reuß [23] themselves. As a result, we believe the difference is due to the computational algorithm proposed by Lim *et al.* [11] as already discussed. For more details, reference is made to Van Impe [19].

If $U_{MAX} = 0.05$ L/h, $C_{S,MAX} = 4$ g/L and $C_{X,MAX} = 23$ g/L then the optimal control results are: $\alpha_{growth} = 339$ g, $t_2 = 24.668$ h, $t_f = 181.407$ h and $P(t_f) = 61.015$ g (see Figure 11.3). In other words the yield $\mathcal{Y}$ decreases with 4.43 %, while the productivity $\mathcal{P} \equiv P(t_f)/t_f$ decreases from $\mathcal{P} = 0.484$ g/h to $\mathcal{P} = 0.336$ g/h, i.e., -30.445 %.

On the other hand, if we optimize the productivity $\mathcal{P}$ (which is a performance index of type (11.5)), the optimal values are (see Figure 11.4): $S_0 = 950$ g, $t_2 = 33.430$ h, $t_f = 71.414$ h and $P(t_f) = 49.721$ g. The corresponding optimal productivity is $\mathcal{P} = 0.696$ g/h. Clearly, the optimal values of yield and productivity do not occur at the same value of S_0. This is an illustration of the well-known *yield-productivity conflict* in fed-batch reactors. A realistic process optimization strategy should consist of searching for the optimal initial substrate amount S_0, which results in the best *trade-off* between *yield* and *productivity*. This represents also a trade-off between *yield* and *total fermentation time*.

11.6 Conclusion

In this chapter we have analyzed the optimization of fed-batch bioreactors with respect to the volumetric substrate feeding rate. As an example, we have derived the optimal control sequence for processes with monotonic growth rate and non-monotonic production rate. The deduced control sequences have been justified based on microbiological/biochemical insight in the process. A more rigorous mathematical verification can be found in Van Impe [19]. The resulting straightforward computational algorithm has some advantages over the algorithm proposed by Lim *et al.* [11]. First, there is no need for costate variables –which have in general no clear physical interpretation– independent of the performance measure considered. Second, as a result this algorithm proves to be numerically more reliable.

In addition, we have presented for the first time a *complete solution in the realistic case of both substrate and biomass inequality constraints.* The optimal solution can be computed using basically the same algorithm.

In Chapter 13 and References [17][18][19] it is shown how a detailed analysis of the singular arc leads in a natural way to the design of model-independent feedback control strategies which combine nearly optimal performance with excellent robustness properties: *optimal adaptive control.*

11.7 Nomenclature

t	: time	[h]
X	: amount of cell mass	[g DW]
P	: amount of product	[g]
S	: amount of substrate	[g]
V	: fermentor volume	[L]
C_X	: cell mass concentration	[g/L]
C_P	: product concentration	[g/L]
C_S	: substrate concentration	[g/L]
$C_{S,in}$	: substrate concentration in feed	[g/L]
u	: volumetric substrate feed rate	[L/h]
m	: maintenance constant	[g/g DW h]
k	: specific product decay rate	[1/h]
$Y_{P/S}$	: product on substrate yield	[g/g]
$Y_{X/S}$	: cell mass on substrate yield	[g DW/g]
μ	: specific growth rate	[1/h]
π	: specific production rate	[g/g DW h]
σ	: specific substrate consumption rate	[g/g DW h]
α	: total amount of substrate available	[g]

Acknowledgments

This chapter presents research results of NFWO Project G.0141.95, NFWO Project G.0286.96, Project OT/95/20 of the Research Council of the Katholieke Universiteit Leuven, and of the Belgian Programme on Interuniversity Poles of Attraction, initiated by the Belgian State, Prime Minister's Office for Science, Technology and Culture. The scientific responsibility rests with its author.

11.8 Bibliography

1. S.J. Parulekar and H.C. Lim 1985. Modeling, optimization and control of semi-batch bioreactors. *Adv. Biochemical Eng. Biotechnol.*, 32, 207-258

2. A. Johnson 1987. The control of fed-batch fermentation processes: a survey. *Automatica*, 23, 691-705

3. V.M. Fishman and V.V. Biryukov 1974. Kinetic model of secondary metabolite production and its use in computation of optimal conditions. *Biotechnol. Bioeng. Symp.*, 4, 647-662

4. H. Ohno, E. Nakanishi and T. Takamatsu 1976. Optimal control of a semibatch fermentation. *Biotechnol. Bioeng.*, 18, 847-864

5. W.A. Weigand, H.C. Lim, C.C. Gregan and R.D. Mohler 1979. Optimization of a repeated fed-batch reactor for maximum cell productivity. *Biotechnol. Bioeng. Symp.*, 9, 335-348

6. W.A. Weigand 1981. Maximum cell productivity by repeated fed-batch culture for constant yield case. *Biotechnol. Bioeng.*, 23, 249-266

7. K.Y. San and G. Stephanopoulos 1984. A note on the optimality criteria for maximum biomass production in a fed-batch fermentor. *Biotechnol. Bioeng.*, 26, 1261-1264

8. J. Staniškis and D. Levišauskas 1984. An adaptive control algorithm for fed-batch culture. *Biotechnol. Bioeng.*, 26, 419-425

9. J. Hong 1986. Optimal substrate feeding policy for a fed batch fermentation with substrate and product inhibition kinetics. *Biotechnol. Bioeng.*, 28, 1421-1431

10. J.M. Modak, H.C. Lim and Y.J. Tayeb 1986. General characteristics of optimal feed rate profiles for various fed-batch fermentation processes. *Biotechnol. Bioeng.*, 28, 1396-1407

11. H.C. Lim, Y.J. Tayeb, J.M. Modak and P. Bonte 1986. Computational algorithms for optimal feed rates for a class of fed-batch fermentation: numerical results for penicillin and cell mass production. *Biotechnol. Bioeng.*, 28, 1408-1420

12. T. Yamane, T. Kume, E. Sada and T. Takamatsu 1977. A simple optimization technique for fed-batch culture. *J. Fermentation Technol.*, 55, 587-598

13. R. Guthke and W.A. Knorre 1981. Optimal substrate profile for antibiotic fermentations. *Biotechnol. Bioeng.*, 23, 2771-2778

14. J.M. Modak and H.C. Lim 1989. Simple nonsingular control approach to fed-batch fermentation optimization. *Biotechnol. Bioeng.*, 33, 11-15

15. K.Y. San and G. Stephanopoulos 1989. Optimization of fed-batch penicillin fermentation: a case of singular optimal control with state constraints. *Biotechnol. Bioeng.*, 34, 72-78

16. Ž. Kurtanjek 1991. Optimal nonsingular control of fed-batch fermentation. *Biotechnol. Bioeng.*, 37, 814-823

17. J.F. Van Impe, B. Nicolaï, P. Vanrolleghem, J. Spriet, B. De Moor and J. Vandewalle 1992. Optimal control of the penicillin G fed-batch fermentation: an analysis of a modified unstructured model. *Chem. Eng. Comm.*, 117, 337-353

18. J.F. Van Impe, and G. Bastin 1995. Optimal adaptive control of fed-batch fermentation processes with growth/production decoupling. *Control Engineering Practice*, 3(7), 939-954

19. J.F. Van Impe 1993. *Modeling and optimal adaptive control of biotechnological processes*, PhD thesis, Department of Electrical Engineering, Katholieke Universiteit Leuven, 304 p.

20. A.E. Jr. Bryson and Y.C. Ho 1975. *Applied optimal control.* Hemisphere, Washington

21. R.K. Bajpai and M. Reuß 1980. A mechanistic model for penicillin production. *J. Chem. Tech. Biotechnol.*, 30, 332-344

22. R.K. Bajpai and M. Reuß 1981. Evaluation of feeding strategies in carbon-regulated secondary metabolite production through mathematical modelling. *Biotechnol. Bioeng.*, 23, 717-738

23. M. Reuß, *Personal communication*

Chapter 12

Monitoring and Adaptive Control of Bioprocesses

D. DOCHAIN
Chercheur qualifié FNRS
Université Catholique de Louvain
CESAME Bâtiment Euler
Av. G. Lemaître 4-6 B-1348 Louvain-La-Neuve (BELGIUM)
email: dochain@auto.ucl.ac.be

M. PERRIER
Ecole Polytechnique de Montréal
Département de Génie Chimique
C.P. 6079 Succ. "A" Montreal H3A 3A7 (CANADA)
email: perrier@auto.polymtl.ca

J.F.M. Van Impe et al. (eds.), Advanced Instrumentation, Data Interpretation, and Control of Biotechnological Processes, 347-400.

12.1 Introduction

Industrial-scale biotechnological processes have progressed vigorously over the last decades. Generally speaking, the problems arising from the implementation of these processes are similar to those of more classical industrial processes and the need for monitoring systems and automatic control in order to optimize production efficiency, to improve products quality or to detect disturbances in process operation is obvious. Nevertheless, automatic control of industrial biotechnological processes is clearly developing very slowly.

There are two main reasons for this:

a) The internal working and dynamics of these processes are as yet badly grasped and many problems of methodology in modeling remain to be solved. It is difficult to develop models taking into account the numerous factors which can influence the specific bacterial growth rate and the yield coefficients which characterize microorganism growth. The modeling effort is often tedious and requires a great number of experiments before producing a reliable model. Reproducibility of experiments is often uncertain due to the difficulty in obtaining the same environmental conditions. Moreover, as these processes involve living organisms, their dynamical behavior is strongly non-linear and non-stationary. Model parameters cannot remain constant over a long period. They will vary, e.g., due to metabolic variations of biomass or to random and unobservable physiological or genetic modifications. It should also be noted that the lack of accuracy of the measurements often leads to identifiability problems.

b) Another essential difficulty lies in the absence, in most cases, of cheap and reliable instrumentation suited to real time monitoring. To date, the market offers very few sensors capable of providing reliable on-line measurements of the biological and biochemical parameters required to implement high performance automatic control strategies. The main variables, i.e., biomass, substrate and synthesis product concentrations, generally need determining through laboratory analyses. The cost and duration of the analyses obviously limit the frequency of the measurements.

The classical monitoring and control methods do not prove to be very efficient to tide over these basic difficulties. Therefore, to reconstruct the state of the system from the only on-line available measurements and to control the biological variables such as the biomass, the substrate or the synthesis product, appropriate algorithms have to be developed. The efficiency of any monitoring or control system highly depends on the design of the control and monitoring techniques and the care taken in their design. Indeed, monitoring or control algorithms will prove to be efficient if they are able to incorporate the important well-known information on the process while being able to deal with the missing informations (lack of on-line measurements, uncertainty on the dynamics, ...) in a *robust* way, i.e., such that these missing informations will not significantly deteriorate the control performance of the process. The present generalized use of computers makes the fairly easy design and implementation of more sophisticated (compared to the classical PID controller) controllers possible. These controllers may refer to quite complex theory (non-linear control, adaptive control) but, as it will be shown, their structure and implementation may remain rather simple while including the key features of simple PID's.

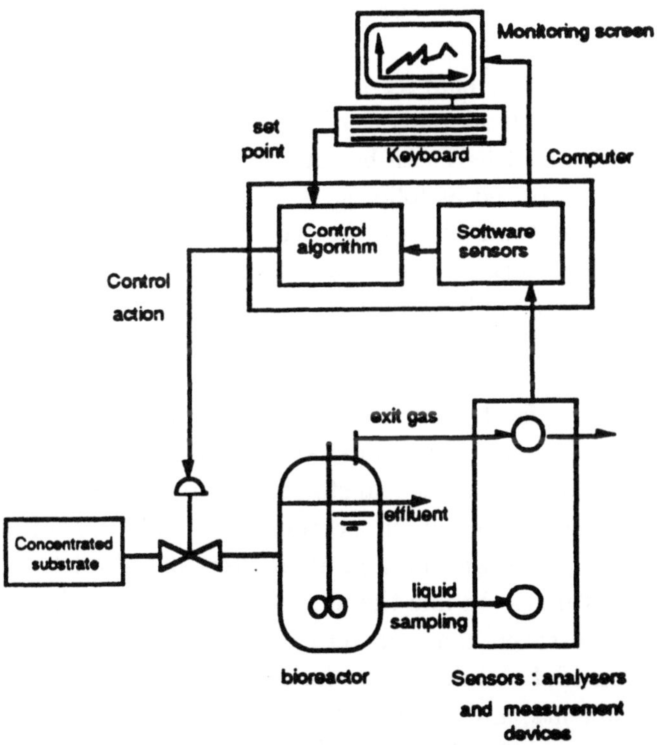

Figure 12.1: Schematic view of a computer controlled bioreactor.

In this chapter, it will be shown how to incorporate the well-known knowledge about the dynamics of biochemical processes (basically, the reaction network and the material balances) in monitoring and control algorithms which are moreover capable of dealing with the process uncertainty (in particular on the reaction kinetics) by introducing, for instance in the control algorithms, an adaptation scheme). Figure 12.1 shows a schematic view of a computer controlled bioreactor.

A key feature of this chapter is to emphasize the central role played by linear algebra which appears to be a very efficient tool in the design of monitoring and control algorithms for apparently complex non-linear systems like bioprocesses, via the use of fairly simple and standard algebraic manipulations. It is also important to draw the attention of the reader to the fact that these control systems are not just the object of academic research but are already used in several applications (see, e.g., [3][9]).

Adaptive as well as non-adaptive linearizing control of bioreactors has been a quite active research area over the last decade. In addition to the works of the authors and co-workers, let us also mention, e.g., [30][25][10][12] [1][33].

A key reference to this chapter is the book by Bastin and Dochain [3] in which a deeper theoretical analysis of most of the monitoring and control methodologies introduced here can be found.

The chapter is organized as follows. In Section 12.2, the general dynamical model for stirred tank bioreactors will be introduced and illustrated with several practical fermentation examples (intracellular production of biodegradable polymers (poly-β-hydroxybutyric acid (PHB)), anaerobic digestion, yeast fermentation, activated sludge process). In this section, it will also be shown how to extend the general dynamical model to multi-tank processes or to non-perfectly mixed reactors like fixed bed reactors. Section 12.3 will concentrate on the dynamical analysis of the stirred tank reactor models by introducing a key state transformation and a general methodology for reducing the order of the model. Sections 12.4 and 12.5 will concentrate on the design of software sensors, i.e., algorithms based on the general dynamical model to estimate on-line the unknown parameters (like the reaction rates) and the unmeasured components from the few available on-line measurements. Section 12.4 will deal with the design of asymptotic observers for the process components, and Section 12.5 will concentrate on the on-line estimation of specific growth rates. In both cases, the proposed methodologies will be illustrated by real-life results on a PHB process and a yeast fermentation, respectively. Finally, Section 12.6 will present the design of adaptive linearizing controllers for bioprocesses based on a reduced order model of the process. The methodology will be illustrated with the anaerobic digestion and the activated sludge examples.

12.2 General Dynamical Model

A biotechnological process can be defined as a set of M biochemical reactions involving N components. The reactions most often encountered in bioprocesses are microbial growth (in which the biomass plays the role of an autocatalyst) and enzyme catalyzed reactions (in which the biomass can be viewed as a simple catalyst); but many other reactions can also take place, like microorganism death, maintenance, ... These reactions can be formalized into reaction schemes, as it is now illustrated.

In the following concentration is made on four process examples which will be used in the rest of the text to illustrate the different themes of this chapter:

1. a biodegradable polymer (PHB) production process;

2. yeast growth, one of the oldest biotechnological processes, which has many applications in the food industry (baker's yeast, wine, ...) and also in some areas, for instance in the pharmaceutical industry (e.g., production of the hepatitis-B vaccine);

3. the two classical biological wastewater treatment processes: the anaerobic digestion and the activated sludge process.

The above choice is somewhat arbitrary (although the chosen processes correspond to typical processes which have been the object of research works over the past ten years and more, and for which estimation and/or control results are available). We could have considered indeed other process examples among which we would like to mention animal cell cultures and penicillin production. Let us mention to the interested reader some references concerning works related to the themes developed here: [13][14][11] about the animal cell cultures, and [52][53] about penicillin G production.

12.2.1 Example #1: PHB production process

Let us first consider the production of poly-β-hydroxybutyric acid (PHB), which is a biodegradable polymer. The PHB can be produced in an aerobic culture of *Alcaligenes eutrophus*, and the production may follow two metabolic pathways:

1. the first metabolic pathway for the production of PHB is growth associated and characterized by a very low yield. The growth takes place with three limiting substrates: oxygen, a source of carbon (e.g., fructose or glucose) and nitrogen (usually under the form of ammonia);

2. the second metabolic pathway for the production of PHB is non-growth associated, where the biomass plays simply the role of a catalyst, and is completely inhibited by nitrogen.

In both cases, CO_2 is a by-product of the reaction. Therefore the production of PHB can be schematized by the following reaction network:

1. Growth associated production:

$$S + N + C \longrightarrow X + P_1 + P_2 \tag{12.1}$$

2. Non-growth associated production:

$$S + C + X \longrightarrow X + P_1 + P_2 \tag{12.2}$$

Where S, C, N, X, P_1 and P_2 represent the carbon source, the oxygen, the nitrogen, the biomass, the PHB and CO_2, respectively.

In the first reaction scheme, the feedback arrow means that, in a growth reaction, the biomass is an autocatalyst, i.e., a product and a catalyst. The presence of X on both sides of the arrow in the second reaction scheme means that X simply plays the role of a catalyst.

Note that in the above reaction network the stoichiometric coefficients have been (deliberately) omitted, since in our view, each reaction scheme simply represents the qualitative mass exchange corresponding to the reaction. The objective of the reaction network here is to use it as a tool for deriving the dynamical model in a general manner. This will be explained in Section 12.2.2.

Note finally that in the following, the symbols used for the components (S, C, N, X, P_1, P_2 in the above example) will either designate the components in the reaction networks or their concentration [g/L] in the dynamical model.

Let us now concentrate on the derivation of the dynamical equations of the PHB process. By considering mass balances for each component in a stirred tank reactor (STR), the following equations can be obtained:

$$\frac{dS}{dt} = -DS + DS_{in} - k_1 r_1 - k_2 r_2 \tag{12.3}$$

$$\frac{dC}{dt} = -DC + Q_{in} - Q_{out} - k_3 r_1 - k_4 r_2 \tag{12.4}$$

$$\frac{dN}{dt} = -DN + DN_{in} - k_5 r_1 \tag{12.5}$$

$$\frac{dX}{dt} = -DX + r_1 \tag{12.6}$$

$$\frac{dP_1}{dt} = -DP_1 + k_6 r_1 + r_2 \tag{12.7}$$

$$\frac{dP_2}{dt} = -DP_2 - Q_2 + k_7 r_1 + k_8 r_2 \tag{12.8}$$

where D is the dilution rate [1/h], S_{in} and N_{in} the influent carbon source and nitrogen concentrations [g/L], Q_{in} and Q_{out} the inlet and outlet gaseous oxygen flow rates [g/h], Q_2 the CO_2 gaseous outflow rate [g/h], r_1 and r_2 the reaction rates of the reactions (12.1) and (12.2) [1/h] respectively and k_i (i = 1 to 8) the yield coefficients.

Note that each reaction rate (r_1 and r_2) has been normalized with respect to one component, the biomass concentration X and the PHB concentration P_1 respectively. Note also that the equations are valid whatever the operating conditions, continuous, fed-batch or batch. The above equations (12.3) to (12.8) can be rewritten in the following matrix form (in which $\Delta Q_{O2} = Q_{in} - Q_{out}$):

$$\frac{d}{dt}\begin{pmatrix} S \\ C \\ N \\ X \\ P_1 \\ P_2 \end{pmatrix} = -D \begin{pmatrix} S \\ C \\ N \\ X \\ P_1 \\ P_2 \end{pmatrix} + \begin{pmatrix} DS_{in} \\ \Delta Q_{02} \\ DN_{in} \\ 0 \\ 0 \\ 0 \end{pmatrix} - \begin{pmatrix} 0 \\ 0 \\ 0 \\ 0 \\ 0 \\ Q_2 \end{pmatrix}$$
$$+ \begin{pmatrix} -k_1 & -k_2 \\ -k_3 & -k_4 \\ -k_5 & 0 \\ 1 & 0 \\ k_6 & 1 \\ k_7 & k_8 \end{pmatrix} \begin{pmatrix} r_1 \\ r_2 \end{pmatrix} \tag{12.9}$$

12.2.2 General dynamical model

The dynamical model (12.9) can be rewritten in the following more compact form:

$$\frac{d\xi}{dt} = -D\xi + Kr + F - Q \tag{12.10}$$

where ξ is the vector of the bioprocess component (dim ξ = N), K is the yield coefficient matrix (dim K = N x M), r is the reaction rate vector (dim r = M), F is the feed rate vector and Q the gaseous outflow rate vector (dim F = dim Q = N).

The model (12.10) has been called the *General Dynamical Model* for stirred tank bioreactors (for further details on the notion scheme and the general dynamical model of bioreactors, see [3]). The derivation of the dynamical model from a reaction network is then straightforward by noting that each component k_{ij} of the yield coefficient matrix

$$K = [k_{ij}] \qquad i = 1 \text{ to } N,\ j = 1 \text{ to } M$$

is representative of the ith component: it is negative if the component is a reactant, it is positive if it is a product and it is equal to zero if the component does not intervene in the reaction.

12.2.3 Example #2: anaerobic digestion

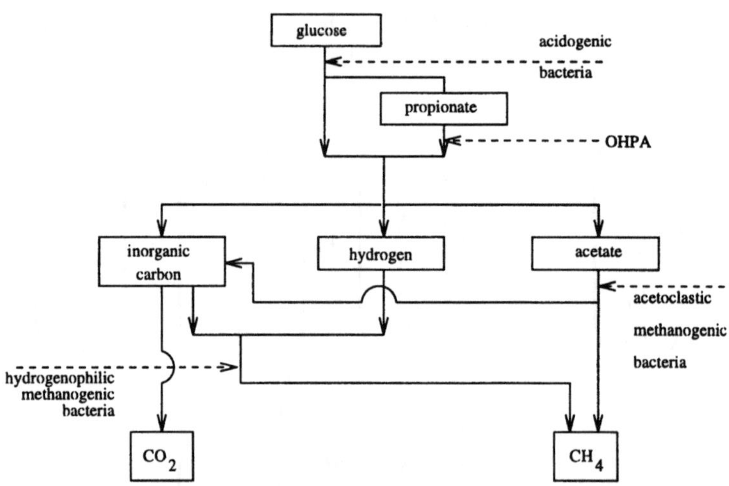

Figure 12.2: Scheme of the anaerobic digestion.

Anaerobic digestion is a biological wastewater treatment process which produces methane. Four metabolic paths [41] can be identified in this process: two for acidogenesis and two for methanization (Figure 12.2).

In the first acidogenic path (Path 1), glucose is decomposed into fatty volatile acids (acetate, propionate), hydrogen and inorganic carbon by acidogenic bacteria. In the second acidogenic path (Path 2), OHPA (Obligate Hydrogen Producing Acetogens) decompose propionate into acetate, hydrogen and inorganic carbon. In a first methanization path (Path 3), acetate is transformed into methane and inorganic carbon by acetoclastic methanogenic bacteria, while in the second methanization path (Path 4), hydrogen combines with inorganic carbon to produce methane under the action of hydrogenophilic methanogenic bacteria. The process can then be described by the following reaction network:

$$S_1 \longrightarrow X_1 + S_2 + S_3 + S_4 + S_5 \tag{12.11}$$

$$S_2 \longrightarrow X_2 + S_3 + S_4 + S_5 \tag{12.12}$$

$$S_3 \longrightarrow X_3 + S_5 + P_1 \tag{12.13}$$

$$S_4 + S_5 \longrightarrow X_4 + P_1 \tag{12.14}$$

where S_1, S_2, S_3, S_4, S_5, X_1, X_2, X_3, X_4 and P_1 are respectively glucose, propionate, acetate, hydrogen, inorganic carbon, acidogenic bacteria, OHPA, acetoclastic methanogenic bacteria, hydrogenophilic methanogenic bacteria and methane. The dynamical model of the anaerobic digestion process ($N = 10, M = 4$) in a stirred tank reactor can be described within the above formalism (12.10) by using the following definitions:

$$\xi = \begin{pmatrix} X_1 \\ S_1 \\ X_2 \\ S_2 \\ X_3 \\ S_3 \\ X_4 \\ S_4 \\ S_5 \\ P_1 \end{pmatrix}, \quad K = \begin{bmatrix} 1 & 0 & 0 & 0 \\ -k_{21} & 0 & 0 & 0 \\ 0 & 1 & 0 & 0 \\ k_{41} & -k_{42} & 0 & 0 \\ 0 & 0 & 1 & 0 \\ k_{61} & k_{62} & 0 & -k_{63} \\ 0 & 0 & 0 & 1 \\ k_{81} & k_{82} & 0 & -k_{84} \\ k_{91} & k_{92} & k_{93} & -k_{94} \\ 0 & 0 & k_{03} & k_{04} \end{bmatrix} \tag{12.15}$$

$$F = \begin{pmatrix} 0 \\ DS_{in} \\ 0 \\ 0 \\ 0 \\ 0 \\ 0 \\ 0 \\ 0 \\ 0 \end{pmatrix}, \quad Q = \begin{pmatrix} 0 \\ 0 \\ 0 \\ 0 \\ 0 \\ 0 \\ 0 \\ Q_1 \\ Q_2 \\ Q_3 \end{pmatrix}, \quad r = \begin{pmatrix} r_1 \\ r_2 \\ r_3 \\ r_4 \end{pmatrix} = \begin{pmatrix} \mu_1 X_1 \\ \mu_2 X_2 \\ \mu_3 X_3 \\ \mu_4 X_4 \end{pmatrix} \tag{12.16}$$

where μ_1, μ_2, μ_3, μ_4 are the specific growth rates [1/h] of reactions (12.11), (12.12), (12.13) and (12.14), respectively, and S_{in}, Q_1, Q_2 and Q_3 represent respectively the influent glucose concentration [g/L] and the gaseous outflow rates of H_2, CO_2 and CH_4 [g/L h].

12.2.4 Example #3: yeast growth

Yeast (*Saccharomyces cerevisiae*) growth is usually characterized by the following three metabolic reactions (see, e.g., [49]):

1. respiratory growth on glucose;
2. fermentative growth on glucose;
3. respiratory growth on ethanol.

These can be formalized by the following reaction schemes:

$$S + C \longrightarrow X + P \tag{12.17}$$

$$S \longrightarrow X + E + P \tag{12.18}$$

$$E + C \longrightarrow X + P \tag{12.19}$$

where S, C, X, P and E are, respectively, glucose, oxygen, yeast, carbon dioxide and ethanol. In absence of growth, substrate and oxygen may be consumed for maintenance, which may be formalized by the following reaction scheme:

$$S + C \longrightarrow P \tag{12.20}$$

The dynamical model of the yeast fermentation (N = 5, M = 4) deduced from material balances can be formalized within the general dynamical model framework (12.10) by considering the following definitions:

$$\xi = \begin{pmatrix} S \\ C \\ X \\ P \\ E \end{pmatrix}, \quad K = \begin{bmatrix} -k_1 & -k_2 & 0 & -k_{10} \\ -k_3 & 0 & -k_4 & -1 \\ 1 & 1 & 1 & 0 \\ k_5 & k_6 & k_7 & k_{11} \\ 0 & k_8 & -k_9 & 0 \end{bmatrix} \tag{12.21}$$

$$F = \begin{pmatrix} DS_{in} \\ \Delta Q_{O_2} \\ 0 \\ 0 \\ 0 \end{pmatrix}, \quad Q = \begin{pmatrix} 0 \\ 0 \\ 0 \\ Q_1 \\ 0 \end{pmatrix}, \quad r = \begin{pmatrix} r_1 \\ r_2 \\ r_3 \\ r_4 \end{pmatrix} = \begin{pmatrix} \mu_o X \\ \mu_r X \\ \mu_e X \\ q_m X \end{pmatrix} \tag{12.22}$$

where μ_o, μ_r and μ_e are the specific gowth rates of the respiratory growth on glucose (12.17), the fermentative growth on glucose (12.18) and the respiratory growth on ethanol (12.19), respectively [1/h], q_m is the specific maintenance rate [1/h], k_i (i = 1 to 9) are yield coefficients, S_{in} is the influent glucose concentration [g/L], ΔQ_{O_2} the gaseous oxygen feed rate and Q_1 the gaseous outflow rate of carbon dioxide P [g/L h].

12.2.5 Example #4: activated sludge process

In the preceding examples, D is a scalar, but it will be a vector if there is more than one tank, as it is illustrated in the activated sludge process example.

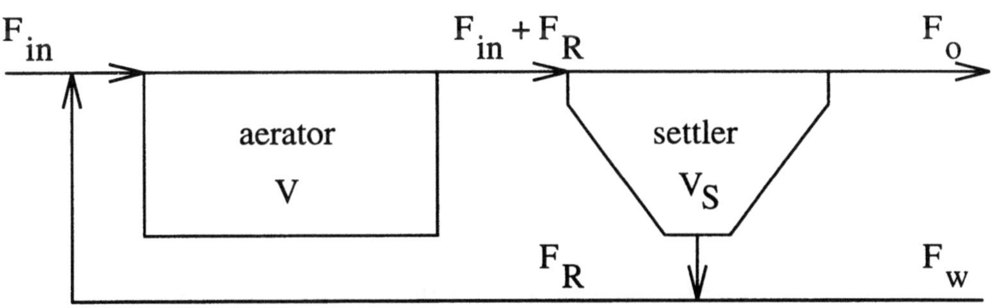

Figure 12.3: Schematic view of an activated sludge process.

The activated sludge process is another classical biological (but aerobic) wastewater treatment process. It is usually operated in two sequential tanks (see Figure 12.3): an aerator (in which the degradation of the pollutants S takes place) and a settler (which is used to recycle part of the biomass X to the aerator). The reaction in the aerator is usually described by a simple microbial growth (see, e.g., [29][39]).

$$S + C \xrightarrow{\curvearrowleft} X \tag{12.23}$$

while the dynamics of the settler are described by the following mass balance equation:

$$\frac{dX_R}{dt} = \frac{F_{in} + F_R}{V_S} X - \frac{F_R + F_W}{V_S} X_R \tag{12.24}$$

where X_R is the concentration of the recycled biomass [g/L], F_{in}, F_R and F_W are the influent, recycle and waste flow rates [g/L h], respectively, and V_S the settler volume [L]. By considering the volume V [L] of the aerator and defining:

$$D_{in} = \frac{F_{in}}{V},\ D_2 = \frac{F_R}{V},\ D_1 = D_{in} + D_2,\ D_3 = \frac{F_{in} + F_R}{V_S},\ D_4 = \frac{F_R + F_W}{V_S} \tag{12.25}$$

The dynamical equations of the process (N = 4, M = 1) can be rewritten in the formalism of the general dynamical model (12.10) with the following definitions:

$$\xi = \begin{pmatrix} S \\ C \\ X \\ X_R \end{pmatrix},\ K = \begin{pmatrix} -k_1 \\ -k_2 \\ 1 \\ 0 \end{pmatrix},\ F = \begin{pmatrix} D_{in}S_{in} \\ \Delta Q_{O_2} \\ 0 \\ 0 \end{pmatrix} \tag{12.26}$$

$$D = \begin{bmatrix} D_1 & 0 & 0 & 0 \\ 0 & D_2 & 0 & 0 \\ 0 & 0 & D_1 & -D_2 \\ 0 & 0 & -D_3 & D_4 \end{bmatrix},\ r = \mu X,\ Q = 0 \tag{12.27}$$

Note that D is now a matrix. In activated sludge processes, the oxygen feed rate ΔQ_{O_2} term in the dynamical equation of the dissolved oxygen is usually set equal to the liquid-gas oxygen transfer rate:

$$\Delta Q_{O_2} = k_L a (C_S - C) \tag{12.28}$$

where k_La is the oxygen mass transfer coefficient and C_S the saturation constant. In line with [32], it is also considered in the following that k_La is a linear function of the air flow rate W:

$$k_La = a_0 W, \; a_0 > 0 \tag{12.29}$$

12.2.6 Fixed bed reactors

Let us now see how to extend the General Dynamical Model (12.10) to non-completely mixed reactors. As a matter of example, we shall concentrate in this section on fixed bed reactors with axial diffusion.

Since the reactor is no longer in completely mixed conditions, the mass balance has to be computed on a thin section dz of the reactor (and not on the whole volume of the reactor anymore) (see Figure 12.4).

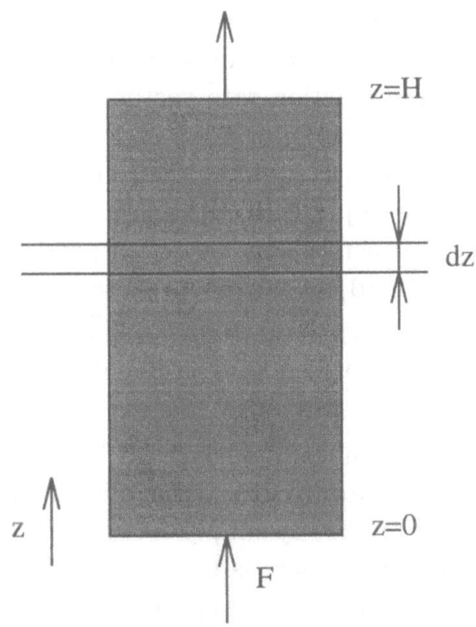

Figure 12.4: Schematic view of a fixed bed reactor.

Let us assume that among the N process components, N_e are microorganisms entrapped or fixed on some support and which remain within the reactor. The N_f other reactants (essentially substrates and products) flow through the reactor. For simplicity, the cross-section of the bioreactor is also considered to be constant and equal to A. From mass balance considerations on a section dz, the following dynamical model can be deduced:

$$\frac{\partial \xi_e}{\partial t} = K_e r(\xi_e, \xi_f) \tag{12.30}$$

$$\frac{\partial \xi_f}{\partial t} = -\frac{F}{A}\frac{\partial \xi_f}{\partial z} + D_{am}\frac{\partial^2 \xi_f}{\partial z^2} + K_f r(\xi_e, \xi_f) \tag{12.31}$$

with the following boundary conditions:

$$D_{am}\frac{\partial \xi_f}{\partial z} = -\frac{F}{A}(\xi_{f,in} - \xi_f) \qquad at\ z = 0 \tag{12.32}$$

$$\frac{\partial \xi_f}{\partial z} = 0 \qquad at\ z = H \tag{12.33}$$

In the above equations, ξ_e is the biomass concentration vector (dim $\xi_e = N_e$), ξ_f is the other reactant concentration vector (dim $\xi_f = N_f$), $\xi_{f,in}(t)$ is the influent concentration of ξ_f (which is different from zero only for external substrates), $r(\xi_e, \xi_f)$ is the reaction rate vector (dim r = M), K_e and K_f are the yield coefficient matrices (dim $K_e = N_e \times$ M; dim $K_f = N_f \times$ M), F the hydraulic flow rate $[m^3/h]$, D_{am} is the axial dispersion coefficient $[m^2/s]$ and z is the space variable [m] (z ∈]0, H]).

Note the similarity between this model (12.30) and (12.31) and the General Dynamical Model (12.10) of stirred tank reactors. The main differences appear in the hydrodynamics term $(-\frac{F}{A}\frac{\partial \xi_f}{\partial z} + D_{am}\frac{\partial^2 \xi_f}{\partial z^2}$ instead of $-D\xi + F - Q)$ and in the presence of boundary conditions (12.32) and (12.33).

Remark: note that the dynamical model of the fixed bed reactor in absence of diffusion, i.e., of the plug flow reactor, is readily obtained from the Equations (12.30), (12.31) and (12.32) by simply setting the dispersion coefficient D_{am} to zero ($D_{am} = 0$). Since there is then only a first order derivative of the state variable ξ_f with respect to the space variable z, only the boundary condition at the reactor input ($z = 0$) is kept (12.32) and the boundary condition at the reactor output (12.33) is dropped.

Comment: the rest of the chapter will mainly concentrate on stirred tank reactors. It is suggested to the interested reader to look at the following papers which deal with the adaptive linearizing control for fixed bed reactors [16][23] and fluidized bed reactors [2], and with the approximation of fixed bed reactor models [19] and the control analysis [6] via singular perturbation techniques.

12.3 Dynamical Analysis of Stirred Tank Bioreactor Models

12.3.1 A key state transformation

The key result of this section is the use of a state transformation by which part of the dynamical model (12.10) becomes independent of the reaction kinetics r (see [3][8]). This transformation will play a very important role in the design of asymptotic observers (Section 4) and adaptive linearizing controllers (Section 12.6). The proposed transformation indeed readily derives from the notion of invariants in reaction systems (see, e.g., [27][24]).

The transformation is defined as follows. Let us denote rank K = R and consider a state partition:

$$\xi = \begin{bmatrix} \xi_a \\ \xi_b \end{bmatrix} \tag{12.34}$$

where ξ_a contains R (arbitrarily chosen) process variables and ξ_b the others, but such that the corresponding submatrix K_a is full rank (rank K_a = R). Let us define the state transformation ζ (dim ζ = N−R):

$$\zeta = C_a\xi_a + C_b\xi_b \tag{12.35}$$

where C_a and C_b are solutions of the matrix equation:

$$C_aK_a + C_bK_b = 0 \tag{12.36}$$

In the particular (but quite general) situation of independent irreversible reactions, then R = M and C_b may be chosen to be a full rank square matrix. Let us derive the dynamics of ζ from (12.10) and the definition (12.35).

1) Single reactor: D = scalar

$$\frac{d\zeta}{dt} = -D\zeta + C_a(F_a - Q_a) + C_b(F_b - Q_b) \tag{12.37}$$

2) Multi-reactor: D = matrix (like in the activated sludge process example)

$$\begin{aligned}\frac{d\zeta}{dt} = & -(C_bD_{bb} + C_aD_{ab})C_b^{-1}\zeta + C_a(F_a - Q_a) + C_b(F_b - Q_b) \\ & +[(C_bD_{bb} + C_aD_{ab})C_b^{-1} - C_bD_{ba} - C_aD_{aa}]\xi_a\end{aligned} \tag{12.38}$$

with:

$$D = \begin{bmatrix} D_{aa} & D_{ab} \\ D_{ba} & D_{bb} \end{bmatrix} \tag{12.39}$$

Note that the dynamical equations of ζ (12.37) and (12.38) are independent of the reaction kinetics $r(\xi)$.

12.3.2 Model order reduction

The examples of bioprocesses presented in the preceding sections have shown that a bioreactor dynamical model may be fairly complex in some instances and involve a large number of differential equations. But there are many practical applications where a simplified reduced order model is sufficient from an engineering viewpoint. Model simplification can be achieved by using the singular perturbation technique, which is a technique for transforming a set of $n+m$ differential equations into a set of n differential equations and a set of m algebraic equations. This technique is suitable when neglecting the dynamics of substrates and of products with low solubility in the liquid phase. The method will be illustrated with one specific example (low solubility product) before stating the general rule for order reduction.

12.3.2.1 Singular perturbation technique for low solubility products

Let us consider a biochemical reaction described by the following reaction scheme:

$$S \longrightarrow P \tag{12.40}$$

where P is a volatile product which can be given off in gaseous form and has low solubility in the liquid phase. The dynamical model is as follows:

$$\frac{dS}{dt} = -r - DS + DS_{in} \tag{12.41}$$

$$\frac{dP}{dt} = kr - DP - Q \tag{12.42}$$

The consistency of this model requires that the product concentration P would be lower than a saturation concentration representative of the product solubility, which is expressed as:

$$P = \Pi P_{sat}, \qquad 0 \leq \Pi(t) \leq 1 \tag{12.43}$$

where P_{sat} is the saturation concentration which is constant in a stable physico-chemical environment. The model (12.41) and (12.42) is rewritten in the standard singular perturbation form, with $\epsilon = P_{sat}$:

$$\frac{dS}{dt} = -r - DS + DS_{in} \tag{12.44}$$

$$\epsilon\frac{d\Pi}{dt} = kr - \epsilon D\Pi - Q \tag{12.45}$$

If the solubility is very low, a reduced order model is obtained by setting $\epsilon = 0$ and replacing the differential equation (12.45) by the algebraic one:

$$Q = kr \tag{12.46}$$

12.3.2.2 A general rule for order reduction

The above example shows that the rule for model simplification is actually very simple and that an explicit singular perturbation analysis is not really needed. Consider that, for some i, the dynamics of the component ξ_i are to be neglected. The dynamics of ξ_i are described by Equation (12.10):

$$\frac{d\xi_i}{dt} = -D\xi_i + K_i r + F_i - Q_i \tag{12.47}$$

where K_i is the line of K corresponding to the component ξ_i. The simplification is then achieved by setting ξ_i and $d\xi_i/dt$ to zero, i.e., by replacing the differential equation (12.47) by the following algebraic equation:

$$K_i r = -F_i + Q_i \tag{12.48}$$

It has been shown that the above model order reduction rule is not only valid for products having a low solubility but also for bioprocesses with fast and slow reactions. Then the above order reduction rule (12.48) applies to substrates of fast reactions (as long as they intervene only in fast reactions) (see [51] for further details).

12.3.2.3 Example: the anaerobic digestion

Let us see how to apply the above model order reduction rule (12.48) to a specific example, the anaerobic digestion. First of all, it is well-known that methane is a badly soluble product. Therefore the above procedure applies. Furthermore, assume that the second methanization path (hydrogen consumption) is limiting, i.e., that the first three reactions (12.11), (12.12) and (12.13) are fast and the fourth one (12.14) is slow. Then the model order reduction rule (12.48) can be applied to the glucose concentration S_1, the propionate concentration S_2, the acetate concentration S_3 and the dissolved methane concentration P_1. By setting their values and their time derivatives to zero

$$S_1 = S_2 = S_3 = P_1 \tag{12.49}$$

$$\frac{dS_1}{dt} = \frac{dS_2}{dt} = \frac{dS_3}{dt} = \frac{dP_1}{dt} = 0 \tag{12.50}$$

their differential equations are reduced to the following set of algebraic equations:

$$\begin{pmatrix} -k_{21} & 0 & 0 & 0 \\ k_{41} & -k_{42} & 0 & 0 \\ k_{61} & k_{62} & -k_{63} & 0 \\ 0 & 0 & k_{03} & k_{04} \end{pmatrix} \begin{pmatrix} r_1 \\ r_2 \\ r_3 \\ r_4 \end{pmatrix} = \begin{pmatrix} -DS_{in} \\ 0 \\ 0 \\ Q_3 \end{pmatrix} \tag{12.51}$$

By inverting the submatrix of the yield coefficients of the left hand side of (12.51), the reaction rates r_1, r_2, r_3 and r_4 can be expressed as functions of the feed rate DS_{in} and of the gaseous methane outflow rate Q_3:

$$r_1 = \frac{1}{k_{21}} DS_{in} \tag{12.52}$$

$$r_2 = \frac{k_{41}}{k_{21}k_{42}} DS_{in} \tag{12.53}$$

$$r_3 = \frac{k_{41}k_{62} + k_{42}k_{61}}{k_{21}k_{42}k_{63}} DS_{in} \tag{12.54}$$

$$r_4 = \frac{1}{k_{04}} Q_3 - \frac{k_{03}}{k_{04}} \frac{k_{41}k_{62} + k_{42}k_{61}}{k_{21}k_{42}k_{63}} DS_{in} \tag{12.55}$$

Let us replace the reaction rates r_1, r_2 and r_4 by their above expressions (12.52), (12.53) and (12.55) in the dynamical equation of the hydrogen concentration S_4, which is then rewritten as follows:

$$\frac{dS_4}{dt} = -DS_4 - Q_1 - k_1 Q_3 + k_2 DS_{in} \tag{12.56}$$

where k_1 and k_2 are defined as follows:

$$k_1 = \frac{k_{84}}{k_{04}}, \; k_2 = \frac{k_{81}}{k_{21}} + \frac{k_{41}k_{82}}{k_{21}k_{42}} + \frac{k_{03}k_{84}}{k_{04}} \frac{k_{41}k_{62} + k_{42}k_{61}}{k_{21}k_{42}k_{63}} \tag{12.57}$$

Equation (12.56) will be used for the design of an adaptive linearizing controller of the hydrogen concentration in Section 12.5. Note that the coefficients k_1 and k_2 are non-linear combinations of the yield coefficients k_{ij}.

12.4 Monitoring of Bioprocesses

A key question in bioprocess control is how to monitor reactant and product concentrations in a reliable and cost effective manner. However, it appears that, in many practical applications, only some of the concentrations of the components involved and critical for quality control are available for on-line measurement. For instance, dissolved oxygen concentration and gaseous flow rates are available for on-line measurement while the values of the concentrations of biomass, substrates and/or synthesis products are often available via off-line analysis. An interesting alternative which circumvents and exploits the use of a model in conjunction with a limited set of measurements are the use of Luenberger or Kalman observers. In these techniques, a model, which includes states that are measured as well as states that are not measured, is used in parallel with the process and the model states may then be used for feedback. This configuration may be used to reduce the effect of noise on measurements as well as to reconstruct the states that are not measured. An introduction to these ideas can be found in, e.g., [35]. These concepts were originally developed for linear problems. Because of the non-linear characteristics of the bioprocess dynamics, it is of interest to extend these concepts and exploit particular structures for biochemical engineering application problems. Linearized versions (the linearized tangent model) of the process dynamics are computed from a Taylor's series expansion of a state space model around some equilibrium point and the observer theory referred to above can be applied. This idea is developed in [3]. These modified observers, particularly the extended Kalman filter (EKF), has found applications in some biochemical processes (e.g., [50][36][54][7]).

One of the reasons for the popularity of the EKF is that it is easy to implement since the algorithm can be derived directly from the state space model. However, since (as the extended Luenberger observer) it is based on a linearized model of the process, the stability and convergence properties are essentially local and valid around some equilibrium point, and it is rather difficult to guarantee its stability over wide ranges of operation. Ljung [37] shows that the EKF for state and parameter estimation of linear systems may give biased estimates or even diverge if it is not carefully initialized.

Let us also point out that the derivation of the EKF is based on some stochastic assumptions on the measurement and process noises, which might be questionable in practice.

One reason for the problem of convergence of EKF is that, in order to guarantee the (arbitrarily chosen) exponential convergence of the observer, the process must be locally observable, i.e., the linearized tangent model must be observable and fulfill the classical observability rank condition. This condition, as it turns out, is restrictive in many practical situations and may account for the failure of EKF to find widespread application (e.g., [3][4][20]).

Another problem is that the theory for the extended Luenberger and Kalman observers is developed using a perfect knowledge of the system parameters, in particular of the process kinetics: it is difficult to develop error bounds and there is often a large uncertainty on these parameters.

It appears from the above remarks that there is a clear incentive to develop new methodologies for the on-line estimation of the unmeasured concentration variables in biochemical reaction systems that do not rely on the explicit use of kinetic models. Indeed, the objective of this section is to propose an alternative to EKF and use process physics in a more direct manner to develop a non-linear observer applicable to the estimation problem of stirred tank reactors (STR). The proposed observer is based on the well-known non-linear model of the process without the knowledge of the process kinetics being necessary. In order to advance the application of this method, its stability and convergence properties are discussed. We would like to emphasize that the presented results are global (i.e., independent of the initial conditions) as opposed to the local properties for EKF. The approach (called *asymptotic observers*) proved to be very successful when applied to bioreactors (see, e.g., [3]). The proposed asymptotic observer for STR's is an intermediate method between the *classical* observers (EKF or extended Luenberger observer) which require a full process model knowledge and the adaptive observers [42][38] which include state and parameter estimation within the same estimation scheme. A review of the application of adaptive observers to biochemical processes is given by Dochain [15].

This section is organized as follows. First, the general methodology for single tank bioprocesses will be presented and its theoretical convergence properties and the practical implementation aspects will be discussed. Then a real life application on a PHB producing process will be presented. Finally, the extension to fixed bed reactors will be introduced.

12.4.1 Asymptotic observers for single tank bioprocesses

The derivation of the asymptotic observer equations are based on the Key State Transformation (12.35) and (12.36) introduced in Section 12.3.1 and on the following assumptions:

1. p ($\geq$ M (the number of reactions)) components are measured on-line.
2. The feed rates F, the gaseous outflow rates Q and the dilution rate D are known either by measurement or by choice of the user.
3. The yield coefficient matrix K is known.
4. The reaction rate vector r is unknown.
5. The M reactions are irreversible and independent, i.e., rank K = R = M.

From assumption 1, the following state partition can be defined:

$$\xi = \begin{bmatrix} \xi_1 \\ \xi_2 \end{bmatrix} \tag{12.58}$$

where ξ_1 and ξ_2 hold for the measured component concentrations and the unmeasured ones, respectively.

Let us consider one (arbitrarily chosen) transformation ζ defined by (12.35) and (12.36). The variable ζ can be rewritten as a linear combination of the measured and unmeasured states ξ_1 and ξ_2, i.e.:

$$\begin{aligned} \zeta &= C_a\xi_a + C_b\xi_b && (12.59) \\ &= A_1\xi_1 + A_2\xi_2 && (12.60) \end{aligned}$$

Recall that the dynamics of ζ are independent of the reaction rate $r(\xi)$:

$$\frac{d\zeta}{dt} = -D\zeta + C_a(F_a - Q_a) + C_b(F_b - Q_b) \tag{12.61}$$

The Equations (12.60),(12.61) are the basis for the derivation of the asymptotic observer. The dynamical equations of ζ are used to calculate an estimate of ζ on-line, which is used, via Equation (12.60) and the on-line data of ξ_1, to derive an estimate of the unmeasured component ξ_2. Let us further assume that the matrix A_2 is (left) invertible.

Case #1: p = M

In this case, the asymptotic observer is written as follows:

$$\frac{d\hat{\zeta}}{dt} = -D\hat{\zeta} + C_a(F_a - Q_a) + C_b(F_b - Q_b) \quad (12.62)$$

$$\hat{\xi}_2 = A_2^{-1}[\hat{\zeta} - A_1\xi_1] \quad (12.63)$$

Comment: if we consider the most simple and straightforward choice for the state transformation ζ, i.e., with $\xi_1 = \xi_a$ and $\xi_2 = \xi_b$, and with $C_b = I_{N-p}$, then we have:

$$A_2 = I_{N-p} \quad (12.64)$$

$$A_1 = -K_2K_1^{-1} \quad (12.65)$$

Therefore the condition on the invertibility of A_2 is in fact a condition on the invertibility of the submatrix K_1 (i.e., K_1 is full rank or rank K_1 = M).

Case #2: p > M

Assume that the number of measured components is larger than the number which is strictly necessary one needed, i.e., p > M. Then the asymptotic observer is modified as follows:

$$\frac{d\hat{\zeta}}{dt} = -D\hat{\zeta} + C_a(F_a - Q_a) + C_b(F_b - Q_b) \quad (12.66)$$

$$\hat{\xi}_2 = A_2^{+}[\hat{\zeta} - A_1\xi_1] \quad (12.67)$$

where A_2^+ is a left inverse of A_2.

The observer (12.62) and (12.63) or (12.66) and (12.67) is completely independent of the process kinetics and can be implemented without the knowledge of the reaction rates $r(\xi)$ being required.

12.4.1.1 Theoretical convergence of the asymptotic observer

The convergence properties can be summarized in the following theorem.

Theorem

If the dilution rate D is a persistently exciting signal, i.e., if there exist positive constants δ and β such that:

$$\delta = \int_t^{t+\delta} D(\tau)d\tau \quad (12.68)$$

then:

$$\lim_{t\to\infty}(\xi_2 - \hat{\xi}_2) = 0 \quad (12.69)$$

Proof

The proof of the theorem is immediate if one observes that, from (12.37), (12.62), (12.63) and (12.67), the dynamics of the estimation error are equal to:

$$\frac{d(\xi_2 - \hat{\xi}_2)}{dt} = -D(\xi_2 - \hat{\xi}_2) \qquad (12.70)$$

QED

Remark #1: the persistence-of-excitation condition on D simply requires that D is not equal to zero for too long. This condition is clearly easily fulfilled in fed-batch and continuous reactors.

Remark #2 (reversible reactions): note that, in presence of reversible reactions, the matrix K will not be full column rank because it will contain two identical columns. However, a simple way to treat the asymptotic observation problem of reversible reactions is to consider each reversible reaction as *one global reaction* (whose rate may then be positive or negative) and therefore characterized by *only one* column in the matrix . This means that if the *forward* and *backward* reactions are characterized by a reaction rate r_1 and r_2 respectively, we consider, for the observation, one global reaction characterized by the same stoichiometric coefficients but one global reaction $r_3 = r_1 - r_2$.

Remark #3: the general formulation of the estimation algorithm (12.66) and (12.67) with the introduction of the left inverse allows for larger flexibility in the use of the asymptotic observer since it permits to take into account the possible presence of a number of measured variables larger than M. Let us illustrate this in the following example.

12.4.1.2 Example: PHB producing process

Let us first define one state transformation ζ, e.g.:

$$\xi_a = \begin{bmatrix} X \\ P_1 \end{bmatrix}, \qquad \xi_b = \begin{bmatrix} S \\ C \\ N \\ P_2 \end{bmatrix} \qquad (12.71)$$

with:

$$K_a = \begin{bmatrix} 1 & 0 \\ k_6 & 1 \end{bmatrix}, \qquad K_b = \begin{bmatrix} -k_1 & -k_2 \\ -k_3 & -k_4 \\ -k_5 & 0 \\ k_7 & k_8 \end{bmatrix} \qquad (12.72)$$

Therefore if C_b is chosen as an identity matrix ($C_b = I_4$), then C_a is equal to:

$$C_a = -K_b K_a^{-1} = \begin{bmatrix} k_1 - k_2k_6 & k_2 \\ k_3 - k_4k_6 & k_4 \\ k_5 & 0 \\ -k_7 + k_6k_8 & -k_8 \end{bmatrix} \tag{12.73}$$

The dynamics of ζ are here equal to:

$$\frac{d}{dt}\begin{pmatrix} \zeta_1 \\ \zeta_2 \\ \zeta_3 \\ \zeta_4 \end{pmatrix} = -D\begin{pmatrix} \zeta_1 \\ \zeta_2 \\ \zeta_3 \\ \zeta_4 \end{pmatrix} + \begin{pmatrix} DS_{in} \\ \Delta Q_{02} \\ DN_{in} \\ -Q_2 \end{pmatrix} \tag{12.74}$$

Note that since $F_a = Q_a = 0$, the dynamics of ζ are also independent of the yield coefficients. This has been called a *nice partition* in [3].

Case #1: X and P_1 are measured on-line

Then $A_1 = C_a$ and $A_2 = C_b = I_4$. Therefore, S, C, N and P_2 can be estimated by using the asymptotic observer via the dynamical equation of ζ (12.74) as follows:

$$\hat{S} = \hat{\zeta}_1 + (k_2k_6 - k_1)X - k_2P_1 \tag{12.75}$$

$$\hat{C} = \hat{\zeta}_2 + (k_4k_6 - k_3)X - k_4P_1 \tag{12.76}$$

$$\hat{N} = \hat{\zeta}_3 - k_5X \tag{12.77}$$

$$\hat{P}_2 = \hat{\zeta}_4 + (k_7 - k_6k_8)X + k_8P_1 \tag{12.78}$$

Case #2: C and P_2 are measured on-line

This choice corresponds to the components which are probably a priori the easiest ones to be measured on-line. Then we have a state partition for ξ_1 and ξ_2 which differs from the one used for ζ:

$$\xi_1 = \begin{bmatrix} C \\ P_2 \end{bmatrix}, \quad \xi_2 = \begin{bmatrix} S \\ N \\ X \\ P_1 \end{bmatrix} \tag{12.79}$$

and A_1 and A_2 are then equal to:

$$A_1 = \begin{pmatrix} 0 & 0 \\ 1 & 0 \\ 0 & 0 \\ 0 & 1 \end{pmatrix}, \quad A_2 = \begin{pmatrix} 1 & 0 & k_1 - k_2k_6 & k_2 \\ 0 & 0 & k_3 - k_4k_6 & k_4 \\ 0 & 1 & k_5 & 0 \\ 0 & 0 & -k_7 + k_6k_8 & -k_8 \end{pmatrix} \tag{12.80}$$

It is straightforward to check that the matrix A_2 is invertible if $k_3k_8 \neq k_4k_7$. In practice, this means that the respiratory quotient must be different from 1 ($RQ \neq 1$). But the situation $RQ = 1$ is the one usually encountered in PHB producing processes (for further details, see [3]). This example shows that not any choice of M measured components is valid for the implementation of the asymptotic observer: the submatrix K_1 must be full rank, this means that the measured components must be independent (here, C and P_2 are not independent if $RQ = 1$) or that the measured components have to take part in all the reactions (at least one in each reaction) in order to avoid a submatrix K_1 with (a) column(s) filled with zeros (A_2 will not be full rank if $\xi_1 = [N\ X]^T$, for instance).

Case #3: C, X and P_1 are measured on-line
Now, the number of measured variables is larger than rank K. In this case, the vectors ξ_1 and ξ_2 are equal to:

$$\xi_1 = \begin{bmatrix} C \\ X \\ P_1 \end{bmatrix}, \quad \xi_2 = \begin{bmatrix} S \\ N \\ P_2 \end{bmatrix} \tag{12.81}$$

and the matrices A_1 and A_2 are then defined as follows:

$$A_1 = \begin{pmatrix} 0 & k_1 - k_2k_6 & k_2 \\ 1 & k_3 - k_4k_6 & k_4 \\ 0 & k_5 & 0 \\ 0 & -k_7 + k_6k_8 & -k_8 \end{pmatrix}, \; A_2 = \begin{pmatrix} 1 & 0 & 0 \\ 0 & 0 & 0 \\ 0 & 1 & 0 \\ 0 & 0 & 1 \end{pmatrix} \tag{12.82}$$

A_2 is not a square matrix anymore. Then its left pseudo-inverse can be chosen as its left inverse:

$$A_2^+ = (A_2^T A_2)^{-1} A_2^T \tag{12.83}$$

which takes the following form in our example:

$$A_2^+ = \begin{pmatrix} 1 & 0 & 0 & 0 \\ 0 & 0 & 1 & 0 \\ 0 & 0 & 0 & 1 \end{pmatrix} \tag{12.84}$$

12.4.1.3 Implementation aspects: choice of the sampling period

Stability of the discrete-time asymptotic observer

Practical computer implementation of the asymptotic observer (12.62) and (12.63) or (12.66) and (12.67) requires that it can be rewritten in a discrete-time form. This can be done simply by replacing the time derivative of ζ by a finite difference (using a first order Euler approximation):

$$\frac{d\hat{\zeta}}{dt} \longrightarrow \frac{\hat{\zeta}_{t+1} - \hat{\zeta}_t}{T} \tag{12.85}$$

where T is the sampling period and t and $t+1$ are time indices. The asymptotic observer is then written as follows for the general case p $\geq$ M:

$$\hat{\zeta}_{t+1} = \hat{\zeta}_t - TD_t\hat{\zeta}_t + TC_a(F_{at} - Q_{at}) + C_b(F_{bt} - Q_{bt}) \tag{12.86}$$

$$\hat{\xi}_{2,t+1} = A_2^+[\hat{\zeta}_{t+1} - A_1\xi_{1,t+1}] \tag{12.87}$$

For the discrete-time equation, the value of the sampling period plays a role in the stability. In fact, if the dilution rate D is bounded as follows:

$$0 \leq D(t) \leq D_{max} \tag{12.88}$$

then Equation (12.86) will remain stable as long as T is smaller than $2/D_{max}$:

$$T \leq \frac{2}{D_{max}} \tag{12.89}$$

This can be shown by considering the following positive definite decrescent Lyapunov function:

$$W_t = (\zeta_t - \hat{\zeta}_t)^T(\zeta_t - \hat{\zeta}_t) \tag{12.90}$$

the time difference of which is equal to:

$$W_{t+1} - W_t = [TD_t(TD_t - 2)](\zeta_t - \hat{\zeta}_t)^T(\zeta_t - \hat{\zeta}_t) \tag{12.91}$$

which is non-positive definite, decrescent as long as inequality (12.89) holds.

Dynamics of the discrete-time asymptotic observer

Equation (12.89) gives a condition for the stability of the discrete-time version of the asymptotic observer. But even if the sampling period T is chosen so as to fulfill condition (12.89), large values of T may introduce oscillations in the estimation. As a matter of fact, if it is assumed that D is constant, the dynamics of ζ are characterized by a (discrete-time) pole equal to (1 - TD). It will be negative if T is larger than $1/D$ ($T > 1/D$) and then correspond to the oscillating (underdamped) dynamics for ζ. Therefore, in order to avoid (undesirable) oscillations, condition (12.89) can be replaced by:

$$T \leq \frac{1}{D_{max}} \tag{12.92}$$

Remark: note that different sampling periods may be used for the computation of the variables ζ and the calculation of the observation of ξ_2 , e.g., a *fast* computation of the variables ζ and a slower computation of the estimated values of ξ_2. This choice may depend on the measurement sampling interval which may be different from one variable to another.

12.4.2 Application to a PHB producing process

One interesting aspect of this example is that it shows that by using extra process information, the monitoring algorithm can be simplified. Indeed it has been shown here above that because $RQ = 1$, the measurements of oxygen and carbon dioxide are not independent and cannot be used to implement the asymptotic observer for the two-reaction PHB process.

However, we have not used yet the fact that the non-growth associated production of PHB is completely inhibited by nitrogen, which on the other hand is a limiting substrate for growth. And in practice the process is operated in two sequential steps: a first step for growth *in presence of nitrogen*, and a second step for non-growth associated production *in absence of nitrogen*. Therefore the process can be viewed as a sequence of two reactions and not simply a two-reaction process. This means that just one measured component is necessary to reconstruct the state of the bioreactor. Here oxygen has been chosen, whose data appeared to be more reliable. Therefore an asymptotic observer has been designed for each step. The first step in the design is the rewriting of the process dynamics, which are indeed the dynamical model (12.9), but once with $r_2 = 0$, and the other time with $r_1 = 0$ and $N = 0$.

1) Growth without production

$$\frac{d}{dt}\begin{pmatrix} S \\ C \\ N \\ X \\ P_1 \\ P_2 \end{pmatrix} = -D\begin{pmatrix} S \\ C \\ N \\ X \\ P_1 \\ P_2 \end{pmatrix} + \begin{pmatrix} DS_{in} \\ \Delta Q_{02} \\ DN_{in} \\ 0 \\ 0 \\ 0 \end{pmatrix} - \begin{pmatrix} 0 \\ 0 \\ 0 \\ 0 \\ 0 \\ Q_2 \end{pmatrix} + \begin{pmatrix} -k_1 \\ -k_3 \\ -k_5 \\ 1 \\ k_6 \\ k_7 \end{pmatrix} r_1 \tag{12.93}$$

2) Production without growth

$$\frac{d}{dt}\begin{pmatrix} S \\ C \\ X \\ P_1 \\ P_2 \end{pmatrix} = -D\begin{pmatrix} S \\ C \\ X \\ P_1 \\ P_2 \end{pmatrix} + \begin{pmatrix} DS_{in} \\ \Delta Q_{02} \\ 0 \\ 0 \\ 0 \end{pmatrix} - \begin{pmatrix} 0 \\ 0 \\ 0 \\ 0 \\ Q_2 \end{pmatrix} + \begin{pmatrix} -k_2 \\ -k_4 \\ 0 \\ 1 \\ k_8 \end{pmatrix} r_2 \quad (12.94)$$

Let us concentrate on the observation of the fructose concentration S, the nitrogen concentration N, the biomass concentration X and the PHB concentration P_1 from the measurement of the dissolved oxygen C.
Let us consider the following variable ζ for both steps:

$$\zeta = \xi_2 - K_2 K_1^{-1} \xi_1 \quad (12.95)$$

which corresponds to the following choices of vectors and matrices:

$$\xi_a = \xi_1, \qquad \xi_b = \xi_2 \quad (12.96)$$

$$C_b = I, \qquad C_a = -K_2 K_1^{-1} \quad (12.97)$$

This means that for each step, we have:
1) Growth without production

$$\xi_1 = C, \qquad \xi_2 = \begin{pmatrix} S \\ N \\ X \\ P_1 \end{pmatrix} \quad (12.98)$$

$$\zeta_1 = S - \frac{k_1}{k_3} C \quad (12.99)$$

$$\zeta_2 = N - \frac{k_5}{k_3} C \quad (12.100)$$

$$\zeta_3 = X + \frac{1}{k_3} C \quad (12.101)$$

$$\zeta_4 = P_1 + \frac{k_6}{k_3} C \quad (12.102)$$

2) Production without growth

$$\xi_1 = C, \qquad \xi_2 = \begin{pmatrix} S \\ X \\ P_1 \end{pmatrix} \quad (12.103)$$

$$\zeta_5 = S - \frac{k_2}{k_4}C \tag{12.104}$$

$$\zeta_6 = X \tag{12.105}$$

$$\zeta_7 = P_1 + \frac{1}{k_4}C \tag{12.106}$$

Since N is not measured on-line, the passage from step 1 to step 2 is based on the value of the estimate of the nitrogen concentration, $\hat{N}$, given by the asymptotic observer in step 1. The asymptotic observer in its discrete-time has then the following form:

1) Growth without production: whenever $\hat{N}_t > 0$

$$\hat{\zeta}_{1,t+1} = \hat{\zeta}_{1,t} - TD_t\hat{\zeta}_{1,t} + TD_tS_{in,t} - \frac{k_1}{k_3}\Delta Q_{O2,t} \tag{12.107}$$

$$\hat{\zeta}_{2,t+1} = \hat{\zeta}_{2,t} - TD_t\hat{\zeta}_{2,t} + TD_tN_{in,t} - \frac{k_5}{k_3}\Delta Q_{O2,t} \tag{12.108}$$

$$\hat{\zeta}_{3,t+1} = \hat{\zeta}_{3,t} - TD_t\hat{\zeta}_{3,t} + \frac{1}{k_3}\Delta Q_{O2,t} \tag{12.109}$$

$$\hat{\zeta}_{4,t+1} = \hat{\zeta}_{4,t} - TD_t\hat{\zeta}_{4,t} + \frac{k_6}{k_3}\Delta Q_{O2,t} \tag{12.110}$$

$$\hat{S}_{t+1} = \hat{\zeta}_{1,t+1} + \frac{k_1}{k_3}C_{t+1} \tag{12.111}$$

$$\hat{N}_{t+1} = \hat{\zeta}_{2,t+1} + \frac{k_5}{k_3}C_{t+1} \tag{12.112}$$

$$\hat{X}_{t+1} = \hat{\zeta}_{3,t+1} - \frac{1}{k_3}C_{t+1} \tag{12.113}$$

$$\hat{P}_{1,t+1} = \hat{\zeta}_{4,t+1} - \frac{k_6}{k_3}C_{t+1} \tag{12.114}$$

2) Production without growth: when $\hat{N}_t = 0$

$$\hat{\zeta}_{5,t+1} = \hat{\zeta}_{5,t} - TD_t\hat{\zeta}_{5,t} + TD_tS_{in,t} - \frac{k_2}{k_4}\Delta Q_{O2,t} \tag{12.115}$$

$$\hat{\zeta}_{6,t+1} = \hat{\zeta}_{6,t} - TD_t\hat{\zeta}_{6,t} \tag{12.116}$$

$$\hat{\zeta}_{7,t+1} = \hat{\zeta}_{7,t} - TD_t\hat{\zeta}_{7,t} + \frac{1}{k_4}\Delta Q_{O2,t} \tag{12.117}$$

$$\hat{S}_{t+1} = \hat{\zeta}_{5,t+1} + \frac{k_2}{k_4}C_{t+1} \tag{12.118}$$

$$\hat{X}_{t+1} = \hat{\zeta}_{6,t+1} \tag{12.119}$$

$$\hat{P}_{1,t+1} = \hat{\zeta}_{7,t+1} - \frac{1}{k_4}C_{t+1} \tag{12.120}$$

The above asymptotic observer has been implemented on a 20 liter pilot fedbatch reactor of the Solvay company, Belgium (see [18]). The values of the yield coefficients are equal to:

$$k_1 = 2.5, \quad k_2 = 2.13, \quad k_3 = 1.23, \quad k_4 = 1, \quad k_5 = 0.11, \quad k_6 = 10 \quad (12.121)$$

Figure 12.5 shows the data of dissolved oxygen C, inlet and outlet oxygen gaseous flow rates $Q_{O2,in}$ and $Q_{O2,out}$, volume V, and inlet fructose and nitrogen concentration S_{in} and N_{in}. Figure 12.6 shows the estimation results for S, N, X, and P: note the good correspondence of the software observation with the off-line analyses. Note that the switching from step 1 to step 2 is efficiently driven by the estimate of nitrogen $\hat{N}$.

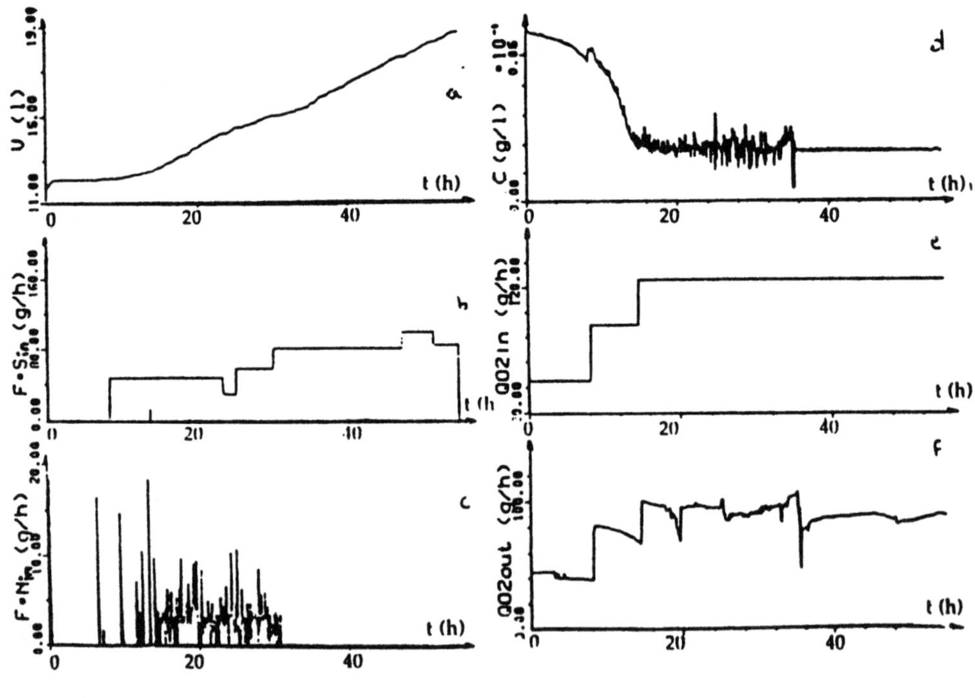

Figure 12.5: On-line data of the asymptotic observer of the PHB process.

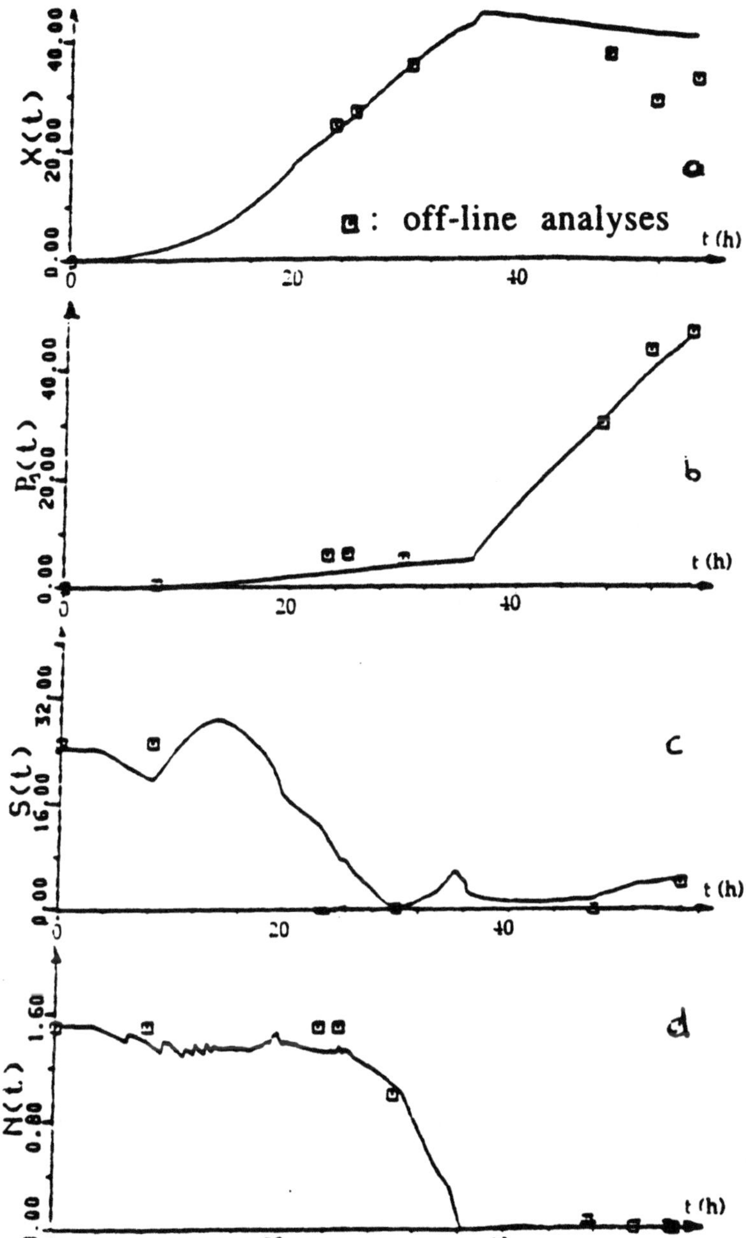

Figure 12.6: Estimation results of the asymptotic observer of the PHB process.

Remark: the design of the above asymptotic observer can be easily extended to multi-reactor processes, since the definition of the transformation ζ is the same. The main question is the stability of the dynamics of the auxiliary variables ζ (12.38). This point, which is largely discussed in [8], will not be developed here. We shall simply concentrate on one example (activated sludge process) in Section 12.6, and discuss its stability.

12.5 On-line Estimation of Reaction Rates

In this section, the problem of estimating the reaction rates from on-line knowledge of the state variables (knowledge available either from measurements or from state estimation) is addressed. The statement of the estimation problem is presented first followed by the development of an observer-based estimator. Finally, an example on the estimation of specific growth rates for baker's yeast is described in detail.

12.5.1 Statement of the estimation problem

A biotechnological process described by the general state space model (12.10) is considered:

$$\frac{d\xi}{dt} = \mathbf{K}r(\xi) - D\xi - Q + F \tag{12.122}$$

It is assumed that:

a) the matrix $\mathbf{K}$ of yield coefficients is known,

b) the dilution rate D, the feed rates F and the gaseous outflow rates Q are measured on-line,

c) the vector of state variables ξ is known either by measurement or by estimation using an asymptotic observer (as described in Section 12.4.1).

It is further assumed that the vector $r(\xi)$ of reaction rates is partially unknown and written as follows:

$$r(\xi) = \begin{bmatrix} \mathbf{H}(\xi)\rho(\xi) \\ h(\xi) \end{bmatrix} \tag{12.123}$$

where $\mathbf{H}(\xi)$ is a diagonal matrix of known functions of the state and $\rho(\xi)$ a vector of unknown functions of ξ with $\dim \rho(\xi) = n_u$. The known reaction rates are given by vector $h(\xi)$ with $\dim h(\xi) = M - n_u$.

Using Equation (12.123), the general dynamical model is rewritten as:

$$\frac{d\xi}{dt} = \mathbf{K_u H}(\xi)\rho(\xi) + \mathbf{K_k} h(\xi) - D\xi - Q + F \tag{12.124}$$

where $\mathbf{K_u}$ and $\mathbf{K_k}$ are matrices of yield coefficients associated with the unknown and known reaction rates respectively.

12.5.2 Observer-based estimator

A state observer is used to provide on-line information for updating the estimate of $\rho(\xi)$. The estimation algorithm is written as follows:

$$\frac{d\hat{\xi}}{dt} = \mathbf{K_u H}(\xi)\hat{\rho} + \mathbf{K_k} h(\xi) - D\xi - Q + F - \mathbf{\Omega}(\xi - \hat{\xi}) \tag{12.125}$$

$$\frac{d\hat{\rho}}{dt} = [\mathbf{K_u H}(\xi)]^T \mathbf{\Gamma}(\xi - \hat{\xi}) \tag{12.126}$$

The update of the parameter vector $\hat{\rho}$ is driven by the deviation $(\xi - \hat{\xi})$ which reflects the mismatch between $\hat{\rho}$ and ρ.

The matrices $\mathbf{\Omega}$ and $\mathbf{\Gamma}$ are tuning parameters for adjusting the rate of convergence of the algorithm. A common choice is:

$$\mathbf{\Omega} = diag(-\omega_i), \quad \mathbf{\Gamma} = diag(-\gamma_j), \quad \omega_i, \gamma_j > 0 \tag{12.127}$$

With this choice, the stability condition of definite negativeness of the matrix $\mathbf{\Omega^T \Gamma} + \mathbf{\Gamma \Omega}$ is automatically satisfied.

The tuning procedure may be simplified if the state equations are first decoupled using the following transformation:

$$\Psi = \mathbf{K_u^{-1}}\xi \tag{12.128}$$

Applying the estimation algorithm to the transformed state equations yields:

$$\frac{d\hat{\Psi}}{dt} = \mathbf{H}\hat{\rho} + \mathbf{K_u^{-1} K_k} h - D\Psi + \mathbf{K_u^{-1}}(F - Q) - \mathbf{\Omega}(\Psi - \hat{\Psi}) \tag{12.129}$$

$$\frac{d\hat{\rho}}{dt} = \mathbf{H\Gamma}(\Psi - \hat{\Psi}) \tag{12.130}$$

12.5.3 Application to the baker's yeast fed-batch process

12.5.3.1 Process model

A modified version of the process model given in Section 12.2.4 is considered. From a global point of view, the baker's yeast fed-batch process can only be in an ethanol production regime or in an ethanol consumption regime. The process model is divided in two partial models to represent the two regimes. The first partial model corresponds to the ethanol production regime and is denoted as the respiro-fermentative (RF) partial model:

$$\frac{d\xi}{dt} = \mathbf{K}_{RF} r_{RF} - D\xi - Q + F \tag{12.131}$$

where

$$\mathbf{K}_{RF} = \begin{bmatrix} -k_1 & -k_2 & -k_{10} \\ -k_3 & 0 & -1 \\ 1 & 1 & 0 \\ k_5 & k_6 & k_{11} \\ 0 & k_8 & 0 \end{bmatrix}, \quad r_{RF} = \begin{bmatrix} \mu_o X \\ \mu_r X \\ q_m X \end{bmatrix} \tag{12.132}$$

The second partial model, the respirative (R) partial model, corresponds to the ethanol consumption regime where oxidation of both glucose and ethanol may occur. The mass balance for this regime is written as:

$$\frac{d\xi}{dt} = \mathbf{K}_R r_R - D\xi - Q + F \tag{12.133}$$

where

$$\mathbf{K}_R = \begin{bmatrix} -k_1 & 0 & -k_{10} \\ -k_3 & -k_4 & -1 \\ 1 & 1 & 0 \\ k_5 & k_7 & k_{11} \\ 0 & -k_9 & 0 \end{bmatrix}, \quad r_R = \begin{bmatrix} \mu_o X \\ \mu_e X \\ q_m X \end{bmatrix} \tag{12.134}$$

12.5.3.2 Specific growth rate estimation

The first step in the estimation procedure is to identify the measured variables. The concentration of ethanol, dissolved oxygen, and carbon dioxide are available for measurement. However, it was shown in [46] that those three measurements are not linearly independent. The two independent measured state variables selected are dissolved oxygen and carbon dioxide concentration. Two specific growth rate estimation algorithms are needed, one for the estimation of $[\mu_o\ \mu_r]$ in the ethanol production regime and one for the estimation of $[\mu_o\ \mu_e]$ in the ethanol consumption regime. The maintenance coefficient q_m is assumed to be known [46]. Only the derivation for the ethanol production partial model is given below, the procedure being identical for the other partial model. The mass balance equations for the measured state variables are:

$$\frac{dC}{dt} = -DC - k_3\mu_o X + q_m X + \Delta Q_{O2} \tag{12.135}$$

$$\frac{dP}{dt} = -DP + k_5\mu_o X + k_6\mu_r X + k_{11}X - Q_1 \tag{12.136}$$

The biomass concentration is unknown and will be estimated by an asymptotic observer as described later. With reference to Equation (12.124), the following vector and matrices can be defined:

$$\mathbf{K_u} = \begin{bmatrix} -k_3 & 0 \\ k_5 & k_6 \end{bmatrix}, \mathbf{H} = \begin{bmatrix} \hat{X} & 0 \\ 0 & \hat{X} \end{bmatrix}, \mathbf{K_k} = \begin{bmatrix} -1 \\ k_{11} \end{bmatrix}$$

$$\xi_1 = \begin{bmatrix} C \\ P \end{bmatrix}, \rho = \begin{bmatrix} \mu_o \\ \mu_r \end{bmatrix}, h = [q_m \hat{X}]$$

A linear transformation is applied to the measured state variables set (ξ_1) to decouple the equations in terms of the specific growth rates:

$$\Psi = \mathbf{K_u}^{-1}\xi_1 \tag{12.137}$$

The matrices of tuning parameters $\mathbf{\Omega}$ and $\mathbf{\Gamma}$ in Equations (12.129) and (12.130) are chosen as follows:

$$\mathbf{\Omega} = -\mathbf{H}\mathbf{C_1} \quad \mathbf{\Gamma} = \mathbf{C_2} \tag{12.138}$$

where $\mathbf{C_1}$ and $\mathbf{C_2}$ are diagonal matrices. The elements of these two matrices are chosen to ensure constant dynamics of the estimation error throughout the fermentation. With $\mathbf{C_1}$ and $\mathbf{C_2}$ chosen as:

$$\mathbf{C_1} = \begin{bmatrix} C_{11} & 0 \\ 0 & C_{11} \end{bmatrix} = \begin{bmatrix} 2\alpha/\hat{X} & 0 \\ 0 & 2\alpha/\hat{X} \end{bmatrix}$$
$$\mathbf{C_2} = \begin{bmatrix} C_{11}^2/4 & 0 \\ 0 & C_{11}^2/4 \end{bmatrix} \tag{12.139}$$

the poles of the error dynamics are all located at $-\alpha$. A single tuning parameter (α) is thus needed.

The non-measured biomass concentration appears in the equations of the estimation algorithm. A biomass concentration observer is thus needed for the application of the estimation algorithm. First, the process model is partitioned in two subsets. The first subset includes the equations associated to the measured state variables ($\xi_1 = [C \ P]^T$), while the second subset is associated to the non-measured state variable ($\xi_2 = X$):

$$\frac{d\xi_1}{dt} = -D\xi_1 + \mathbf{K_u}\mathbf{H}\rho + \mathbf{K_k}h + F_1 - Q_1 \tag{12.140}$$

$$\frac{d\xi_2}{dt} = -D\xi_2 + \mathbf{K_2}\mathbf{H}\rho \tag{12.141}$$

Applying the following linear transformation between the unmeasured and measured state variables

$$Z = \xi_2 - \mathbf{K_2}\mathbf{K_u^{-1}}\xi_1 \tag{12.142}$$

we obtain:

$$\begin{aligned} \frac{dZ}{dt} = & -(D + \mathbf{K_2}\mathbf{K_u^{-1}}\mathbf{K_k})Z \\ & -\mathbf{K_2}\mathbf{K_u^{-1}}(F_1 - Q_1 + \mathbf{K_k}\mathbf{K_2}\mathbf{K_u^{-1}}\xi_1) \end{aligned} \tag{12.143}$$

By substituting $\hat{Z}$ for Z, the biomass concentration observer is then:

$$\begin{aligned} \frac{d\hat{Z}}{dt} = & -(D + \mathbf{K_2}\mathbf{K_u^{-1}}\mathbf{K_k})\hat{Z} \\ & -\mathbf{K_2}\mathbf{K_u^{-1}}(F_1 - Q_1 + \mathbf{K_k}\mathbf{K_2}\mathbf{K_u^{-1}}\xi_1) \end{aligned} \tag{12.144}$$

$$\hat{X} = \hat{Z} + \mathbf{K_2}\mathbf{K_u^{-1}}\xi_1 \tag{12.145}$$

The procedure proposed so far is based on the use of the proper partial model algorithm set according to the process state. The problems that remain to be solved are the detection of the proper regime and the transition between the two estimation algorithms.

The first problem is resolved by looking at the values of the specific growth rate estimates. If the process is not in the corresponding regimestimate directly related to ethanol (μ_r in the respiro-fermentative model and μ_e in the respirative model) will have a negative value. The criterion for the transition between partial model algorithm sets is given by the transition between positive and negative values of these estimates.

The transition procedure between the two sets of algorithms is more complex. In each algorithm set, five variables have to be monitored in time: the transformed variable estimate in the biomass observer ($\hat{Z}$), the two transformed measured state estimates (Ψ) and the two specific growth rate estimates. It is important to monitor the time course of these variables also when the estimation algorithm does not correspond to the process regime to obtain appropriate values of these variables at transition time.

For the transformed variable in the biomass observer, the following technique is suggested. When the process regime changes, the transformed variable and the estimated biomass concentration issued from the other partial model algorithm set are used to provide a value of $\hat{Z}$ for the new partial model. Then the biomass observer algorithm of the new partial model is used. As an example, if the transition between respiro-fermentative model to respirative model occurs, then:

$$\hat{X}_{RF} = \hat{Z}_{RF} - \mathbf{K_2}\mathbf{K_u^{-1}}_{RF}\xi_{1RF} \qquad (12.146)$$

$$\hat{Z}_{R} = \hat{X}_{RF} + \mathbf{K_2}\mathbf{K_u^{-1}}_{R}\xi_{1R} \qquad (12.147)$$

For the specific growth rate estimates, distinct treatment is needed depending if the specific growth rate is related to ethanol or not. In the case of the estimation of μ_o, the time trajectory is still followed during the period when the partial model algorithm set is not used. Experiences have shown that the estimation error is small and that convergence following the transition is so rapid that a more sophisticated treatment is not required. For ethanol related specific growth rates, μ_e or μ_r, the estimated value is forced to zero when the partial model does not correspond to the process state.

For the transformed state variables (Ψ) of the specific growth rate estimator, the time trajectories have also to be followed during partial model and process mismatch, but the estimated value for biomass concentration used during this period is the one issued from the valid partial model observer. Also, the zero value of the ethanol related specific growth rate estimate is used in the prediction equations. This technique is required to avoid a too strong perturbation of the estimation algorithm at transition time.

12.5.3.3 Experimental verification

Experimental conditions

The experimental fermentations have been achieved in a 20 L BioEngineering fermentor. The yeast strain, *Saccharomyces cerevisiae* and the carbon source (a mix of cane and beat molasses) were supplied by an industrial baker's yeast producer. The operating conditions were chosen to reproduce the industrial fermentations.

The fermentations were carried out under ethanol concentration regulation with different non-linear adaptive control laws [45]. The ethanol regulation keeps the process near the boundary between the ethanol production and ethanol consumption state. Setpoint changes, agitation speed and aeration rate perturbations have been applied to test the estimation algorithm. These perturbations create a diversity of process conditions and some of these can be considered as extreme conditions which do not appear in normal operation. The dissolved oxygen concentration was not controlled and the agitation speed (700 rpm) and the aeration rate (2 L/(L·min)) were kept constant except at the time of perturbation. The pH was kept constant at 5.0. At this low pH there is no influence on the equilibrium of the different forms of dissolved carbon dioxide.

The ethanol concentration was measured in the exit gas with a Figaro sensor (TGS822) and calibrated as a function of the liquid ethanol concentration in the fermentor. Dissolved oxygen concentration was measured with an Ingold probe. Carbon dioxide concentration was presumed to be directly proportional to the carbon dioxide content of the exit gas. Oxygen transfer rate (ΔQ_{O2}) and carbon dioxide transfer rate (Q_1) were evaluated with off-gas analysis performed by a magnetic sector mass spectrometer (VG-MM8-80). The molasses feed rate was controlled by a variable speed peristaltic pump (Watson-Marlow 501U/R).

Results and Discussion

Figure 12.7a shows the biomass concentration estimates produced by the alternating use of the two sets of algorithms issued from the two partial models. The comparison of this time profile with the measured biomass concentration values from different samples is also shown. The precision of the biomass concentration estimate is within the precision of the measured values.

Figure 12.7b shows the estimated values of specific growth rate associated with the sugar oxidation (μ_o). The estimate of μ_o remains quite constant during the non-limiting oxygen transfer condition ($t \in [1,2]\ h$) and corresponds to the maximum specific growth rate achievable without ethanol production. At $t = 2\ h$ the ethanol setpoint was increased from 0.28 g/L to 1.88 g/L. The estimate of μ_r increased accordingly. The setpoint was brought back to 0.26 g/L at $t = 3.3$ h. The feed was decreased resulting in a decrease of μ_o and in an increase of μ_e as expected. The agitation speed was decreased from 700 rpm to 600 rpm at $t = 4.8\ h$. The process went into oxygen transfer limitation as confirmed by the sharp decrease in μ_o and in the specific oxygen uptake rate (not shown). In this case, the sugar uptake saturates the oxidation capacity of the yeast and the overflow of sugar is directed to the anaerobic pathway with ethanol production. When it is below, the sugar did not use all the oxidation capacity and allowed the ethanol oxidation.

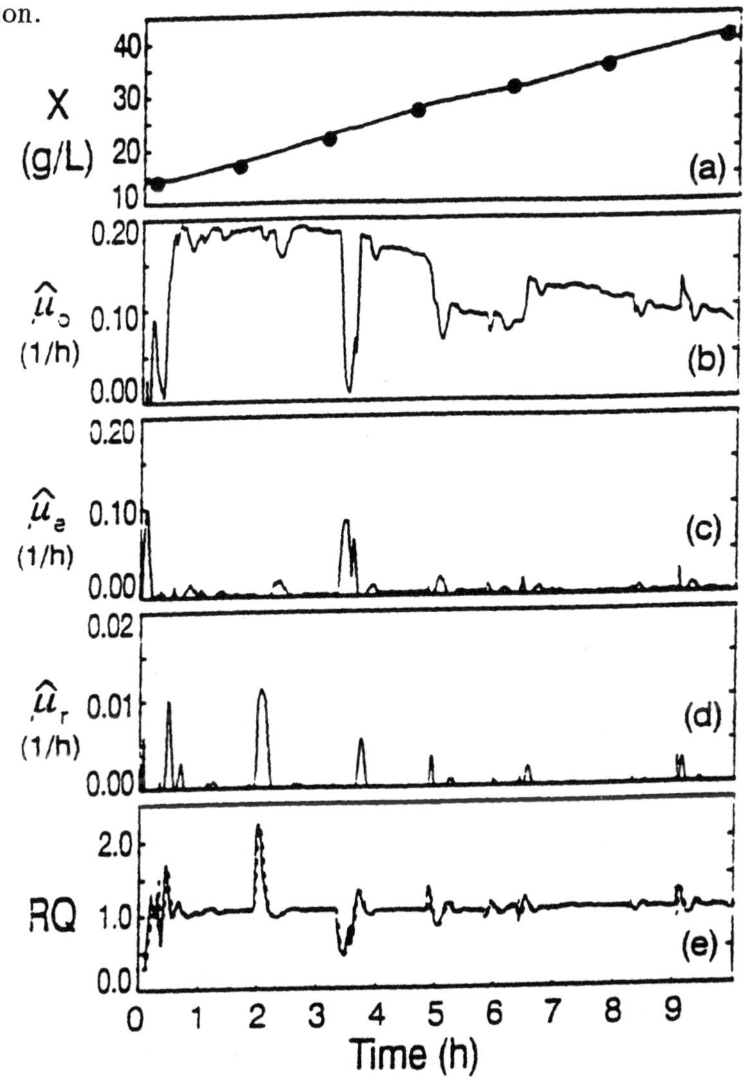

Figure 12.7: Estimation of biomass and specific growth rates for the baker's yeast example.

This is confirmed by the values of the estimates of the two other specific growth rates (μ_e and μ_r) presented in Figures 12.7c and 12.7d. This behavior corresponds quite well to the hypothesis of the limiting oxidation capacity of the yeast [49]. The agitation speed was returned to 700 rpm at $t = 6.4\ h$ resulting in an increase in μ_o. The air flow rate was decreased at $t = 8.2\ h$ and increased to its previous value at $t = 9.0\ h$. The estimate of μ_o decreased and increased accordingly.

The estimated values of the three specific growth rates show rapid fluctuations. To determine whether these fluctuations actually correspond to process fluctuations or are simply artefacts of the overall estimation procedure, a criterion to evaluate the accuracy of the three specific growth rates is proposed. The criterion is to compare the estimated respiratory quotient ($\widehat{RQ}$) obtained from the estimates of the three specific growth rates

$$\widehat{RQ} = \frac{k_5\hat{\mu}_o + k_6\hat{\mu}_r + k_7\hat{\mu}_e}{k_3\hat{\mu}_o + k_4\hat{\mu}_e} \tag{12.148}$$

to the value evaluated from experimental data. The values of each of the three specific growth rate estimates have to be accurate in order to produce a good estimate of the respiratory quotient. Figure 12.7e shows the comparison between the experimental and the estimated respiratory quotient. A respiratory quotient over 1.06 indicates an ethanol production state and below 1.06 an ethanol consumption state. The agreement between the two curves is very good in spite of the inherent lag of the estimation procedure. This comparison allows us to be confident in the proposed estimation procedure.

12.6 Adaptive Linearizing Control of Bioprocesses

12.6.1 Design of the adaptive linearizing controller

Now concentration is made on the design of model-based controllers for bioreactors. The key idea of the control design is here again to take advantage of what is well known about the dynamics of bioprocesses (reaction network and mass balances) which are summarized in the General Dynamical Model (12.10) while taking into account the model uncertainty (mainly the kinetics). Since the model is generally non-linear, the model-based control design will result in a *linearizing* control structure, in which the on-line estimation of the unknown variables (component concentrations) and parameters (reaction rates and yield coefficients) are incorporated. The resulting controller will be an *adaptive linearizing controller.* After the introduction of the general control design formulation, two typical adaptive linearizing control designs will be presented:

1. Control of anaerobic digestion processes;

2. Control of activated sludge processes.

One interesting aspect of the control design in the first example will be to show how to eliminate the (unknown) kinetics terms by incorporating the gas measurements into the controller. The first example is concerned with the SISO (Single Input Single Output) control case, while the second one will concentrate on the design of a MIMO (Multi-Input Multi-Output) adaptive linearizing controller. The design of the control algorithm is based on the General Dynamical Model (12.10) or on a reduced-order form of (12.10). In the following, two control design examples will be presented: one based on the dynamical model (12.10) (activated sludge process), and one based on a reduced-order version of (12.10) (anaerobic digestion). In each case, the control design follows the same line of reasoning described below. Let us first define the control problem.

The objective is to control the concentration of some reactant components (one in example #1, two in example #2) by acting on flow rates (the air flow rate and the recycle flow rate in example #2, the dilution rate in example #1) under the following practical constraints:

C1. the controlled components are assumed to be measured on-line;
C2. the concentrations of the other components (particularly of the biomass) are not available for on-line measurement;
C3. the kinetics $r(\xi)$ are unknown;
C4. most of the yield coefficients are unknown (only the ratio k_1/k_2 will be assumed to be known, in example #2);
C5. the mass transfer coefficients are known;
C6. the feed rates F, the dilution rate D and the gaseous outflow rates Q are known (either by user's choice or by measurement).

By defining y as the controlled component(s), the dynamics of y are simply the equation(s) of y in model (12.10) and can be rewritten as follows:

$$\frac{dy}{dt} = -Dy + K_y r + F_y - Q_y \qquad (12.149)$$

where the index y holds for the rows of K, F and Q corresponding to the controlled output y. By considering the above control problem and defining u the control input, the output equation (12.149) can be rewritten as follows:

$$\frac{dy}{dt} = f(F_y, Q_y, K_y r) + g(F_y, y)u \tag{12.150}$$

Assume now that we wish to have a linear stable closed loop (process + controller) dynamical behavior, i.e.:

$$\frac{dy}{dt} = C_1(y^* - y), \; C_1 > 0 \tag{12.151}$$

By combining Equations (12.150) and (12.151), the control law is readily obtained:

$$u = g(F_y, y)^{-1}[C_1(y^* - y) - f(F_y, Q_y, K_y r)] \tag{12.152}$$

Since the kinetics and most of the yield coefficients are assumed to be unknown, they are replaced by on-line estimates of selected parameters:

$$u = g(F_y, y)^{-1}[C_1(y^* - y) - f(F_y, Q_y, \hat{K}_y \hat{r})] \tag{12.153}$$

The above controller (12.153) is also known as the model reference adaptive linearizing control law (see, e.g., [3]). As the examples will show, the unknown parameters appear linearly in the equations and will therefore be estimated by using linear regression techniques (example #2) or via a Lyapunov design estimation approach (example #1). Moreover, the unknown components that may appear explicitly in the output equation (12.150) via the reaction rate r will be replaced either by an auxiliary variable easy to compute (example #2) or by gaseous outflow rates (example #1) (in the latter instance, the effect will also make disappear the kinetics terms).

Comment. Note that the adaptive linearizing controller contains all the important features of a PI controller: the proportional term $C_1(y^* - y)$ and the integral action via the parameter adaptation (see [44] for an explicit analysis). Moreover, the other terms can be viewed as feed forward action terms.

12.6.2 Example #1: anaerobic digestion

The importance of implementing efficient control systems for anaerobic digestion processes clearly appears from the following two points:

1. Anaerobic digestion is intrinsically a very unstable process: variations of the input variables (hydraulic flow rate, influent organic load) may easily lead the process to a wash-out, i.e., a state where the bacterial life has disappeared. This phenomenon takes place under the form of acid accumulation in the reactor (see [5][26]). It is therefore essential to implement controllers which are capable of stabilizing the process via a carefully designed control strategy.

2. If the process is used for wastewater treatment purposes, the control objective consists of maintaining the output pollution at a prescribed level despite the fluctuations of the input pollution (organic load).

From the above two comments, it is clear that control strategies should particularly concentrate on the control of the substrate concentration (which characterize the pollution level and the presence of acids). However, intricate difficulties inherent to the process make the control problem very hard to solve. Anaerobic digestion is a very complex process in which many different bacterial populations intervene. Its kinetics are basically non-linear and non-stationary, and they are far from being fully understood. Moreover, the concentrations of the different bacterial populations are not available from any direct measurement, even from off-line analyses.

Finally, there remains the problem of choosing an appropriate substrate to be controlled. COD (Chemical Oxygen Demand) is a priori a very good candidate: an adaptive linearizing controller has been designed, theoretically analyzed and experimentally validated on a pilot anaerobic digester (for further details, see [17][15][48]). But the industrial applicability of this control solution may appear to be limited by the need of on-line COD (or equivalent substrate concentration) measurements.

Therefore, there is a clear incentive to look for alternative substrate candidates. In this context, the use of hydrogen as a controlled variable appears to be very promising. As it has been recently emphasized [43], hydrogen plays an important role in the kinetics and stability of anaerobic digestion, particularly when the organic substrate is mainly composed of glucose (e.g., waste from the sugar industry). Finally, hydrogen is easy to measure on-line [43].

Let us now deal with the controller design. One specific aspect of the anaerobic digestion example is that the design of the controller is based on a *reduced-order* form of the general dynamical model (12.15) and (12.16). The order reduction has been done in Section 12.3.2.3 and resulted in Equation

(12.56) for the dissolved oxygen concentration:

$$\frac{dS_4}{dt} = -DS_4 - Q_1 - k_1 Q_3 + k_2 D S_{in} \tag{12.154}$$

Let us further motivate the use of a reduced order model. The design of a controller based on a reduced order model will result in a simpler controller structure but will still contain the important process features if the model reduction has been correctly performed (here the controller will incorporate the influent organic matter concentration S_{in} which can be viewed as a feed forward term and the methane gas production rate Q_3 which gives an important information about the state of the process). Recall that the order reduction is based on the notion of fast and slow reactions. It is difficult to know a priori which reactions in the anaerobic digestion reaction network will be, in a broad range of operating conditions, fast or slow. On the other hand, one may wish to avoid undesirable effects such as accumulation of propionate or hydrogen: then, in that situation, the corresponding reaction (reaction 2 or 4) can be considered as the limiting, i.e., *slow*, reaction, and the others as *fast* reactions. For instance, assume that the main problem is to avoid accumulation of hydrogen, then reaction 4 is the slow reaction and reactions 1, 2 and 3 are the fast reactions.

Let us go back to our control problem. If it is considered that the control input is the dilution rate D, the expressions of f and g are here equal to:

$$f = -Q_1 - k_1 Q_3, \; g = k_2 S_{in} - S_4 \tag{12.155}$$

and the controller (12.153) is written here as follows:

$$D = \frac{C_1(S_4^* - S_4) + Q_1 + \hat{k}_1 Q_3}{\hat{k}_2 S_{in} - S_4} \tag{12.156}$$

The unknown parameters are now k_1 and k_2. They can be estimated on-line by using an appropriate updating, for instance here a *Lyapunov design* estimation equation:

$$\frac{d\hat{k}_i}{dt} = C_{2i}(S_4^* - S_4), \qquad C_{2i} > 0, \qquad i = 1, 2 \tag{12.157}$$

A typical simulation is shown in Figure 12.8 (see also [44][21][22]). In this simulation, the hydrogen gaseous outflow rate Q_1 has been assumed to be negligible, and the parameter k_2 has been assumed to be known. Therefore, only k_1 has been estimated on-line.

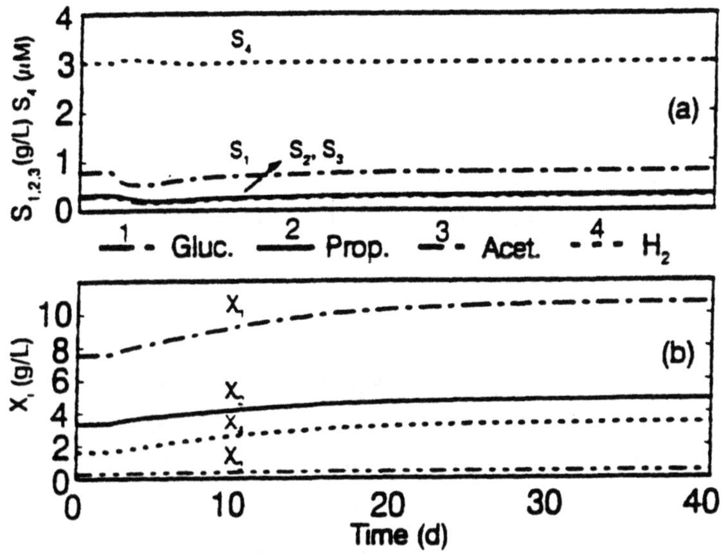

Figure 12.8: Control of the hydrogen concentration in an anaerobic digestion reactor.

The anaerobic process has been simulated by integration of the mass balance equations with the following yield coefficients, specific growth rate expressions, and initial conditions:

Yield coefficients

$$\begin{array}{llllll} k_1 = 3.2 & k_2 = 1.5 & k_3 = 0.7 & k_4 = 12 & k_5 = 0.27 & k_6 = 0.53 \\ k_7 = 1.15 & k_8 = 0.6 & k_9 = 0.1 & k_{14} = 0.3 & k_{15} = 0.08 & \end{array} \quad (12.158)$$

Specific growth rate expressions

$$\mu_1 = \frac{\mu_{max1} S_1}{1 + (S_2 + S_3)/K_{I_1}} \quad (12.159)$$

$$\mu_2 = \frac{\mu_{max2} S_2}{K_{S_2}(1 + S_3/K_{I_2})} \quad (12.160)$$

$$\mu_3 = \frac{\mu_{max3} S_3}{K_{S_3} + S_3 + S_3^2/K_{I_3}} \qquad (12.161)$$

$$\mu_4 = \frac{\mu_{max4} S_4}{K_{S_4} + S_4 + S_4/K_{I_4}} \qquad (12.162)$$

with:

$$\begin{array}{llllll} \mu_{max_1} = 0.2 & K_{I_1} = 10 & & \\ \mu_{max_2} = 0.5 & K_{S_2} = 0.4 & K_{I_2} = 0.5 \\ \mu_{max_3} = 0.4 & K_{S_3} = 0.5 & K_{I_3} = 4 \\ \mu_{max_4} = 0.5 & K_{S_2} = 4 & K_{I_4} = 3 \end{array} \qquad (12.163)$$

Initial steady state conditions

$$\begin{array}{ll} S_1(0) = 0.79\ g/L & S_2(0) = 0.28\ g/L \\ S_3(0) = 0.31\ g/L & S_4(0) = 3.0\ \mu M \\ X_1(0) = 7.6\ g/L & X_2(0) = 3.3\ g/L \\ X_3(0) = 0.29\ g/L & X_4(0) = 1.6\ g/L \end{array} \qquad (12.164)$$

The performance of the controller was evaluated for an influent substrate concentration (S_{in1}) change from 25 to 35 g/L applied at $t = 2$ days. The value of the design parameter C_1 was chosen to be equal to 1 1/day. The parameter estimation dynamics are imposed by C_2. As for the on-line estimation of the specific growth rates (Section 12.5), in order to have a closed loop dynamics independent of the (time-varying) value of the regressor Q_3, the design parameter C_2 was chosen to be inversely proportional to the regressor, i.e.:

$$C_2 = C_1/(4Q_3^2) \qquad (12.165)$$

The concentration of each substrate and bacterial population is shown in Figures 12.8a and 12.8b, respectively, for dissolved hydrogen control. The controller proves to be able to maintain the hydrogen concentration at the desired setpoint. The concentration of bacterial populations slowly reaches new levels while the other substrates go back to their initial values.

Remark: a similar approach has been used to design an adaptive linearizing controller for propionate as well as for COD (as already mentioned). It is worth noting that the three control designs result in similar controller structures [44][21]. The controllers of propionate and of COD have been successfully applied to a pilot-scale anaerobic digestion reactor [48][47].

12.6.3 Example #2: activated sludge process

The following example is treated in details in [21]. The control problem considered now can be formulated as follows. Like in any aerobic fermentation process, proper aeration is crucial to process efficiency, and adequate control of the dissolved oxygen concentration in the aerator is very important. But it is also important to limit load variations and substrate concentration variations by acting on them. As suggested by many authors (e.g., [39][31]), load variations can be expected to be smoothed by using the recycle flow rate as a control input. The above control algorithm will be considered under the following conditions:

1. The controlled outputs are the effluent BOD (Biological Oxygen Demand) and the dissolved oxygen concentrations, which are assumed to be measured on-line (see [34] for an on-line measuring device of the BOD in activated sludge processes).

2. The control inputs are the recycle flow rate and the air flow rate.

Therefore the input u and the output y are defined as follows:

$$u = \begin{pmatrix} F_R \\ W \end{pmatrix}, \quad y = \begin{pmatrix} S \\ C \end{pmatrix} \tag{12.166}$$

and the control problem is multi-variable (2 inputs, 2 outputs). The expressions of f and g in (12.150) are readily obtained from model (12.26):

$$f = \begin{pmatrix} -k_1\mu X + D_{in}(S_{in} - S) \\ -k_2\mu X - D_{in}C \end{pmatrix}, \quad g = \begin{pmatrix} -\frac{S}{V} & 0 \\ -\frac{C}{V} & a_0(C_S - C) \end{pmatrix} \tag{12.167}$$

g is invertible (i.e., g^{-1} exists) as long as S and $C_S - C$ are different from zero. Under these physically realistic conditions, the linearizing controller (12.153) can be applied. The remaining problem is how to deal with the *unknown* parameters k_1, k_2 and μ, and the variable X. The solution proceeds as follows. Let us first define the auxiliary variables ζ (as introduced in Section 12.3.1):

$$z = \begin{pmatrix} \zeta_1 \\ \zeta_2 \\ \zeta_3 \end{pmatrix} = \begin{pmatrix} k_1 \\ k_2 \\ 0 \end{pmatrix} X + \begin{pmatrix} 1 & 0 & 0 \\ 0 & 1 & 0 \\ 0 & 0 & k_2 \end{pmatrix} \begin{pmatrix} S \\ C \\ X_R \end{pmatrix} \tag{12.168}$$

The dynamical equation of ζ is derived from model (12.26) and definitions (12.168):

$$\frac{d}{dt}\begin{pmatrix}\zeta_1\\ \zeta_2\\ \zeta_3\end{pmatrix} = \begin{pmatrix}-D_1 & 0 & D_2 k_1/k_2\\ 0 & -D_1 & D_2\\ 0 & D_3 & -D_4\end{pmatrix}\begin{pmatrix}\zeta_1\\ \zeta_2\\ \zeta_3\end{pmatrix} + \begin{pmatrix}D_{in}S_{in}\\ \Delta Q_{O_2}\\ -D_3 C\end{pmatrix} \tag{12.169}$$

with D_i (i = 1 to 4) as defined in Section 12.2.5. One important feature of the above dynamics (12.169) is that these are independent of the (unknown) kinetics. Equation (12.169) is a dynamical system linear-in-the-state-ζ. It can be checked that its state matrix is asymptotically stable if $F_{in} - F_R > 0$ and $F_W > 0$. Therefore Equation (12.169) can be used to compute the value of ζ on-line from the knowledge of the flow rates F_{in}, F_R and F_W, the feed rates $F_{in}D_{in}$ and ΔQ_{O_2}, the dissolved oxygen concentration C and the ratio k_1/k_2. Let us now rewrite the specific growth rate μ as follows:

$$\mu = \alpha S C \tag{12.170}$$

where α is an (unknown) positive function of the process components: Equation (12.170) simply implies that there is no growth in absence of one of the limiting substrate. By introducing (12.168) and (12.170), the function f in (12.167) becomes:

$$f = \begin{pmatrix} -\alpha SC(\zeta_1 - S) + D_{in}(S_{in} - S)\\ -\alpha SC(z_2 - C) - D_{in}C \end{pmatrix} \tag{12.171}$$

This expression for f is used in the computation of the control law (12.153):

$$F_R = -\frac{V}{S}[C_{11}(S^* - S) - D_{in}(S_{in} - S) + \hat{\alpha}_1 SC(\zeta_1 - S)] \tag{12.172}$$

$$\begin{aligned} W &= \frac{1}{a_0(C_S - C)}[C_{12}(C^* - C) - \frac{C_{11}(S^* - S)C}{S} \\ &\quad + \frac{D_{in}S_{in}C}{S} + \hat{\alpha}_2 C(S\zeta_2 - C\zeta_1)] \end{aligned} \tag{12.173}$$

C_{11} and C_{12} (> 0) are the control design parameters (a decoupling diagonal matrix C_1 has been considered here). The variables ζ_1 and ζ_2 are computed via the use of Equations (12.169) (which only require the knowledge of the ratio k_1/k_2) and the parameter α is estimated on-line by using, e.g., a recursive least square (RLS) algorithm. In discrete-time, if α is estimated via the substrate concentration equation S (α_1, to be used in Equation (12.172)) or via the dissolved oxygen concentration C (α_2, to be used in Equation (12.173)), the RLS algorithm is written as follows:

$$\begin{aligned} \hat{\theta}_{i,t+1} &= \hat{\theta}_{i,t} + g_{i,t}\phi_{i,t}e_{i,t+1} & i = 1,2 \qquad (12.174)\\ g_{i,t} &= \frac{g_{i,t-1}}{\gamma_i + g_{i,t-1}\phi_{i,t}^2} & 0 < \gamma_i \leq 1 \qquad (12.175) \end{aligned}$$

with:

$$\begin{aligned} \hat{\theta}_{1,t} &= \hat{\alpha}_{1,t},\ \phi_{1,t} = TS_tC_t(S_t - z_{1,t}) & (12.176)\\ e_{1,t+1} &= S_{t+1} - S_t + TD_{1,t}S_t - TD_{in,t}S_{in,t} - \hat{\alpha}_{1,t}\phi_{1,t} & (12.177)\\ \hat{\theta}_{2,t} &= \hat{\alpha}_{2,t},\ \phi_{2,t} = TS_tC_t(C_t - z_{2,t}) & (12.178)\\ e_{2,t+1} &= C_{t+1} - C_t + TD_{1,t}C_t - Ta_0W_t(C_s - C_t) - \hat{\alpha}_{2,t}\phi_{2,t} & (12.179) \end{aligned}$$

and t the time index, and γ_i (i = 1, 2) a forgetting factor. Note that now because of the variables ζ, the biomass X does not appear explicitly anymore in the controller (12.172) and (12.173).

A typical simulation result is shown in Figure 12.9 (see also [21]). The activated sludge process has been simulated by numerical integration of the basic dynamical model equations (12.26) (using a 4th order Runge-Kutta method) with the following (Monod type) model for the specific growth rate:

$$\mu = \mu_{max}\frac{S}{K_S + S}\frac{C}{K_C + C} \qquad (12.180)$$

and a decay rate ($-k_dX$) which has been added to the biomass equation (in order to simulate biomass mortality). The specific growth rate model and the decay rate term are (obviously) completely ignored by the control algorithm. The decay rate term can be viewed as an unmeasured disturbance. The model parameters have been set to the following values (inspired from literature data, in particular from [31][40]):

$$k_1 = 1.2, \quad k_2 = 0.565, \quad \mu_{max} = 0.2\ 1/\text{h}, \quad K_S = 75\ \text{mg/L}, \quad K_C = 2\ \text{mg/L}$$

$$a_0 = 0.018\ \text{m}^{-3},\ C_S = 10\ \text{mg/L},\ V = 100\ \text{m}^3,\ V_S = 50\ \text{m}^3,\ k_d = 0.001\ 1/\text{h}$$

In Figure 12.9, a square wave signal of the influent BOD concentration (disturbance input) S_{in} (from 150 mg/L to 200 mg/L) (in order to simulate the periodical variation of the pollutant load) has been applied over a period of 20 days (480 hours). The sampling period has been set to 3 minutes following the constraints of the commercially available BOD measuring device proposed by [34]. The following initial process conditions have been considered in the simulation:

$$S = 5 \text{ mg/L}, \quad C = 6 \text{ mg/L}, \quad X = 1225 \text{ mg/L}, \quad X_R = 2333 \text{ mg/L}$$

The controller parameters have been set to the following values:

$$S* = 5 \text{ mg/L}, \quad C* = 6 \text{ mg/L}, \quad C_{11} = 1 \text{ 1/h}, \quad C_{12} = 10 \text{ 1/h}$$

The auxiliary variables ζ, the unknown parameters and the tuning estimation variables have been initialized as follows:

$$\zeta_1 = 1400 \text{ mg/L}, \quad \zeta_2 = 750 \text{ mg/L}, \quad \zeta_3 = 1400 \text{ mg/L}, \quad g_{1,0} = g_{2,0} = 10^{-3}$$

$$\gamma_1 = \gamma_2 = 0.9, \quad \hat{\alpha}_{1,0} = \hat{\alpha}_{2,0} = 0.00025 \text{ L}^2/\text{mg}^2 \cdot \text{h}$$

Note the ability of the controller to maintain the controlled outputs S and C close to their desired values in spite of the unknown disturbance.

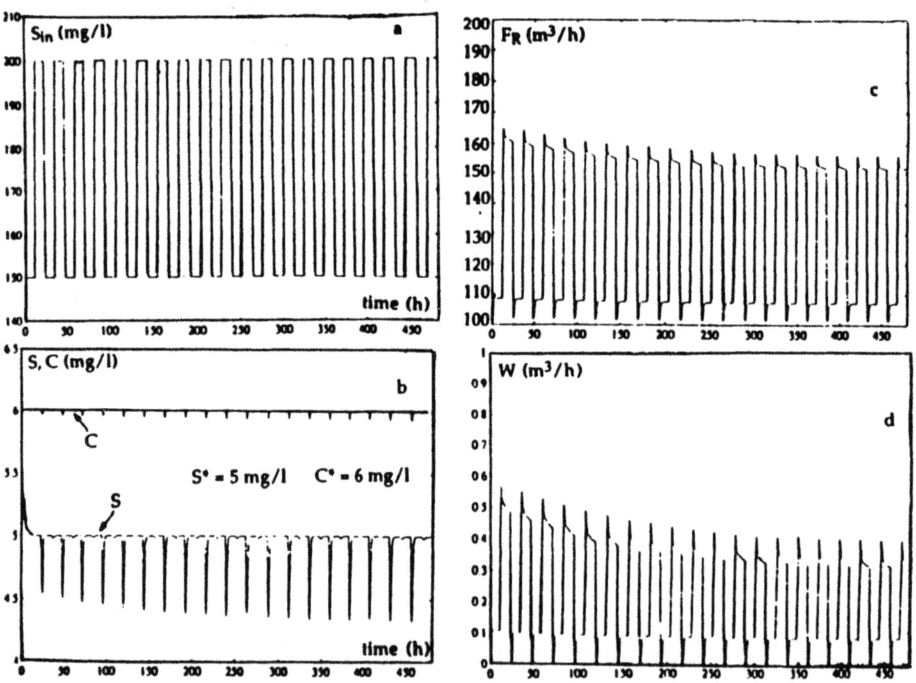

Figure 12.9: Control of an activated sludge process in presence of an unknown biomass mortality.

12.7 Conclusions

The objective of this chapter was to present a survey about recent approaches of model-based monitoring and control of bioreactors. The proposed results cover the whole range from theory (dynamical modeling, dynamical analysis, monitoring and control design) to practice (experimental results). They apply to different types of reactors, mainly stirred tank reactors and fixed bed reactors, but could also be extended to, e.g., fluidized bed reactors. The dynamical model of the bioprocesses is based on material balances and is formalized into a General Dynamical Model framework which serves as a basis for dynamical analysis and for monitoring and control algorithm design. Monitoring was found to be a key question in bioprocess applications. The design of asymptotic observers for the concentrations of the process components has been developed, analyzed and illustrated. Due to the usually large uncertainty of some process parameters, namely, e.g., the process kinetics, on-line estimation of the uncertain parameters has been considered either for monitoring purposes or for being included in an adaptive model-based control scheme. In the latter case, the incorporation of *physical* parameters presents the double advantage of introducing an integral action in the controller while giving extra information about the process behavior and performance.

Acknowledgments

This chapter presents results of the Belgian Programme on Interuniversity Poles of Attraction initiated by the Belgian State, Prime Minister's Office, Science Policy Programming. The scientific responsibility rests with its authors.

12.8 Bibliography

1. J. Alvarez and J. Alvarez 1988. Analysis and control of fermentation processes by optimal and geometric methods. *Proc. ACC*, 2, 1112-1117
2. H. Aoufoussi, M. Perrier, J. Chaouki, C. Chavarie and D. Dochain 1992. Feedback linearizing control of a fluidized-bed reactor. *Can. J. Chem. Eng.*, 70, 356-367
3. G. Bastin and D. Dochain 1990. *On-line Estimation and Adaptive Control of Bioreactors.* Elsevier, Amsterdam
4. G. Bastin and J. Levine 1990. On state reachability of reaction systems. *Proc. 29th CDC*, 2819-2824

5. R. Binot, T. Bol, H. Naveau and E. Nyns, 1983. Biomethanation by immobilized fluidised cells. *Wat. Sci. Tech.*, 15, 103-115
6. B. Bouaziz and D. Dochain 1993. Control analysis of fixed bed reactors: a singular perturbation approach. *Proc. ECC'93*, 1741-1745
7. G. Caminal, F. Lafuente, J. Lopez-Santin, M. Poch and C. Sola 1987. Application of the extended Kalman filter to identification of enzymatic deactivation. *Biotechnol. Bioeng.*, 24, 366-369
8. L. Chen 1992. *Modelling, Identifiability and Control of Complex Biotechnological Processes.* PhD thesis, Université Catholique de Louvain, Belgium
9. L. Chen, G. Bastin and V. Van Breusegem 1991. Adaptive nonlinear regulation of fed-batch biological reactors: an industrial application. *Proc. 30th IEEE CDC*, 2130-2135
10. M. Chidambaram and Y. Malleswararao 1991. Nonlinear control of a mixed-culture fed-batch bioreactor. *Ind. Chem. Eng.*, 33(2), 53-57
11. V. Chotteau and G. Bastin 1992. Identification of a reaction mechanism for a class of animal cell cultures. *Proc. ICCAFT 5/IFAC-BIO 2*, 215-218
12. B. Dahhou, J. Bordeneuve and J. Babary 1991. Multivariable long-range predictive control algorithm applied to a continuous flow fermentation process. *Proc. IFAC World Congress*, 393-397
13. M. DeTremblay and M. Perrier 1992. Optimisation of fed-batch culture of hybridoma cells using dynamic programming: single and multi-feed cases. *Bioprocess Eng.*, 7, 229-234
14. M. DeTremblay, M. Perrier, C. Chavarie and J. Archambault 1993. Fed-batch culture of hybridoma cells: comparison of optimal control approach and closed-loop strategies. *Bioprocess Eng.*, 9, 13-21
15. D. Dochain 1986. *On-line Parameter Estimation, Adaptive State Estimation and Adaptive Control of Fermentation Processes.* PhD thesis, Université Catholique de Louvain, Belgium
16. D. Dochain, J. Babary and M. Tali-Maamar 1992. Modelling and adaptive control of nonlinear distributed parameter bioreactors via orthogonal collocation. *Automatica*, 68, 873-883
17. D. Dochain and G. Bastin, 1984. Adaptive identification and control algorithms for nonlinear bacterial growth systems. *Automatica*, 20, 621-634
18. D. Dochain and G. Bastin 1987. Convention de recherche Solvay - UCL, rapport final. *Technical report*, UCL, Belgium
19. D. Dochain and B. Bouaziz 1993. Approximation of the dynamical model of fixed bed reactors via a singular perturbation approach. *Proc. IMACS Int. Symp. MIM-S2'93*, 34-39
20. D. Dochain and L. Chen 1992. Local observability and controllability of stirred tank reactors. *J. Process Control*, 2(3), 139-144
21. D. Dochain and M. Perrier 1992. Adaptive linearizing control of activated sludge processes. *Proc. Control Systems'92*, 211-215

22. D. Dochain, M. Perrier and A. Pauss 1991. Adaptive control of the hydrogen concentration in anaerobic digestion. *Ind. Eng. Chem. Res.*, 30, 129-136
23. D. Dochain, N. Tali-Maamar and J. Babary 1994. Design of adaptive linearizing controllers for fixed bed reactors. *Proc. ACC.*
24. M. Fjeld, O. Asbjornsen and K. Astrom 1974. Reaction invariants and their importance in the analysis of eigenvectors, state observability and controllability of the continuous stirred tank reactor. *Chem. Eng. Sci.*, 29, 1917-1926
25. J. Flaus, A. Cheruy and J. Engasser 1991. An adaptive controller for batch feed bioprocess. applicaton to lysine production. *J. Proc. Cont.*, 1, 271-281
26. J. Fripiat, T. Bol, R. Binot, H. Naveau and E. Nyns 1984. A strategy for the evaluation of methane production from different types of substrate biomass. In: R. Buvet, M.F. Fox and D.J. Picker (Eds.), *Biomethane, Production and uses*, 95-105, Roger Bowskill Ltd, Exeter (UK)
27. G. Gavalas 1968. *Nonlinear Differential Equations of Chemically Reacting Systems.* Springer Verlag, Berlin
28. M. Golden, B. Pangrie and B. Ydstie 1986. Nonlinear adaptive optimization of a continuous bioreactor. *Proc. AIChE 1986 National Meeting*, Pap.125b
29. R. Hamalainen, A. Halme and A. Gyllenberg 1975. A control model for activated sludge wastewater treatment process. *Proc. 6th IFAC World Congress, Boston, Paper 61:6*
30. M. Henson and D. Seborg 1992. Nonlinear control strategies for continuous fermenters. *Chem. Eng. Sci.*, 47(4), 821-835
31. A. Holmberg 1983. A microprocessor-based estimation and control system for the activated sludge process. In: A. Halme (Ed.), *Modelling and Control of Biotechnical Processes*, 111-120, Pergamon
32. A. Holmberg and J. Ranta 1982. Procedures for parameter and state estimation of microbial growth process models. *Automatica*, 18, 181-193
33. K. Hoo and J. Kantor 1986. Linear feedback equivalence and control of an unstable biological reactor. *Chem. Eng. Comm.*, 46, 385-399
34. M. Köhne 1985. Practical experiences with a new on-line BOD measuring device. *Env. Technol. Letters*, 6, 546-555
35. H. Kwakernaak and R. Sivan 1972. *Linear Optimal Control Systems.* John Wiley, New York
36. P. Tsobanakis, S. Lee, J. Phillips and C. Georgakis 1992. Issues in the optimization, estimation and control of fed-batch bioreactors using tendency models. *Proc. ICCAFT 5/IFAC-BIO 2*, 71-76
37. L. Ljung 1979. Asymptotic behavior of the extended Kalman filter as a parameter estimator for linear systems. *IEEE Trans. Aut. Cont.*, 24, 36-50
38. R. Marino 1990. Adaptive observers for single output nonlinear systems. *IEEE Trans. Aut. Cont.*, 35, 1054-1058
39. S. Marsili-Libelli 1984. Optimal control of the activated sludge process. *Trans. Inst. Meas. Control*, 6, 146-152

40. S. Marsili-Libelli 1989. Modelling, identification and control of the activated sludge process. *Adv. Biochem. Eng. Biotechnol.*, 38, 90-148
41. F. Mosey 1983. Mathematical modelling of the anaerobic digestion process: regulatory mechanisms for the formation of short-chain volatile acids from glucose. *Wat. Sci. Technol.*, 15, 209-232
42. K. Narendra and A. Annaswamy 1989. *Stable Adaptive Systems.* Prentice-Hall, Englewood Cliffs, New Jersey
43. A. Pauss, C. Beauchemin, R. Samson and S. Guiot 1990. Continuous measurement of dissolved H_2 in an anaerobic reactor using a commercial probe hydrogen/air fuel cell-based. *Biotechnol. Bioeng.*, 35, 492-501
44. M. Perrier and D. Dochain 1993. Evaluation of control strategies for anaerobic digestion processes. *Int. J. Adaptive Cont. Signal Proc.*, 7(4), 309-321
45. Y. Pomerleau 1990. *Modélisation et contrôle d'un procédé fed-batch de culture des levures à pain Saccharomyces cerevisiae.* PhD thesis, Ecole Polytechnique de Montréal, Canada
46. Y. Pomerleau and M. Perrier 1990. Estimation of multiple specific growth rates in bioprocesses. *AIChE J.*, 36(2), 207-215
47. P. Renard, V. Van Breusegem, N. Nguyen, H. Naveau and E. Nyns 1991. Implementation of an adaptive controller for the start-up and steady-state running of a biomethanation process operated in the CSTR mode. *Biotechnol. Bioeng.*, 38, 805-812
48. P. Renard, D. Dochain, G. Bastin, H. Naveau and E. Nyns 1988. Adaptive control of anaerobic digestion processes: a pilot-scale application. *Biotechnol. Bioeng.*, 31, 287-294
49. B. Sonnleitner and O. Kappeli 1986. Growth of *Saccharomyces cerevisiae* is controlled by its limited respiratory capacity: formulation and verification of an hypothesis. *Biotechnol. Bioeng.*, 28, 927-937
50. G. Stephanopoulos and K.-Y. San 1984. Studies on on-line bioreactor identification. *Biotechnol. Bioeng.*, 26, 1176-1188
51. V. Van Breusegem and G. Bastin 1992. Order reduction in bioprocess modelling: a singular perturbation solution. *Proc. ICCAFT 5/IFAC-BIO 2*, 347-350
52. J. Van Impe 1993. *Modelling and Optimal Adaptive Control of Biotechnological Processes.* PhD thesis, Katholieke Universiteit Leuven, Belgium, 304 p.
53. J. Van Impe, B. Nicolai, P. Vanrolleghem, J. Spriet, B. De Moor and J. Vandewalle 1992. Optimal control of the penicillin G fed-batch fermentation: an analysis of a modified unstructured model. *Chem. Eng. Comm.*, 117, 337-353
54. Y. Yoo, J. Hong and R. Hatch 1985. Sequential estimation of states and kinetic parameters and optimization of fermentation processes. *Proc. ACC*, 2, 866-871

Chapter 13

Optimal Adaptive Control of Fed-Batch Fermentation Processes

J.F. VAN IMPE
Katholieke Universiteit Leuven
Department of Food and Microbial Technology - BioTeC
Kardinaal Mercierlaan 92 B-3001 Leuven (BELGIUM)
email: jan.vanimpe@agr.kuleuven.ac.be

G. BASTIN
Université Catholique de Louvain
CESAME Bâtiment Euler
Av. G. Lemaître 4-6 B-1348 Louvain-La-Neuve (BELGIUM)
email: bastin@auto.ucl.ac.be

ABSTRACT – *This chapter presents a unifying methodology for optimization of biotechnological processes, namely optimal adaptive control, which combines the advantages of both the optimal control and the adaptive control approaches. As an example, the design of a substrate feeding rate controller for a class of biotechnological processes in stirred tank reactors characterized by a decoupling between biomass growth and product formation is considered. More specifically, the most common case is considered of a process with monotonic specific growth rate and non-monotonic specific*

J.F.M. Van Impe et al. (eds.), Advanced Instrumentation, Data Interpretation, and Control of Biotechnological Processes, 401-435.

production rate as functions of substrate concentration. The main contribution is to illustrate how the insight, obtained by preliminary optimal control studies, leads to the design of easy-to-implement adaptive controllers. The controllers derived in this way combine a nearly optimal performance with good robustness properties against modeling uncertainties and process disturbances. Since they can be considered model-independent, they may be very helpful also in solving the model discrimination problem, which often occurs during biotechnological process modeling. To illustrate the method and the results obtained, simulation results are given for the penicillin G fed-batch fermentation process.

KEYWORDS – *Fed-batch Fermentation Processes, Optimal Control, Adaptive Control, Linearizing Control*

13.1 Introduction

The design of high-performance model-based control algorithms for biotechnological processes is hampered by two major problems which call for adequate engineering solutions. First, the process kinetics are most often poorly understood non-linear functions, while the corresponding parameters are in general time-varying. Second, up till now there has been a lack of reliable sensors suited to real-time monitoring of the process variables which are needed in advanced control algorithms. Therefore, the earliest attempts at control of a biotechnological process used no model at all. Successful state trajectories from previous runs which had been stored in the process computer were tracked using open-loop control. Many industrial fermentations are still operated using this method.

During the last two decades, two trends for the design of monitoring and control algorithms for fermentation processes have emerged [4]. In a first approach, the difficulties in obtaining an accurate mathematical process model are ignored. In numerous papers classical methods (e.g., Kalman filtering, optimal control theory, ...) are applied under the assumption that the model is perfectly known. Due to this oversimplification, it is very unlikely that a real-life implementation of such controllers –very often this implementation is already hampered by, e.g., monitoring problems– would result in the predicted simulation results. In a second approach, the aim is to design specific monitoring and control algorithms without the need for a complete knowledge of the process model, using concepts from, e.g., adaptive control and non-linear linearizing control. A comprehensive treatment of these ideas can be found in the text-book by Bastin and Dochain [3] and the references therein.

This chapter shows how to combine the best of both trends into one unifying methodology for optimization of biotechnological processes: *optimal adaptive control.* This is motivated as follows. Model-based *optimal control* studies provide a theoretically realizable optimum. However, the real-life implementation will fail in the first place due to modeling uncertainties. On the other hand, model-independent *adaptive controllers* can be designed, but there is *a priori* no guarantee of the optimality of the results obtained. The gap between both approaches is bridged in two steps. First, heuristic control strategies are developed with nearly optimal performance under all conditions. These suboptimal controllers are based on biochemical knowledge concerning the process and on a careful mathematical analysis of the optimal control solution. In a second step, implementation of these profiles in an adaptive, model-independent way combines excellent robustness properties with nearly optimal performance. As an example, the design of a substrate feeding rate controller for a class of biotechnological processes in stirred tank reactors characterized by a decoupling between biomass growth and product formation is considered. More specifically, the most common case of a process with monotonic specific growth rate and non-monotonic specific production rate as functions of substrate concentration is investigated.

The chapter is organized as follows. In Section 13.2 the concept of *optimal adaptive control* is motivated in detail. Starting from the *optimal control* solution, the *heuristic substrate controllers* described in [13][15] are briefly reconsidered with respect to a real-life implementation. It is indicated that a straightforward implementation is not robust at all. However, as these controllers are the translation of a realistic control objective, namely setpoint control, they can serve as a basis for the development of more reliable, robust, model-independent adaptive control schemes. To do so, they are interpreted within the framework of non-linear linearizing control theory. In this way, a mechanism is incorporated that makes them both stable and robust against disturbances.

The next step is then to cope with the monitoring problem, i.e., how to determine on-line the non-measurable variables needed in the controller, in other words, how to make the non-linear linearizing controller *adaptive.* Sections 13.3, 13.4, and 13.5 present three possible solutions depending on which variables are on-line available. The use of *software sensors* is clearly illustrated. Both a linear regression estimator and a state-observer-based estimator can be used for on-line tracking of the unknown states and specific rates. According to the minimal modeling concept, these specific rates are considered as time-varying parameters. Furthermore, it is possible to take a measurement delay into account. The three solutions are compared with simulation results for the penicillin G fed-batch fermentation process.

A remarkable result is as follows. If there are no state inequality constraints, a substrate concentration reference level is specified only for the second phase, namely the production phase. However, the proposed controllers can be implemented without difficulty from the start of the fermentation on. The first phase, the growth phase, is then mainly used to obtain estimator convergence. Furthermore, the extension to a problem with constraints is straightforward.

Finally, Section 13.6 presents some conclusions. The main contribution of this chapter is to illustrate that the design of an adaptive controller for the considered class of fermentation processes can be successfully based on a preliminary optimal control study. For instance, a substrate concentration reference profile is designed which guarantees nearly optimal performance. As such, this chapter can be considered as an extension of the methods presented in Chapter 5 of the text-book by Bastin and Dochain [3], where the reference substrate concentration is kept constant throughout the whole fermentation process.

13.2 Optimal Adaptive Control: Motivation

13.2.1 Problem statement

This chapter considers the class of fed-batch fermentation processes described by an (unstructured) model of the form:

$$\begin{aligned} \frac{dS}{dt} &= -\sigma X + C_{S,in} u \\ \frac{dX}{dt} &= \mu X \\ \frac{dP}{dt} &= \pi X - kP \\ \frac{dV}{dt} &= u \end{aligned} \tag{13.1}$$

where the state variables S, X, P, and V are respectively the amount of the only limiting substrate in broth [g], the amount of cell mass in broth [g DW] (DW stands for dry weight), the amount of product in broth [g], and the volume of the liquid phase in the fermentor [L]. Dissolved oxygen is considered non-limiting, by maintaining a sufficiently high aeration level. The input u of the system is the volumetric substrate feed rate [L/h]. $C_{S,in}$ (expressed in [g/L]) is the (constant) substrate concentration in the feed stream u, while k [1/h] is the product hydrolysis or degradation constant. σ, μ, and π are respectively the (overall) specific substrate consumption rate [g/g DW h], the (overall) specific growth rate [1/h], and the specific production rate [g/g DW h]. These rates are interrelated by:

$$\sigma = \mu/Y_{X/S} + m + \pi/Y_{P/S} \tag{13.2}$$

with $Y_{X/S}$ the biomass on substrate yield coefficient [g DW/g], $Y_{P/S}$ the product on substrate yield coefficient [g/g], and m the (overall) specific maintenance demand [g/g DW h]. When introducing the concept of *endogenous fractions* [13][15] all kinds of metabolism for biomass survival and product synthesis can be easily described within one unifying frame. The *endogenous fraction* $f_m \in [0,1]$ of the overall specific maintenance demand m, and the *endogenous fraction* $f_P \in [0,1]$ for product synthesis, are defined through the following equations:

$$\mu \triangleq \mu_{substr} - Y_{X/S}(f_m m + f_P \frac{\pi}{Y_{P/S}}) \tag{13.3}$$

$$\sigma \triangleq \frac{\mu_{substr}}{Y_{X/S}} + (1 - f_m)m + (1 - f_P)\frac{\pi}{Y_{P/S}} \tag{13.4}$$

where μ_{substr} is the specific substrate-to-biomass conversion rate. Observe that Equation (13.2) holds true independent of the value of the fractions f_m and f_P. If $f_m = f_P = 1$, then *endogenous* metabolism is assumed: biomass survival and product synthesis are assumed due to combustion of part of the biomass. On the other hand, if $f_m = f_P = 0$, then *maintenance* metabolism is assumed: biomass survival and product synthesis are assumed due to consumption of part of the external substrate. If $f_m \in]0,1[$ and $f_P \in]0,1[$, then a mixed maintenance/endogenous metabolism can be modeled.

If the concentrations C_S, C_X, and C_P (defined as S/V, X/V, and P/V respectively) are used as state variables, the following equivalent model is obtained:

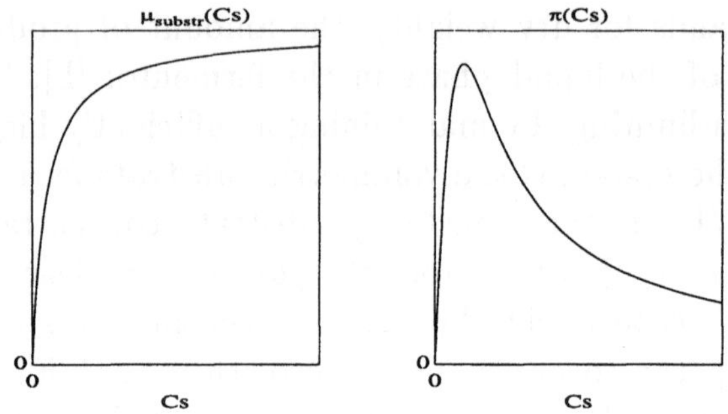

Figure 13.1: Process with monotonic μ_{substr} and non-monotonic π.

$$
\begin{aligned}
\frac{dC_S}{dt} &= -\sigma C_X - C_S\frac{u}{V} + C_{S,in}\frac{u}{V} \\
\frac{dC_X}{dt} &= \mu C_X - C_X\frac{u}{V} \\
\frac{dC_P}{dt} &= \pi C_X - C_P\frac{u}{V} - kC_P \\
\frac{dV}{dt} &= u
\end{aligned}
\tag{13.5}
$$

In this chapter, the *shape* of the specific rates μ_{substr} and π as functions of substrate concentration C_S is assumed as depicted in Figure 13.1: μ_{substr} is a monotonically increasing function, while π is non-monotonic exhibiting a maximum. In other words, the enzyme catalyzed production is *not associated* to the microbial growth. Observe that this model structure represents the most common case of fermentation processes with product formation and growth/production decoupling. Obviously, the specific rates μ_{substr} and π may be functions of other component concentrations as well.

Consider the following optimization problem. *Determine for the set of dynamic equations* (13.1) (or (13.5)) *the optimal volumetric substrate feed rate profile* $u^*(t)$ *which maximizes the final amount of product* $P(t_f)$, *subject to the following constraints:*

1. $t_0 = 0$, $t_f =$ *free*

2. $P(0) = P_0$ and $X(0) = X_0$ are given. $S(0) = S_0$ is *free*. $V(0) = V_0$ follows from:

$$V_0 = V_* + S_0/C_{S,in}$$

with V_* the initial volume without substrate. Remember that substrate is added as a solution with concentration $C_{S,in}$.

3. The total amount of substrate available for fermentation, denoted by α, is *fixed*. In other words, the final reactor volume is fixed to $V(t_f) = V_f$, with V_f given.

Observe that, from a mathematical point of view, the problem of determining both S_0 and $u^*(t)$ is equivalent to determining $u^*(t)$, $t \in [0, t_f]$ with a Dirac input at time $t = 0$.

13.2.2 Case study: the penicillin G fed-batch fermentation

As a test case for all methods presented in this chapter, consider the penicillin G fed-batch fermentation process modeled by Bajpai and Reuß [1][2]. This unstructured model is of the form (13.1) (or equivalently (13.5)), with the specific rates defined as follows:

$$\begin{aligned} \pi &= \pi_m \frac{C_S}{K_P + C_S + C_S^2/K_I} && \text{(Haldane)} \\ \mu_{substr} &= \mu_C \frac{C_S}{K_X C_X + C_S} && \text{(Contois)} \end{aligned}$$

with π_m the specific production constant [g/g DW h], K_P the saturation constant for substrate limitation of product formation [g/L], K_I the substrate inhibition constant for product formation [g/L], μ_C the maximum specific growth rate for Contois kinetics [1/h], and K_X the Contois saturation constant for substrate limitation of biomass production [g/g DW]. Observe that these specific rates have the general shape shown in Figure 13.1.

Bajpai and Reuß assume a completely maintenance metabolism for both biomass survival and product synthesis. In other words, the *endogenous fractions* f_m and f_P, as introduced in definitions (13.3) and (13.4), are both equal to zero: $f_m = f_P = 0$. The extension to a mixed maintenance/endogenous or completely endogenous metabolism assumption has been described by Nicolaï *et al.* [11] and Van Impe *et al.* [13], with the endogenous fractions f_m and f_P modeled as functions of substrate concentration C_S.

In the application considered here, Eqs. (13.3) and (13.4) reduce to:

$$\begin{aligned} \mu &= \mu_{substr} \\ \sigma &= \mu_{substr}/Y_{X/S} + m + \pi/Y_{P/S} \end{aligned}$$

which clearly satisfy relation (13.2).

parameters			
μ_C	0.11	K_X	0.06
π_m	0.004	k	0.01
K_P	0.1	K_I	0.1
$Y_{X/S}$	0.47	$Y_{P/S}$	1.2
m	0.029	$C_{S,in}$	500
initial conditions			
X_0	10.5	S_0	*to be specified*
P_0	0	V_0	$7 + S_0/C_{S,in}$
t_0	0	α	1500

Table 13.1: Parameters and initial conditions used in simulations.

Based on some experimental evidence, Bajpai and Reuß [1][2] preferred Contois kinetics over the following, more commonly used Monod kinetics in modeling the specific substrate-to-biomass conversion rate μ_{substr}:

$$\mu_{substr} = \mu_M \frac{C_S}{K_S + C_S} \qquad \text{(Monod)}$$

with μ_M the maximum specific growth rate for Monod kinetics [1/h], and K_S the Monod saturation constant for substrate limitation of biomass production [g/L]. One reason is that at high cell densities serious diffusional limitations can be expected which would cause the apparent value of the saturation constant K_S in Monod kinetics to be higher than its value at lower cell densities. In Contois kinetics this behavior is modeled by the term $K_X C_X$ in the denominator. However, *both* kinetics have been shown to be valid for fungal growth. From the *mathematical* point of view, it is interesting to consider Monod kinetics as well, as all specific rates then become functions of substrate concentration C_S only. A procedure to calculate the kinetic constants μ_M and K_S from the Contois constants μ_C and K_X can be found in [15]. The numerical value of the original Bajpai and Reuss model parameters –i.e., involving Contois kinetics– is given in Table 13.1, together with the operational and initial conditions used in simulations. The value of K_P and K_X has been adjusted to obtain agreement with recent biochemical knowledge of the penicillin G fermentation [15]. The results of a constant feeding strategy during 100 h are summarized in Table 13.2.

	$S(0)$	t_2	t_f	$P(t_f)$
Constant control	0	-	100.000	19.422
Optimal control	337	25.312	139.533	22.606
Heuristic C_S-control: $C_S(t_2) = (K_P K_I)^{1/2}$	345	25.530	138.478	22.515
Optimal adaptive control using C_S and C_X	345	-	126.072	22.233
Optimal adaptive control using C_S	345	-	137.538	22.484
Optimal adaptive control using CER	345	-	146.275	22.273

Table 13.2: Numerical results for constant control, optimal control, heuristic C_S-control, and optimal adaptive control.

13.2.3 Optimal control strategy

The optimal control solution –in the sense of the Minimum Principle– for a general model (13.1) (or equivalently, (13.5)) has been analyzed by Van Impe [15] (see also Chapter 11). Initial work along the same lines has been reported by Modak *et al.* [9]. Due to the decoupling between biomass growth and product synthesis, this type of fermentation behaves as a *biphasic* process.

The state vector $\mathbf{x}$ is defined as:

$$\mathbf{x}^{\mathbf{T}} \triangleq [\, S \quad X \quad P \quad V \,]$$

while the control input u is the volumetric substrate feeding rate. This chapter is limited to a problem with an unconstrained state vector $\mathbf{x}$ and control input u, and free initial substrate concentration $C_S(0)$. The extension to problems involving one or more constraints is described in [14][15]. The optimal substrate feed rate profile $u^*(t)$ can be characterized as follows (the results of the optimal control strategy for the penicillin G fed-batch fermentation are summarized in Figure 13.2 and Table 13.2):

1. The first phase, the *growth* phase, is a *batch* phase, i.e., $[u^*(t) = 0,\ 0 < t < t_2]$ (see the left plot of Figure 13.2). All the substrate consumed during growth, denoted by α_{growth}, is added all at once at time $t = 0$, thus ensuring the highest possible specific growth rate μ for all $t \in [0, t_2]$, with a low production rate (see the right plot of Figure 13.2).

2. During the second phase, the *production* phase, a *singular control* $[u^*(t) = u_{sing}(t), t_2 < t < t_3]$ (see the left plot of Figure 13.2) forces the process to produce the product as fast as possible. At any time,

there is a balance between glucose feeding and glucose demand for production and possibly maintenance, thus ensuring the lowest possible growth rate (see the right plot of Figure 13.2). When $V(t_3) = V_f$, the fermentation continues in *batch* $[u^*(t) = 0, t_3 < t < t_f]$ until the net product formation rate dP/dt equals zero at $t = t_f$.

This solution is similar to the one reported by San and Stephanopoulos [12] who used C_S as control input. In summary, the two point boundary value problem, which results from the application of the Minimum Principle and involves the state vector $\mathbf{x}$ and the so-called costate vector $\mathbf{p}$, has been reduced to the *two-dimensional optimization of the initial substrate amount $S(0) = S_0$ (or equivalently, $C_S(0)$), and the time t_2 at which the switching from batch to singular control occurs.* It can be shown that for the performance measure considered here *the singular control $u_{sing}(t)$ is a non-linear feedback law of the state variables S, X, P, and V only* [9][15]. In other words, the costate vector $\mathbf{p}$ (which has no clear physical interpretation) has been eliminated completely.

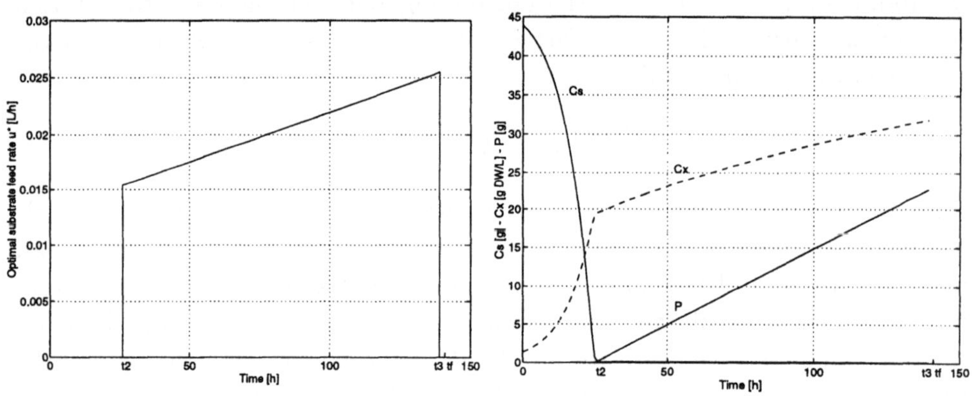

Figure 13.2: Optimal control. Left plot: optimal substrate feed rate profile $u^*(t)$. Right plot: substrate concentration $C_S(t)$, biomass concentration $C_X(t)$, and product amount $P(t)$ profiles.

13.2.4 Heuristic control strategies

The most important drawbacks of the optimal control solution can be summarized as follows.

1. Optimal control is a very *model sensitive* technique. It requires a complete knowledge of the process model, including an analytic expression for all specific rates. Since in biotechnology this assumption is in practice never fulfilled, the optimal profile is generally calculated using a highly simplified model describing the process more or less correctly only from a qualitative view-point. Therefore, the resulting optimal profiles can be used only to increase the insight into both the process and the quality of the model.

2. For the performance measure considered in this chapter (i.e., maximization of final product amount $P(t_f)$), the optimal feed rate profile is obtained in complete state feedback form except for the switching time t_2 between the batch growth phase and the singular production phase (Figure 13.2). In general, t_2 must be determined numerically in advance.

3. Necessary and sufficient conditions can be derived for which t_2 also becomes a function of state variables only [15]. However, even if a perfect process model could be available which satisfies all conditions to obtain the complete optimal solution in closed loop, real-life implementation is still hampered by the lack of *reliable sensors* suited to real-time monitoring of the process variables needed in the controller. Besides a perfect analytical knowledge of all specific rates and corresponding parameters, the control during the singular phase $u_{sing}(t)$ requires on-line measurements of all state variables S, X, P, and V [15].

Therefore, it is very useful to construct *suboptimal* strategies that do not suffer from the above difficulties, at the expense of as small as possible a decrease in performance. In [13][15] suboptimal *heuristic* controllers for both the substrate concentration C_S (*heuristic C_S-control*) and the overall specific growth rate μ (*heuristic μ-control*) are designed. As an example, in this chapter heuristic C_S-control is considered, which can be motivated from both the *microbiological* and *mathematical* point of view.

13.2.4.1 Microbiological and experimental motivation

The construction of a suboptimal profile for the type of biotechnological processes under consideration can be based on the concept of a *biphasic* fermentation.

1. *Growth phase* $[0, t_2]$. During the growth phase the specific substrate-to-biomass conversion rate μ_{substr} is focused. For the control needed reference is made to the optimal control results: in the case of an unbounded input u, an unconstrained state vector $\mathbf{x}$, and a free initial substrate concentration $C_S(0)$, the growth phase is a *batch* phase. In the case of a constraint on the input or the state and/or a fixed initial state, some minor modifications are required. A general strategy is that *the fraction α_{growth} of the total amount of substrate available α, which is consumed for biomass accumulation during the growth phase, must be added as fast as possible in order to obtain the highest possible value of μ_{substr}.*

2. *Production phase* $[t_2, t_3]$. During production the specific production rate π is focused. As shown in Figure 13.1, π exhibits a maximum as a function of the substrate concentration C_S. So, it is a reasonable control objective to keep the substrate concentration during the production phase constant at the level $C_{S,\pi}$ which maximizes π. For instance, in the case of Haldane kinetics, $C_{S,\pi}$ equals $(K_P K_I)^{1/2}$. Therefore, as soon as $C_S(t)$ equals $C_{S,\pi}$, the feed rate switches from $u(t) = 0$ to:

$$u_{production} = \frac{\sigma C_X V}{C_{S,in} - C_S} \tag{13.6}$$

   which keeps substrate concentration C_S constant during production. This can be readily seen using model equations (13.5). Controller (13.6) is shut off when all substrate α has been added at time $t = t_3$, or equivalently, when $V(t_3) = V_f$. As in the case of optimal control, the fermentation continues in *batch* $[u(t) = 0, t_3 < t < t_f]$ until the net product formation rate dP/dt equals zero at $t = t_f$.

Obviously, the switching time t_2 (Figure 13.2) is known in closed loop: the production phase starts when the substrate concentration C_S becomes equal to $C_{S,\pi}$. As a result, the optimization problem has been reduced to the *one-dimensional optimization of the initial substrate concentration $C_S(0)$, or more generally, of the fraction α_{growth} of the total substrate amount available.*

A further refinement of this strategy consists of optimizing the value of the substrate concentration level during production (denoted by C_S^*, which plays the role of a *setpoint*). In other words, during production C_S is kept constant, but not necessarily at the value $C_{S,\pi}$ which maximizes π. As in the case of optimal control, optimization of final product amount reduces to a *two-dimensional optimization problem.* The degrees of freedom are *the initial substrate concentration* $C_S(0)$ *–or more generally, the fraction* α_{growth}*– and the substrate concentration setpoint* C_S^* *during production.*

13.2.4.2 Mathematical justification

The mathematical motivation of the proposed heuristic C_S-controller is based on the following considerations. If it is assumed that μ_{substr} and π (and thus σ through relation (13.2)) are functions of substrate concentration C_S only –consider for instance the above model for penicillin G fermentation with μ_{substr} modeled by Monod kinetics– then the *optimal* feed rate during the singular production phase $[t_2, t_3]$ is given by [15]:

$$u_{sing}(t) = \frac{\sigma C_X V}{C_{S,in} - C_S} + k \frac{p_3 V(\pi' X - \mu' P)}{X(C_{S,in} - C_S)(p_1\sigma'' - p_2\mu'' - p_3\pi'')} \qquad (13.7)$$

where a prime denotes derivation with respect to substrate concentration, and p_i is the *costate* associated with component x_i of the state vector $\mathbf{x}$. Note that this expression is linear in the specific product decay rate k, and is a feedback law of state variables only (it can be shown that the costates p_1 and p_2 depend linearly on p_3). Furthermore, the second term requires knowledge of an analytical expression of the derivatives of all specific rates up to second order. It is shown in [15] that the proposed heuristic C_S-controller reduces to the optimal profile if (and only if) (i) the performance index is independent of final time t_f, (ii) the specific rates μ_{substr} and π are functions of C_S only, (iii) $k = 0$, and (iv) the production phase starts when the substrate concentration C_S reaches the level which maximizes the ratio π/σ. In cases where (some of) these conditions are not satisfied, the proposed heuristic C_S-controller is at least a very good approximation of the optimal solution.

Note that the specific rates μ_{substr} and π can be allowed to be functions of other component concentrations as well (e.g., C_X), provided these specific rates as functions of substrate concentration C_S have the general shape shown in Figure 13.1. As an example, the results given in Table 13.2 are obtained for the penicillin G fermentation process with μ_{substr} modeled by Contois kinetics.

13.2.5 Linearizing control

With respect to a real-life implementation, the heuristic controller (13.6) has the following advantages over the optimal controller (13.7). First, the switching time t_2 between growth and production (and thus the complete control) is known in closed loop as a function of the state: $C_S(t_2) = C_S^*$. Second, as for the modeling uncertainty problem, only the specific substrate consumption rate σ is required. Third, as for the on-line monitoring problem, the number of state variables to be measured on-line has been reduced by one: there is no need for a measurement of the product P. This is an important advantage in cases where the product remains (almost) completely in the liquid phase of the reactor. Finally, the most important advantage is that the given optimal control problem –namely, optimization of the final product amount $P(t_f)$ at some unknown final time t_f– has been replaced by a more common *regulator problem* –namely, regulation of substrate concentration C_S to some setpoint C_S^* for all time t during production– for which feedback control loops can be developed.

However, a real-life implementation is still far away. Two important problems remain to be solved.

> *Problem 1: The monitoring problem.* Although the number of unknowns has been reduced, the heuristic C_S-controller still needs on-line measurements –or at least reliable estimates– of substrate S, biomass X, volume of the liquid phase V, and of the specific substrate consumption rate σ.
>
> *Problem 2: The stability problem.* The closed loop stability is not guaranteed *a priori*. From general model (13.5) the closed loop dynamics during production for substrate concentration C_S when using controller (13.6) are simply:
>
> $$\frac{dC_S}{dt} = 0$$
>
> Clearly, even a small disturbance can move substrate concentration irreversibly away from its desired value C_S^*, resulting in performance degradation.

In the following it is illustrated how to design controllers based on the heuristic approach that do not suffer from the above drawbacks.

REMARK

It is emphasized that the primary goal of a substrate feedback controller for a fed-batch fermentation process is *not to stabilize the process globally*, but rather to *optimize* it while *keeping an inherently unstable type of behavior under control* (see also [3]). As an example, consider the growth phase of a fed-batch fermentation process with a substrate inequality constraint. The optimal strategy then consists of keeping C_S at its maximum value, say $C_{S,MAX}$, using a control of the form (13.6) until all substrate available for growth α_{growth} has been added [15]. The closed loop dynamics of biomass are then, using (13.1) and (13.5):

$$\begin{aligned}\frac{dX}{dt} &= \mu(C_{S,MAX})X \\ \frac{dC_X}{dt} &= [\mu(C_{S,MAX}) - \frac{\sigma(C_{S,MAX})}{C_{S,in} - C_{S,MAX}}C_X]C_X\end{aligned}$$

During growth, in general the specific growth rate μ is much larger than the dilution rate $D = u/V$. As a result, both the absolute amount X and the concentration C_X of biomass increase in an exponential way. Clearly, the substrate controller (13.6) *does not stabilize* the growth phase. It *optimizes* growth by keeping substrate concentration at its optimal value. ■

13.2.6 The stability problem

The second problem is considered first. When replacing the optimal controller (13.7) by the heuristic controller (13.6) the control objective becomes more realistic, namely *setpoint control* or more generally *tracking of a reference profile*. The heuristic controller (13.6) performs well if there are no disturbances, measurement errors, ..., and if the switch from growth to production occurs exactly when $C_S(t_2) = C_S^*$. As in general these assumptions are not fulfilled, some mechanism must be incorporated in control law (13.6) which controls the *tracking error* in presence of disturbances, ... At this point the principle of *linearizing control* can be used. An introduction and several applications in bioreactor control can be found in [3] and the references therein.

1. In the application considered in this chapter, the control variable is the volumetric feed rate u, while the controlled variable is the substrate concentration C_S. So an *input/output model* for this case is simply the first differential equation of (13.5):

$$\frac{dC_S}{dt} = -\sigma C_X - C_S\frac{u}{V} + C_{S,in}\frac{u}{V} \tag{13.8}$$

This input/output model (which is linear in the control u) is of *relative degree one*: the control u appears explicitly in the first derivative (with respect to time t) of the controlled variable C_S.

2. A *linear stable* (λ is a strictly positive given number) *reference model* for the tracking error is then:

$$\frac{d(C_S - C_S^*)}{dt} = -\lambda(C_S - C_S^*) \qquad (13.9)$$

Note that the reference model is of the same degree as the input/output model. At this point, the reference signal C_S^* may be time-varying.

3. A *non-linear linearizing controller* is obtained by eliminating dC_S/dt between (13.8) and (13.9):

$$u_0 = \frac{\sigma C_X + \frac{dC_S^*}{dt} - \lambda(C_S - C_S^*)}{C_{S,in} - C_S} V$$

In the application considered here, a *constant* substrate concentration during the production phase is desired, so:

$$u_0 = \frac{\sigma C_X - \lambda(C_S - C_S^*)}{C_{S,in} - C_S} V$$

In most practical situations the control action (i.e., the feeding pump capacity) is bounded. The resulting controller during the production phase is then:

$$u_{production} = \begin{cases} u_0 & \text{if } \; 0 \leq u_0 \leq U_{MAX} \\ 0 & \text{if } \; u_0 \leq 0 \\ U_{MAX} & \text{if } \; u_0 \geq U_{MAX} \end{cases} \qquad (13.10)$$

REMARKS

1. Note that it is preferred to apply non-linear linearizing control theory. Since a fermentation process is known to be inherently non-linear, it is reasonable to expect that better control would result by exploiting from the outset the non-linearities in the model in the design of a non-linear adaptive control algorithm.

2. Obviously, this controller reduces to heuristic controller (13.6) if the tracking error $(C_S - C_S^*)$ due to disturbances, measurement errors, ..., equals zero.

3. An important advantage as compared with the heuristic controller (13.6) is the following. Controller (13.10) *can be implemented from $t = 0$ on, by considering C_S^* as the setpoint from $t = 0$ on.* This idea can be motivated as follows. If there are no inequality constraints on the state $\mathbf{x}$ and the initial substrate concentration is free, the control sequence is as shown in Figure 13.2 with u_{sing} replaced by the above non-linear linearizing controller. The growth phase is a *batch* phase, as all substrate consumed during growth α_{growth} is added at time $t = 0$ in order to maximize the specific growth rate μ. In other words, there is no reference profile for substrate concentration during the growth phase. In general, the optimal initial substrate amount S_0 is sufficiently high so that during growth $C_S(t) >> C_S^*$, with C_S^* the desired substrate concentration level during production. As a result, the tracking error $(C_S(t) - C_S^*)$ is a very large positive number. Consequently, the control calculated using (13.10) is set equal to $u = 0$, i.e., a *batch* phase as required. Furthermore, the feed rate switches *automatically* to positive values as soon as $C_S \to C_S^*$, so controller (13.10) can indeed be implemented from $t = 0$ on. As in the case of optimal and heuristic control, this is a *two-dimensional optimization problem.* The degrees of freedom are *the initial substrate concentration $C_S(0)$ –or more generally, the fraction α_{growth}– and the reference substrate concentration during production C_S^*.*

The extension to the case of a substrate inequality constraint $(C_S(t) \leq C_{S,MAX})$ during growth is straightforward. The optimal initial substrate concentration is then [15] $C_S(0) = C_{S,MAX}$. During growth, a controller of the form (13.10) tries to keep C_S at $C_{S,MAX}$ ($C_{S,MAX}$ then plays the role of the reference value) until at some time instant $t = t_1$ the total amount of substrate reserved for growth, α_{growth}, is added. At $t = t_1$ the reference level switches to the level C_S^* which is optimal for production. Since at $t = t_1$ the actual substrate concentration C_S is much larger than C_S^*, controller (13.10) switches to zero –i.e., batch mode– as required. If $C_S \to C_S^*$, controller (13.10) switches *automatically* to positive values to keep C_S around C_S^*. Controller (13.10) can again be implemented from $t = 0$ on. For this case the two degrees of freedom in the optimization are *the time t_1 –or equivalently, the amount of substrate reserved to growth α_{growth}– and again the reference substrate concentration level during production C_S^*.*

Both cases are illustrated for the penicillin G fed-batch fermentation in Figure 13.3 where it is assumed that the measurements are all perfect and that there are no disturbances. Observe that the non-linear linearizing controller (13.10) becomes positive *before* C_S reaches C_S^*. This guarantees a smooth transition in the substrate concentration profile at the start of the production phase. This behavior can be obtained using a small value of λ (e.g., $\lambda = 1$) in the reference model (13.9).

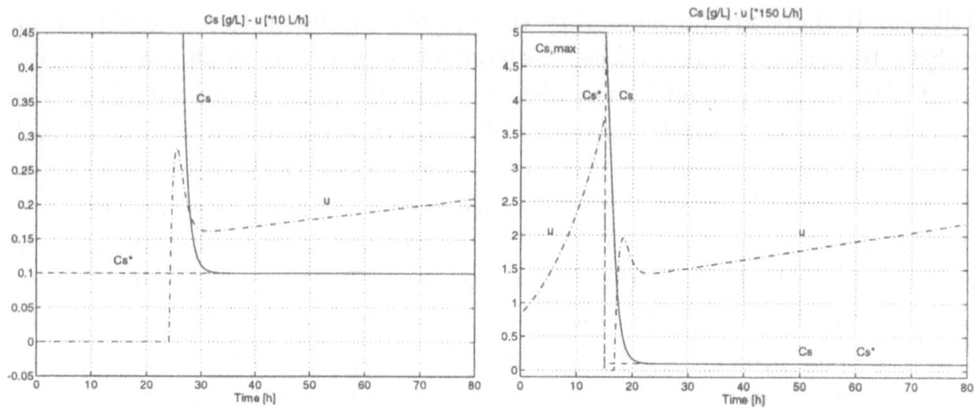

Figure 13.3: Non-linear linearizing control. Left plot: no constraints. Right plot: inequality constraint on C_S. Legend: — actual substrate concentration C_S, – – reference profile C_S^*, – · control action u.

4. On the other hand, a large value of λ increases the stability margin, tracking behavior and disturbance rejection. This can be easily seen when calculating the closed loop response of substrate concentration C_S. Since λ is at the disposal of the user, it can be used to search for an optimal *trade-off.* ■

13.2.7 The monitoring problem

In the following sections the first problem is considered, i.e., monitoring of all variables required in controller (13.10). Three solutions depending on which measurements are available on-line are presented. The remaining variables are then estimated on-line using *software sensors.* The three algorithms proposed are all based on the *minimal modeling concept* introduced in [3]. In this approach no assumption is made concerning the exact analytical structure of the specific rates required in the control law, thus circumventing the modeling and corresponding parameter identification problem. Instead, they are treated as *time-varying parameters* which are estimated on-line. By doing so, the non-linear linearizing controller (13.10) is made *adaptive* and can be implemented *independently* of the –usually unknown– analytical expression for the specific rates. Simulation results for the penicillin G fed-batch fermentation process as described above illustrate each of the variants proposed.

13.3 Optimal Adaptive Control: On-line Measurements of C_S and C_X

13.3.1 Mathematical description

The following assumptions are made.

1. Both substrate concentration C_S and biomass concentration C_X are measured on-line.

2. The results of the on-line measurement devices become available to the controller only after a time delay ΔT, which is assumed equal for both measurements. This delay represents the time required to analyze a sample from the reactor contents.

3. The volume of the liquid phase V is available on-line without time delay.

4. In agreement with the minimal modeling concept, no assumption is made concerning the exact analytical structure of the specific rates σ and μ.

In practice, a control algorithm will be implemented in discrete time. So the differential equations for C_S, C_X, and V (see Equations (13.5)) are discretized first. As an on-line measurement device can take a new sample from the reactor only after finishing the analysis of the previous one, the discretization interval is set equal to the sampling interval ΔT. A first-order forward Euler discretization results in the following equations:

$$\begin{aligned}
C_{S,k+1} - C_{S,k} &= -\sigma_k C_{X,k}\Delta T - C_{S,k}\frac{u_k}{V_k}\Delta T + C_{S,in}\frac{u_k}{V_k}\Delta T \\
C_{X,k+1} - C_{X,k} &= \mu_k C_{X,k}\Delta T - C_{X,k}\frac{u_k}{V_k}\Delta T \\
V_{k+1} - V_k &= u_k\Delta T
\end{aligned} \tag{13.11}$$

The discrete-time version of linearizing controller (13.10) is:

$$u_k = \frac{\sigma_k C_{X,k} - \lambda(C_{S,k} - C_S^*)}{C_{S,in} - C_{S,k}} V_k$$

$$u_{production} = \begin{cases} u_k & \text{if} \quad 0 \leq u_k \leq U_{MAX} \\ 0 & \text{if} \quad u_k \leq 0 \\ U_{MAX} & \text{if} \quad u_k \geq U_{MAX} \end{cases} \tag{13.12}$$

Besides the on-line measurement of V_k, controller (13.12) needs on-line estimates of $C_{S,k}$, $C_{X,k}$, and σ_k. Observe that $C_{S,in}$ and λ are prespecified constants.

Since the discrete-time model equations (13.11) are *linear* in the specific rates σ and μ, these rates can be estimated using a *recursive least squares algorithm with forgetting factor* ζ fed with on-line data of C_S and C_X. A standard text-book formulation of *RLS* can be found in, e.g., [7].

However, due to the measurement delay, at time $t = k\Delta T$ only $C_{S,k-1}$ and $C_{X,k-1}$ are known. An adaptive version of controller (13.12) can then be obtained using the following algorithm (a hat '^' denotes an estimate).

Algorithm 1

Step 1: *Estimation of σ_{k-1} and μ_{k-1} using RLS*
Using Equations (13.11) the *residuals* ε can be written as:

$$\begin{aligned} \varepsilon_{C_S,k-1} &\triangleq C_{S,k-1} - \hat{C}_{s,k-1} \\ &= C_{S,k-1} + \hat{\sigma}_{k-2} C_{X,k-2} \Delta T \\ &- (C_{S,in} - C_{S,k-2}) \frac{u_{k-2}}{V_{k-2}} \Delta T - C_{S,k-2} \\ \varepsilon_{C_X,k-1} &\triangleq C_{X,k-1} - \hat{C}_{x,k-1} \\ &= C_{X,k-1} - \hat{\mu}_{k-2} C_{X,k-2} \Delta T \\ &+ C_{X,k-2} \frac{u_{k-2}}{V_{k-2}} \Delta T - C_{X,k-2} \end{aligned}$$

The gain K is obtained via:

$$\begin{aligned} P_{k-1} &= P_{k-2} / (\zeta + C_{X,k-2}^2 \Delta T^2 P_{k-2}) \\ K_{k-1} &= -C_{X,k-2} \Delta T P_{k-1} \end{aligned}$$

An estimation of σ_{k-1} and μ_{k-1} is then:

$$\begin{aligned} \hat{\sigma}_{k-1} &= \hat{\sigma}_{k-2} + K_{k-1} \varepsilon_{C_S,k-1} \\ \hat{\mu}_{k-1} &= \hat{\mu}_{k-2} - K_{k-1} \varepsilon_{C_X,k-1} \end{aligned}$$

Step 2: *Prediction of $C_{S,k}$ and $C_{X,k}$*
These variables can be calculated by using the estimates $\hat{\sigma}_{k-1}$ and $\hat{\mu}_{k-1}$ from Step 1 in the discrete equations (13.11) rewritten at time k.

Step 3: *Calculation of the controller action u_k*
u_k is calculated by substituting the results of the previous steps in (13.12):

$$u_k = \frac{\hat{\sigma}_{k-1}\hat{C}_{x,k} - \lambda(\hat{C}_{s,k} - C_S^*)}{C_{S,in} - \hat{C}_{s,k}} V_k$$

In the above expression the required estimate $\hat{\sigma}_k$ is replaced by the (available) estimate $\hat{\sigma}_{k-1}$, as σ varies only slowly as compared with the dynamics of the process. ■

REMARKS

1. Following a same line of reasoning as presented in Section 13.2.6 it can be concluded that this controller can also be implemented from $k = 0$ on.
2. Instead of a *linear regression estimator*, an *observer-based estimator* could be used as well for the estimation of the specific rates σ and μ. The use of an observer-based estimator will be illustrated in the following sections.
3. Although the specific growth rate μ is not explicitly required in controller (13.12), an estimate is needed in order to predict the state at time $t = k\Delta T$ in Step 2 of the algorithm. Obviously, this complication is entirely due to the measurement delay ΔT.
4. As already indicated in Section 13.2.6, the parameter λ is at the disposal of the user to search for an optimal *trade-off* between smoothness of the controller action on the one hand, and stability margin and tracking behavior on the other. ■

13.3.2 Simulation results

Consider again the penicillin G fed-batch fermentation described in Section 13.2.2. All simulations are carried out using a *continuous-time process model* and a *discrete-time controller* action. Between two samples the controller action is kept constant.

In addition to the measurement time delay ΔT, the on-line measurements of C_S and C_X are assumed to be corrupted by zero mean white noise. The standard deviation is set equal to **std** $(C_X) = 0.25$ g/L, and **std** $(C_S) = 0.01$ g/L. For a typical value of $C_X = 10$ g/L, this represents a standard deviation of 2.5 %. A typical value of substrate concentration C_S during the production phase is the level $C_{S,\pi}$ which maximizes the specific production rate π. For the set of parameters given in Table 13.1, $C_{S,\pi}$ is equal to $(K_P K_I)^{1/2} = 0.1$ g/L. Under the above assumptions, a standard deviation on C_S of even 10 % is allowed.

For a measurement time delay $\Delta T = 0.1$ h, the following results are obtained. The *RLS* scheme is initialized with $P_{k=0} = 10^9$, $\hat{\sigma}_{k=0} = 0$ [g/g DW h], and $\hat{\mu}_{k=0} = 0$ [1/h]. The forgetting factor ζ is set equal to $\zeta = 0.98$, while the parameter λ is set equal to $\lambda = 10$. The substrate concentration setpoint during production C_S^* is set equal to $C_S^* = C_{S,\pi} = 0.1$ g/L. When using the same initial substrate amount $S_0 = 345$ g as given in Table 13.2 for the heuristic C_S-controller, a final product amount $P(t_f) = 22.233$ g is obtained at $t_f = 126.072$ h, which comes very close to the optimal value $P_{opt}(t_f) = 22.606$ g.

The convergence of the estimation of the specific rates σ and μ is illustrated in the left plot of Figure 13.4. The right plot shows the regulation of the actual substrate concentration $C_S(t)$ towards its setpoint C_S^*, and the corresponding adaptive control action u_k. From both plots it can be concluded that the algorithm has converged shortly after the beginning of the production phase.

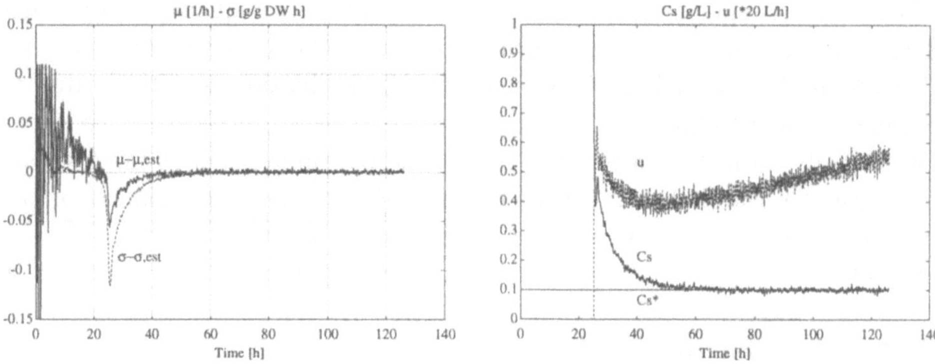

Figure 13.4: Optimal adaptive control: on-line measurements of C_S and C_X. Left plot: estimation errors of σ and μ. Right plot: regulation of C_S and adaptive control action u_k.

For $\Delta T = 1$ h the algorithm did *not* converge. Remember that ΔT represents both the sampling interval in discretizing the model equations and the measurement delay. As for Step 1 in the algorithm, it can be easily verified that there always exist an initial value $P_{k=0}$ and a forgetting factor ζ such that the estimations σ_{k-1} and μ_{k-1} and the corresponding gain K_{k-1} remain bounded if biomass concentration C_X is strictly positive for all t and if the control input u is bounded. As a result, the instability at too large values of ΔT is completely due to the increasing inaccuracy of the predictions made in Step 2. It can be expected –and this has been confirmed during simulations– that if the measurement delay ΔT becomes too large, especially the prediction of substrate concentration C_S around the transition from growth to production becomes worse.

13.4 Optimal Adaptive Control: On-line Measurements of C_S

13.4.1 Mathematical description

The following assumptions are made.

1. Besides the volume V, substrate concentration C_S is the only available on-line measurement.

2. In this and the next section, a measurement time delay is not explicitly taken into account. If the measurement delay cannot be ignored, an additional prediction step must be incorporated in the discrete-time implementation of the proposed algorithms. An example is given in Section 13.3. Therefore, from now on ΔT only represents the discretization interval.

3. In agreement with the minimal modeling concept, no assumption is being made concerning the exact analytical structure of the specific rate σ.

An *adaptive* implementation of controller (13.10) can be obtained as follows. Since $C_{S,in}$ and λ are known constants, we only need an on-line estimate of both the specific substrate consumption rate σ and the biomass concentration C_X. More precisely, only an estimate of the *product* of σ and C_X is needed. Therefore, we define the rate β –with dimension [g/L h]– as follows:

$$\beta \triangleq \sigma C_X$$

By doing so, the only unknown is exactly the rate β, which is considered as a time-varying parameter. In the following continuous-time algorithm, β is estimated using a *state-observer-based parameter estimator.*

Algorithm 2

Step 1: *Estimation of β*

$$\begin{aligned}\frac{d\hat{C}_S}{dt} &= -\hat{\beta} + (C_{S,in} - C_S)\frac{u}{V} + \omega(C_S - \hat{C}_S)\\ \frac{d\hat{\beta}}{dt} &= -\gamma(C_S - \hat{C}_S)\end{aligned}$$

Step 2: *Calculation of the controller action u*

$$\begin{aligned}u_0 &= \frac{\hat{\beta} - \lambda(C_S - C_S^*)}{C_{S,in} - C_S}V\\ u &= \begin{cases} u_0 & \text{if } 0 \le u_0 \le U_{MAX}\\ 0 & \text{if } u_0 \le 0\\ U_{MAX} & \text{if } u_0 \ge U_{MAX}\end{cases}\end{aligned}$$

■

REMARKS

1. Tuning of the state-observer-based parameter estimator proposed in Step 1 reduces to the calibration of the (positive) constants ω and γ.

2. Just like the controller proposed in Section 13.3, this controller does not need any *a priori* information either, such as, e.g., yield coefficients, ... Moreover, treating β (and thus σ) as a time-varying parameter makes it robust against modeling uncertainties.

3. During simulations, a continuous-time process model and a discrete-time version of the above estimator and controller have been used. Between sampling instants –at distance ΔT– the controller action u_k is kept constant. It can be easily shown that convergence of the estimator is guaranteed if the following inequalities are satisfied:

$$\begin{aligned}|1 - \omega\Delta T| &\le 1\\ |1 - \omega\Delta T + \gamma\Delta T^2| &\le 1\end{aligned}$$

Observe that these constraints are independent of the design parameter λ in the controller action u_k. ■

13.4.2 Simulation results

Some simulation results for the penicillin G model presented in Section 13.2.2 are shown in Figure 13.5. The sampling interval ΔT is set equal to $\Delta T = 0.1$ h. The initial substrate amount S_0 is set equal to $S_0 = 345$ g,

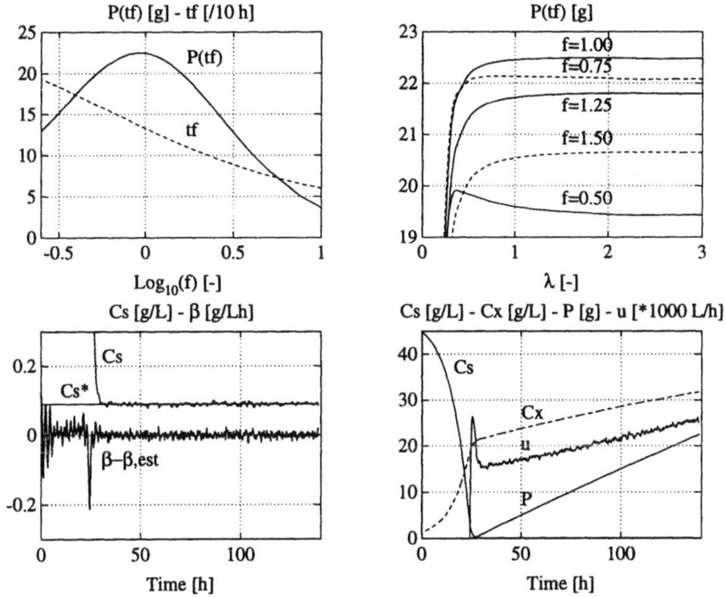

Figure 13.5: Optimal adaptive control: on-line measurements of C_S. Upper left plot: optimization of $P(t_f)$ as function of f. Upper right plot: influence of f and λ on $P(t_f)$. Lower plots: time profiles for the setpoint $C_S^* = 0.9(K_P K_I)^{1/2}$ g/L.

which is optimal for heuristic C_S-control (Table 13.2). Tuning of the estimator, initialized with $\hat{\beta}_{k=0} = 0.25$ [g/L h], leads to the following efficient values: $\omega = 1$ and $\gamma = 10$.

Define a factor f as:

$$f \triangleq \frac{C_S^*}{C_{S,\pi}} \equiv \frac{C_S^*}{(K_P K_I)^{1/2}}$$

The results of an optimization with respect to this factor f –in other words, with respect to the setpoint during production–, and the controller parameter λ are shown in the upper plots of Figure 13.5. The upper left plot shows the final production $P(t_f)$ and the final time t_f as functions of the factor f, for the controller parameter λ equal to $\lambda = 1$. Observe that the optimal value does *not* occur at $f = 1$ –in other words, at $C_S^* = C_{S,\pi}$–, but at:

$$C_{S,opt}^* = 0.9\ C_{S,\pi}$$

Due to the shape of the function $P(t_f)$ as function of f, it is clear that in practice three experiments should suffice to optimize the process. In addition, f can be used to search for a *trade-off* between $P(t_f)$ and t_f. The upper right plot illustrates the influence of the controller parameter λ on the final product amount $P(t_f)$, for different choices of the factor f. It can be seen that values of λ larger than 1 have little influence upon the final product amount.

The lower plots show the time profiles for the optimal values $f = 0.9$ and $\lambda = 1$, while assuming a zero mean white noise on the measurements of C_S with standard deviation $\mathtt{std}(C_S)$=0.01 g/L. This represents an admissible standard deviation of 10 % for substrate concentration within the order of magnitude $C_S = \mathcal{O}[(K_P K_I)^{1/2}] = \mathcal{O}[0.1]$ g/L during production. The lower left plot illustrates the convergence of the estimator for the rate β, and the regulation of the actual substrate concentration $C_S(t)$ towards its setpoint $C_S^* = 0.09$ g/L. As in the previous section, the proposed algorithm has converged shortly after the beginning of the production phase. The lower right plot shows the adaptive control action u_k, together with the corresponding profiles for the state variables. The final product amount is $P(t_f) = 22.484$ g at $t_f = 137.538$ h, which comes again very close to the optimal value of $P_{opt}(t_f) = 22.606$ g (Table 13.2).

13.5 Optimal Adaptive Control: On-line Measurements of *CER*

13.5.1 Mathematical description

In the algorithms of Sections 13.3 and 13.4, the major bottleneck is the *accuracy of the on-line substrate concentration measurements.* Kleman *et al.* [8] reported a control algorithm maintaining C_S as tight as 0.49 ± 0.04 g/L during growth of *E. coli.* Using the parameter values of Table 13.1, the optimal value for substrate concentration during production is in the order of magnitude $C_S = \mathcal{O}[(K_P K_I)^{1/2}] = \mathcal{O}[0.1]$ g/L. Although the proposed algorithms proved robust against standard deviations of even 10 %, the question arises whether such a small, locally determined concentration level can be considered as representative for the whole reactor contents, which is in practice not perfectly mixed.

This problem can be circumvented as follows. In order not to overload the notation, we assume a completely maintenance metabolism, i.e., $f_m = f_P \equiv 0$ in expressions (13.3) and (13.4). In other words, the specific substrate-to-biomass conversion rate μ_{substr} is identical to the specific growth rate μ. If the specific growth rate μ is a monotonically increasing function of substrate concentration C_S (see Figure 13.1), then prespecifying a reference profile for C_S can be replaced by *prespecifying a reference profile for the specific growth rate.* In the case of μ function of C_S only this is even identical. An appropriate reference profile is then: during growth μ should be as high as possible, while during production μ should be kept constant at $\mu = \mu^*$.

Obviously, an on-line estimation of the specific growth rate μ is required. This can be done, e.g., using the easily accessible measurement of CO_2 in the effluent gas from the fermentor. The dissolved carbon dioxide dynamics are given by:

$$\frac{dC_C}{dt} = CER - DC_C - Q_{out}$$

with C_C the dissolved carbon dioxide concentration [L CO_2/L], and Q_{out} the rate of outflow of carbon dioxide from the reactor in gaseous form [L CO_2/L h]. This model is valid only if the dissolved carbon dioxide concentration C_C is lower than the saturation concentration $C_{C,sat}$ representative of the CO_2 solubility:

$$C_C \triangleq \Pi\ C_{C,sat} \qquad \Pi \in [0,1]$$

The differential equation can then be written as

$$C_{C,sat}\frac{d\Pi}{dt} = CER - \Pi D C_{C,sat} - Q_{out} \tag{13.13}$$

In agreement with (13.2) the carbon dioxide evolution rate *CER* [L CO_2/L h] can be described by:

$$CER = [Y_{C/X}\mu + m_C + Y_{C/P}\pi]C_X$$

under suitably controlled conditions for pH. At any time during the fermentation, carbon dioxide arises from (i) growth and associated energy production (yield coefficient $Y_{C/X}$ [L CO_2/g DW]), (ii) maintenance energy (specific rate m_C [L CO_2/g DW h]), and (iii) product biosynthesis and other possible specialized metabolism (yield coefficient $Y_{C/P}$ [L CO_2/g]).

In most applications $C_{C,sat}$ is very low (i.e., CO_2 solubility is very low), which means that carbon dioxide appears almost completely in gaseous form. Thus, letting $C_{C,sat} = 0$ is a meaningful *singular perturbation.* Equation (13.13) reduces to the following algebraic equation:

$$Q_{out} = CER$$

which means that an on-line CO_2 analysis of the effluent gas flow from the fermentor can be used to extract the variables needed. Note that $C_{C,sat}$ is *not* assumed to be equal to zero. It is only assumed that $C_{C,sat}$ is small enough to neglect the terms $C_{C,sat}(d\,\Pi/dt)$ and $\Pi D C_{C,sat}$ in differential equation (13.13). Calam and Ismail [5] reported the following slightly simplified relation in the case of penicillin G fermentation:

$$CER = Y_{C/X}\mu C_X + m_C C_X + k_P \tag{13.14}$$

Based on experimental results, a *constant* value k_P [L CO_2/L h] representing the contribution of product synthesis is proposed, instead of a term involving the penicillin production rate. This can be motivated as follows. First, during the main production period the rate of biosynthesis is remarkably steady. Second, it is known that penicillin production is accompanied by decomposition (see, e.g., the hydrolysis constant k in the penicillin G model of Section 13.2.2). Therefore, it seems possible that as production later appears to slow down, biosynthesis itself may be continuing or may be diverted to non-antibiotic substances.

In the following algorithm the adaptive observer for C_X and μ is inspired by [6]. This is only a *partially* adaptive observer, as some model constants are required *a priori*.

Algorithm 3

Step 1: *Estimation of μ and C_X*

$$\begin{aligned}
\frac{d\hat{C}_X}{dt} &= \hat{\delta} - \hat{C}_X u/V + \omega(CER - \hat{CER}) \\
\frac{d\hat{\delta}}{dt} &= \gamma(CER - \hat{CER}) \\
\hat{CER} &= Y_{C/X}\hat{\delta} + m_C\hat{C}_X + k_P \\
\hat{\mu} &= \frac{\hat{\delta}}{\hat{C}_X} \\
\hat{\sigma} &= \hat{\mu}/Y_{X/S} + m + C
\end{aligned}$$

Step 2: *Calculation of the controller action u*

$$\begin{aligned}
u_0 &= \frac{\hat{\sigma}\hat{C}_X - \lambda(\hat{\mu} - \mu^*)}{C_{S,in}} V \\
u &= \begin{cases} u_0 & \text{if } 0 \le u_0 \le U_{MAX} \\ 0 & \text{if } u_0 \le 0 \\ U_{MAX} & \text{if } u_0 \ge U_{MAX} \end{cases}
\end{aligned}$$

■

REMARKS

1. The time-varying parameter $\hat{\delta}$ can be interpreted as an estimate of the *biomass growth rate* μC_X [g DW/L h]. During estimation of σ in Step 1, the contribution of $\pi / Y_{P/S}$ is replaced by the *constant* term C [g/g DW h]. This can be motivated as follows. During production, the objective is to keep the specific growth rate μ constant. In the case of μ and π functions of C_S only, this corresponds to keeping C_S –and thus also π– constant during production. In any other case this is at least an excellent approximation. Observe that the expression for $\hat{\sigma}$ has now exactly the same form as model equation (13.14) for *CER*.

2. In the denominator of the controller action in Step 2 substrate concentration C_S is considered negligible as compared with $C_{S,in}$.

3. Note that this scheme requires the *a priori* knowledge of the parameters $Y_{C/X}$, m_C, and k_P, and $Y_{X/S}$, m, and C. In addition, if the endogenous fractions f_m and f_P are different from zero, their value should be known as well. Clearly, this is the price to pay for estimating state variables using only on-line measurements of easily accessible auxiliary variables.

4. However, this estimation procedure has an additional benefit over the algorithms presented in Sections 13.3 and 13.4 when scaling up the production from a pilot plant towards an industrial fermentor. When measuring substrate and/or biomass on-line, it becomes very important where to place the sampling devices on such a large reactor (e.g., actual penicillin production on an industrial scale takes place in fermentors of about 150,000 L). Due to an imperfectly mixed reactor, the question is whether a locally determined concentration is representative of the whole reactor contents. On the other hand, an analysis of the effluent gas from the fermentor provides in some sense averaged values of the reactor state which can be used immediately in the feed rate controller.

5. Obviously, this algorithm can also be implemented from time $t = 0$ on by considering the setpoint for production μ^* as the reference from $t = 0$ on. In case of a substrate inequality constraint during growth, the reference profile has the form shown in the right plot of Figure 13.3 with $C_{S,MAX}$ and C_S^* replaced by the corresponding values for μ.

6. As in Sections 13.3 and 13.4, a continuous-time process model and a discrete-time version of the above estimator and controller have been used. Between sampling instants –at distance ΔT– the controller action u_k is kept constant. ■

13.5.2 Simulation results

Some simulation results for the penicillin G model presented in Section 13.2.2 are shown in Figure 13.6.

The model constants required in the estimator of *CER* are set equal to $Y_{C/X} = 0.4$ [L CO_2/g DW], $m_C = 0.01$ [L CO_2/g DW h], and $k_P = 0.3$ [L CO_2/L h] [10]. Using the parameters given in Table 13.1, the maximum value of π is $\pi_{max} = 1.333\ 10^{-3}$ [g/g DW h]. Therefore, the constant C in the estimator of σ, which is an approximation of $\pi/Y_{P/S}$, is set equal to $C = 10^{-3}$ [g/g DW h].

The sampling interval ΔT is set equal to $\Delta T = 0.1$ h. The initial substrate amount S_0 is set equal to $S_0 = 345$ g, which is optimal for heuristic C_S-control (Table 13.2). Tuning of the estimator, initialized with $\hat{C}_{x,k=0} = 1.3$ [g DW/L], $\hat{\delta}_{k=0} = 0.13$ [g DW/L h], and $\hat{CER}_{k=0} = 0.375$ [L CO_2/L h], leads to the following efficient values: $\omega = 5$ and $\gamma = 25$. The controller parameter λ is set equal to $\lambda = 50$.

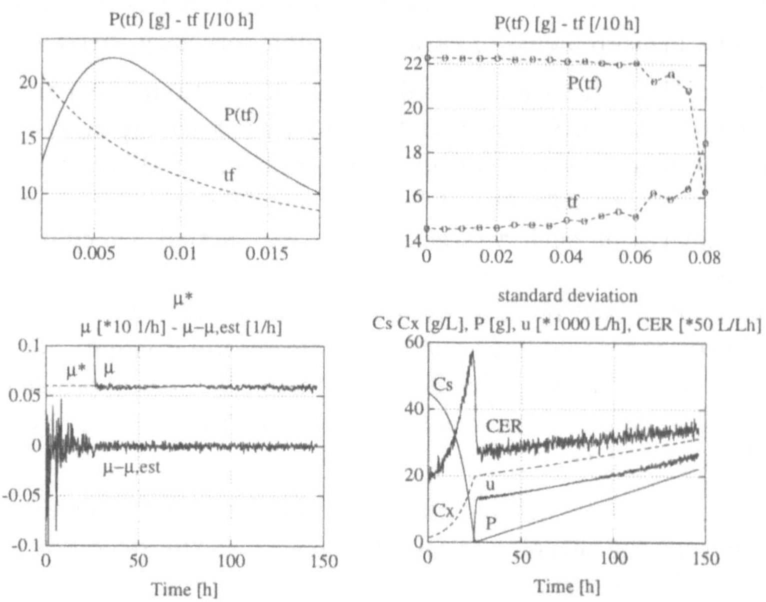

Figure 13.6: Optimal adaptive control: on-line measurements of *CER*. Upper left plot: optimization of $P(t_f)$ as function of μ^*. Upper right plot: sensitivity of optimal values to increasing measurement noise. Lower plots: time profiles for the setpoint $\mu^* = 0.006$ 1/h.

The upper left plot of Figure 13.6 shows the optimization of the final product amount $P(t_f)$ with respect to the setpoint μ^* for the specific growth rate during production, and the corresponding values of the final time t_f. The optimum occurs at $\mu^* = 0.006$ 1/h. As in Section 13.4, a *trade-off* can be made between $P(t_f)$ and t_f with respect to μ^*. Furthermore, from the shape of the function $P(t_f)$ versus μ^* it can be concluded again that in practice three experiments should suffice to optimize the process.

The lower plots show the time profiles for this optimal setpoint $\mu^* = 0.006$ 1/h, while assuming a zero mean white noise on the measurements of *CER* with standard deviation std(*CER*)=0.025 L CO_2/L h. This represents an admissible standard deviation of 5 % for *CER* within the order of magnitude *CER* = $\mathcal{O}[0.5]$ L CO_2/L h during production. The lower left plot illustrates the convergence of the estimator for the specific growth rate μ, and the regulation of the actual specific growth rate $\mu(t)$ towards its setpoint $\mu^* = 0.006$ 1/h. As in Sections 13.3 and 13.4, the proposed algorithm has converged shortly after the beginning of the production phase. The right plot shows the noisy measurements of *CER*, the adaptive control action u_k, together with the corresponding profiles for the state variables. The final product amount is $P(t_f) = 22.273$ g at $t_f = 146.275$ h, which again comes very close to the optimal value of $P_{opt}(t_f) = 22.606$ g (Table 13.2).

The upper right plot illustrates the robustness of the optimal final product amount $P(t_f)$, and the corresponding final time t_f, with respect to increasing standard deviation of the measurement noise on the carbon dioxide evolution rate *CER*. It can be seen that the optimal values are more or less insensitive to this measurement noise, for standard deviations up to std(*CER*)=0.06 L CO_2/L h. For *CER* within the order of magnitude *CER* = $\mathcal{O}[0.5]$ L CO_2/L h during production, this represents an admissible standard deviation of more than 10 %.

13.6 Conclusions

The main contribution of this chapter was to present a unifying methodology for optimization of biotechnological processes, namely *optimal adaptive control*, by combining concepts and techniques from both *optimal control* and *adaptive linearizing control*. As an example, substrate feed rate controllers have been designed for a class of biotechnological processes, characterized by a decoupling between biomass growth and product formation.

optimal control	model sensitive, open loop
↓	microbiological/biochemical process knowledge mathematical analysis of the optimal control solution
heuristic control	model-independent control objective
↓	linearizing control adaptive state and parameter estimation
optimal adaptive control	robust, nearly optimal performance

Table 13.3: Optimal adaptive control: a unifying approach.

It has been illustrated how the information obtained during preliminary optimal control studies leads to the design of easy-to-implement adaptive controllers. The optimal adaptive control procedure is summarized in a schematic way in Table 13.3. The design consists of the following steps.

Step 1 Derivation of the *optimal control solution* to the given optimization problem, under the assumption of a perfectly known process model.

Step 2 Derivation of nearly optimal *heuristic controllers*, based on a careful analysis of the optimal control solution of Step 1 from both the biochemical and the mathematical point of view. This second step itself consists of:

1. Detection of process variables which characterize the optimal control solution, such as a concentration, a specific rate, ...
2. Construction of a *reference profile* for the characteristic process variable as a function of time.

As such, the optimization problem of Step 1 is replaced by a more common tracking control problem, for which feedback control loops are designed in Step 3.

Step 3 Non-linear adaptive implementation of the derived heuristic controller in two steps:

1. Embedding of the heuristic controller within a non-linear linearizing controller.
2. Adaptive estimation of the states and parameters which are not available on-line. According to the minimum modeling principle, no assumption is made concerning the exact analytic nature of the specific rates needed in the control algorithm.

The optimal adaptive controllers derived in this way combine a nearly optimal performance with good robustness properties against modeling uncertainties and process disturbances.

To illustrate the method and the results obtained, simulation results have been given for the penicillin G fed-batch fermentation process. Three possible implementations have been presented, depending on which variables are available by means of on-line measurements. The trade-off between *on-line measurement requirements* (such as accessibility and accuracy) and *a priori information needs* (such as yield and maintenance coefficients) has been clearly illustrated.

Acknowledgments

This chapter presents research results of NFWO Project G.0141.95, NFWO Project G.0286.96, Project OT/95/20 of the Research Council of the Katholieke Universiteit Leuven, and of the Belgian Programme on Interuniversity Poles of Attraction, initiated by the Belgian State, Prime Minister's Office for Science, Technology and Culture. The scientific responsibility rests with its authors.

13.7 Nomenclature

t	: time	[h]
S	: absolute substrate amount in the reactor	[g]
X	: absolute amount of biomass	[g DW (Dry Weight)]
P	: absolute amount of product	[g]
V	: reactor volume	[L]
$C_{S,in}$	: substrate concentration in the influent	[g/L]
C_S	: substrate concentration in the reactor	[g/L]
C_X	: biomass concentration	[g DW/L]
C_P	: product concentration	[g/L]
C_C	: dissolved CO_2 concentration	[L CO_2/L]

CER	: carbon dioxide evolution rate	[L CO_2/L h]
u	: influent volumetric flow rate	[L/h]
f_m	: endogenous fraction of the overall specific oxygen demand	[-]
f_P	: endogenous fraction of product synthesis	[-]
α	: total amount of substrate consumed during fermentation	[g]
σ	: specific substrate consumption rate	[g/g DW h]
μ	: overall specific growth rate	[1/h]
μ_{substr}	: specific substrate-to-biomass conversion rate	[1/h]
μ_C	: maximum specific growth rate for Contois kinetics	[1/h]
K_X	: Contois saturation constant for substrate limitation of biomass production	[g/g DW]
μ_M	: maximum specific growth rate for Monod kinetics	[1/h]
K_S	: Monod saturation constant for substrate limitation of biomass production	[g/L]
π	: specific production rate	[g/g DW h]
π_m	: specific production constant	[g/g DW h]
K_P	: Monod saturation constant for substrate limitation of product formation	[g/L]
K_I	: substrate inhibition constant for product formation	[g/L]
m	: overall specific maintenance demand	[g/g DW h]
m_C	: specific CO_2 production rate in maintenance processes	[L CO_2/g DW h]
k	: product degradation constant	[1/h]
k_P	: CO_2 due to production	[L CO_2/L h]
$Y_{X/S}$	: cell mass on substrate yield coefficient	[g DW/g]
$Y_{P/S}$	: product on substrate yield coefficient	[g/g]
$Y_{C/P}$	: CO_2 on product yield coefficient	[L CO_2/g]
$Y_{C/X}$	: CO_2 on biomass yield coefficient	[L CO_2/g DW]

13.8 Bibliography

1. R.K. Bajpai and M. Reuß 1980. A mechanistic model for penicillin production. *J. Chem. Tech. Biotechnol.*, 30, 332-344

2. R.K. Bajpai and M. Reuß 1981. Evaluation of feeding strategies in carbon-regulated secondary metabolite production through mathematical modelling. *Biotechnol. Bioeng.*, 23, 717-738

3. G. Bastin and D. Dochain 1990. *On-line estimation and adaptive control of bioreactors.* Elsevier Science Publishing Co.

4. G. Bastin and J.F. Van Impe 1995. Nonlinear and adaptive control in biotechnology: a tutorial. *European Journal of Control*, 1, 37-53

5. C.T. Calam and B.A.-K. Ismail 1980. Investigation of factors in the optimisation of penicillin production. *J. Chem. Tech. Biotechnol.*, 30, 249-262

6. C. Di Massimo, A.C.G. Saunders, A.J. Morris and G.A. Montague 1989. Nonlinear estimation and control of mycelial fermentations. *Proceedings of the American Control Conference, June 21-23 1989, Pittsburgh (USA)*, 1994-1999

7. G.C. Goodwin and K.S. Sin 1984. *Adaptive filtering, prediction and control.* Prentice-Hall, Englewood Cliffs, New Jersey

8. G.L. Kleman, J.C. Chalmers, G.W. Luli and W.R. Strohl 1991. A predictive and feedback control algorithm maintains a constant glucose concentration in fed-batch fermentations. *Appl. Environ. Microbiol.*, 57, 910-917

9. J.M. Modak, H.C. Lim and Y.J. Tayeb 1986. General characteristics of optimal feed rate profiles for various fed-batch fermentation processes. *Biotechnol. Bioeng.*, 28, 1396-1407

10. I. Nelligan and C.T. Calam 1983. Optimal control of penicillin production using a mini-computer. *Biotechnology Letters*, 5, 561-566

11. B.M. Nicolaï, J.F. Van Impe, P.A. Vanrolleghem and J. Vandewalle 1991. A modified unstructured mathematical model for the penicillin G fed-batch fermentation. *Biotechnology Letters*, 13 (7), 489-494

12. K.Y. San and G. Stephanopoulos 1989. Optimization of fed-batch penicillin fermentation : a case of singular optimal control with state constraints. *Biotechnol. Bioeng.*, 34, 72-78

13. J.F. Van Impe, B. Nicolaï, P. Vanrolleghem, J. Spriet, B. De Moor and J. Vandewalle 1992. Optimal control of the penicillin G fed-batch fermentation: an analysis of a modified unstructured model. *Chem. Eng. Comm.*, 117, 337-353

14. J.F. Van Impe, B. De Moor and J. Vandewalle 1993. Singular optimal control of fed-batch fermentation processes with growth/production decoupling and state inequality constraints. *Preprints of the 12th World Congress International Federation of Automatic Control IFAC*, 9, 125-128

15. J.F. Van Impe 1993. *Modeling and optimal adaptive control of biotechnological processes*, PhD thesis, Department of Electrical Engineering, Katholieke Universiteit Leuven, 304 p.

Chapter 14

Predictive Control in Biotechnology using Fuzzy and Neural Models

H. TE BRAAKE, R. BABUSKA
Delft University of Technology
Control Laboratory
Mekelweg 4 2628 CD Delft (THE NETHERLANDS)

E. VAN CAN, C. HELLINGA*
Delft University of Technology
Kluyver Laboratory for Biotechnology
Julianalaan 67 2628 BC Delft (THE NETHERLANDS)
* to whom all correspondence should be addressed

ABSTRACT – *New developments in process modeling, identification, measurement and control are likely to cause some major breakthroughs in process control in the next decade. Especially black box modeling techniques based on Artificial Neural Networks and Fuzzy Set theory are opening new horizons for modeling and controlling non-linear processes in biotechnology. The link between accurate dynamic process models and actual process control is provided by the concept of Model-based Predictive Control (MBPC). A model serves here as process output predictor so that the effect of (future) control actions can be evaluated automatically, before the process is activated. This chapter presents a brief introduction to modeling with fuzzy sets and artificial neural nets. To demonstrate the practical applicability, laboratory*

J.F.M. Van Impe et al. (eds.), Advanced Instrumentation, Data Interpretation, and Control of Biotechnological Processes, 437-464.

experiments are described where MBPC was applied to a non-linear pressure control problem in a fermentor. Both fuzzy and neural models were developed and identified for this process and as the results show the fuzzy and neural MPBC outperform the classical PI controller. Controller tuning was very easy compared to classical (linear) techniques.

KEYWORDS – *Predictive Control, Fuzzy Sets, Artificial Neural Networks*

14.1 Introduction

In biotechnological conversion processes living organisms (or enzymes extracted from them) produce desired products in response to the physical and chemical environment they are exposed to. Such processes are multivariable, non-linear and have time varying parameters. Moreover, the number of practically applicable on-line sensors is limited, and thus the process state can in general only partly be monitored. Non-linearities not only stem from the complicated input/output relations of the biocatalysts but also from the fact that most industrial processes are carried out in batch or fed-batch mode. *Optimal control* is not very feasible under such conditions and in practice, processes are controlled with classical PID controllers in Single Input Single Output (SISO) control structures. Optimal or acceptable process conditions are found in laboratory experiments and translated into fixed setpoints for industrial production, that can be maintained relatively easily in decoupled control loops. With an increasing understanding of biocatalyst behavior and improving measurement technologies, new, more intelligent control approaches will become available in the next decade. This will help to meet the ever increasing requirements with respect to processing costs, safety, environmental regulations and product quality. Particularly, the need for overdimensioning plants (for example in wastewater treatment) will decrease with more adequate control techniques.

Various control structures containing fuzzy or neural models have been proposed in literature (e.g., [8][9][13][14][21]). Both fuzzy and neural models can be *retrained* on-line, i.e., the model parameters can be updated recursively when new measurements are available. Many proposed controller structures use this capability to construct adaptive controllers. However, a disadvantage of on-line adaptation is the fact that the parameters are updated according to a local situation, resulting in a loss of global modeling properties. In fact, Model-based Predictive Control (MBPC) with *fixed* fuzzy or neural models is easier to implement and more reliable and attractive for the process operator, because it does not suffer from the *learning (adapting)/forgetting* contradiction. It is more appropriate to adapt the neural or fuzzy model only when this is necessary. With fast *training* algorithms, an *on-line* batch retraining procedure should be possible.

In MBPC, a process model is used to predict process response to alternative controller outputs. There are practically no restrictions with respect to the model structure, so that MBPC can very well deal with process non-linearities. Model-based predictive control has become an important research area of automatic control theory and, moreover, it has been accepted also in industry [4]. A number of successful applications to industrial processes has been reported, see [15] for a survey. The ability to handle constraints is one of the reasons for this success.

Model development and identification are the most time consuming parts in designing such controllers. Therefore, good, efficient modeling and identification methods are most important. This chapter will focus on fuzzy logic and neural network models as new black box methods for (non-linear) process modeling. The models are very promising in combination with MBPC. These techniques are not only capable of modeling non-linear processes, also, these techniques are easy to apply and to implement.

First, some general remarks are made with respect to black and white box modeling. Then fuzzy logic and neural networks are briefly introduced, highlighting the specific features of these two approaches for modeling. Finally, some practical experiments with a pressure control system in a fermentor are presented to demonstrate the power of this method and to indicate what efforts are necessary to realize a practical MBPC implementation based on such black box models.

Black, grey and white box modeling

In their education, (bio-)chemical engineers are brought up with mechanistic, or white box, models. Such models are most valuable tools when studying the *first principles* underlying (bio-)chemical conversions. Model parameters are tightly connected to the physical-chemical reality and have *universal values*: mechanistic models are very well suited for process design, scale-up, scale-down and optimization. Characteristically, conversion processes are described with differential (balance) equations for all considered compounds stating that accumulation equals the sum of transport (to and from the reaction space) and conversion (production and/or consumption). Process dynamics as well as steady state conditions can be evaluated with the same set of equations. It seems difficult not to forget that even such, often complicated, process descriptions are still approximations. Model structure and model parameters are never completely valid for the whole range of relevant subprocesses. It is especially important to note that subprocesses have there own typical time scale. For describing process dynamics, only subprocesses having roughly the same time scale need to be considered. The other processes are either much faster (pseudo steady state) or much slower (frozen) and need not to be taken into account in the dynamic description [16]. When the relevant subprocesses are ordered according to their relative speed, the model structure and parameters become more accurate towards the slower processes, because they have been built and measured primarily for the design purposes. Proper design justifies high investments in model building and experimental verification, especially for large scale operations. A rough description of the process dynamics is often sufficient for design purposes.

For model-based control however, process dynamics must be described much more precisely. It is then very much the question whether it is still useful to stick to mechanistic process descriptions. Although for many cases the mathematically convenient lower order and/or linear(ized) process models are sufficient, these models also are no longer sufficient for use in MBPC. One is primarily interested in an accurate input/output description and not so much in understanding the underlying mechanisms. A flexible mathematical structure and a standard prescription for model design and parameter estimation are believed not only to result in more accurate models, but also in far reduced development time and costs. Artificial neural networks and fuzzy set theory provide frameworks that fulfil the demand for flexible mathematical structures. It was proven that both neural and fuzzy models can approximate any continuous (non-linear) mapping between input and output, i.e., between independent and dependent variables [7][5], which makes them suitable for modeling non-linear process description.

Anticipating the following sections, it should be noted that a disadvantage of black box models is that a rich set of input/output data is needed, that gives a *complete* information about the process dynamics. This is especially true for complex multivariable systems. Extrapolations outside the experimental *space* are not save with pure black box models! Current research in the field of grey box modeling indicates that combinations of white and black box models can contribute to solve this problem. In a recent paper, Thompson and Kramer [17] discuss how *a priori* knowledge can be used when modeling complex systems (choice of inputs/submodels, data preprocessing, selecting constraints for training sets, ...). In an example they show that a *hybrid* model, incorporating both a mechanistic and a neural part, has indeed important advantages for modeling a fed-batch penicillin fermentation, especially with respect to extrapolative power. Also for the fuzzy model structure it will be shown briefly, in the subsequent section, that fuzzy set theory provides an intrinsic framework for combining mechanistic knowledge with a black box description.

14.2 Non-linear Model-based Predictive Control Structure

Predictive controllers are based on the prediction of the future behavior of the process to be controlled (see Figure 14.1). This prediction is obtained from a (non-)linear model of the process that is assumed to be available.

A predictive controller calculates a sequence of future controller outputs $u(k+1), u(k+2), \ldots, u(k+H_p)$ over the prediction horizon H_p, such that the predicted output of the process is as close as possible to a reference trajectory defined by the operator. This control sequence is obtained by the optimization of a certain criterion function which describes the control goals. An example of a criterion function to be minimized is:

$$J[u] = \sum_{i=1}^{H_p} [\hat{y}(k+i) - w(k+i)]^2 + \beta\,[u(k+i-1)]^2 \tag{14.1}$$

with $\hat{y}$ being the predicted process output, w the desired process behavior (reference trajectory) and u the future control signal, weighted by a scalar β. Such a cost function aims simultaneously at the minimization of the output error and the energy spent on manipulating the process input. Parameter β determines the weighting of these two criteria.

Usually, the prediction horizon is as large as possible but at least it is larger than the process delay time. To avoid too many calculations, commonly the control horizon, i.e., the number of future control actions, is defined smaller than the prediction horizon. To construct an *open loop feedback controller*, only the first of the calculated control signals is applied. At every sample time, the optimization is repeated with new measurements and the previously calculated future control actions may be used as initial values for the next optimization.

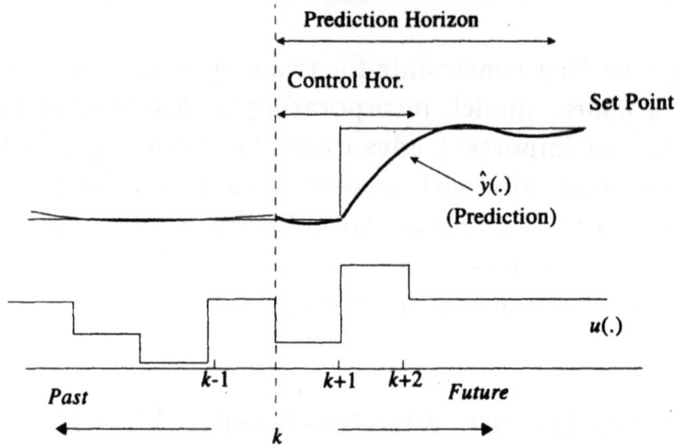

Figure 14.1: Model-based predictive control principle.

It is known that a predictive controller based on a linear process model, in the absence of constraints, results in a stable control law which can be calculated analytically, minimizing a predetermined criterion [18][23]. In the case of MBPC using a neural or fuzzy model, an optimal control action can only be calculated by means of numerical optimization of the criterion function. The resulting controller (optimizer) is a part of the MBPC structure, as depicted in Figure 14.2.

Based on *Non-linear Model 1* (see Figure 14.2), the *Optimizer* finds an optimal solution of the criterion function and the first element of the resulting control sequence can be applied to the process.

Both process inputs and outputs can be subject to operation constraints which should be incorporated in the optimization routine as hard or soft constraints.

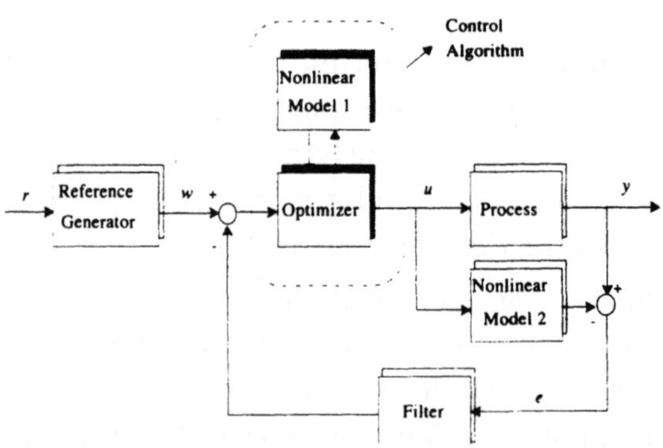

Figure 14.2: Non-linear predictive controller. r denotes the setpoint, w the reference signal (reshaped setpoint), u the calculated control action, and e the error between the process and the non-linear process model.

Suppose that the *Non-linear Model 2*, which is exactly the same as the *Non-linear Model 1* used in the optimization routine, predicts exactly the process output y. Then the error between the predicted (based on the non-linear model) and the measured output will be zero. The control structure becomes a simple feedforward controller. However, due to the recalculation of the controller outputs at every sampling instant, actually a combination of *feedforward* (prediction part) and *feedback* (recalculation at every time instant) action is present, which is able to react on differences in process and reference requirements.

If the predicted and process outputs are not equal (due to noise, disturbances or mismatch between model and process) a feedback loop corrects this error. A *Filter* (see Figure 14.2) is designed to filter out the process noise and to make the loop more stable. Because at every time instant the controller minimizes the error between the setpoint (minus the filtered error e) and process output, an offset-free response is guaranteed. A reference generator can be used to reshape the setpoint to a smoother reference signal, before it is used in the calculation of the control action.

14.3 Modeling with Neural Networks

The original idea of applying artificial neural networks (ANN) was to imitate the way the human brain processes information. For our purposes, however, this *background* will not be elaborated and an ANN will simply be regarded as a convenient way to model a non-linear input-output mapping. The neural network will be denoted by:

$$\mathbf{y} = f_{nn}(\mathbf{x}) \tag{14.2}$$

($\mathbf{x}$ is the input and $\mathbf{y}$ is the output vector). This non-linear mapping capability can be used in the modeling of non-linear dynamic processes. In the next section the structure of a neural network will be presented in more detail. If one is not interested in these details, this section can be skipped without loosing the essence of this chapter.

14.3.1 Neural network structure

The basic entity in a neural network is a neuron (see Figure 14.3). It consists of a summator, which adds the weighted inputs, and a non-linear *activation* function, which maps this sum onto a bounded interval (usually [-1,1] or

[0,1]), using a non-linear *s-shaped* (sigmoidal) function. In our experiments, the tangent hyperbolic function was used.

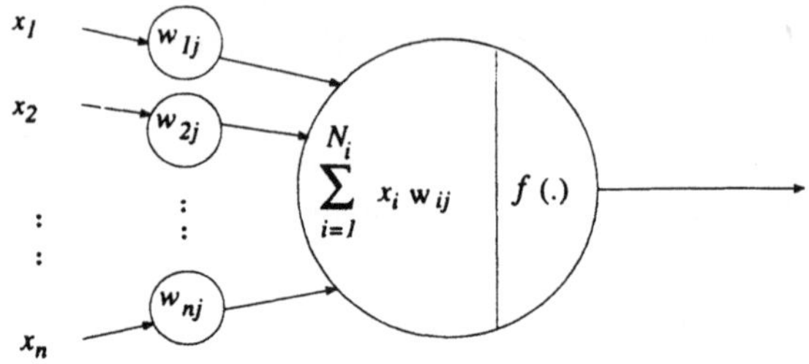

Figure 14.3: Artificial neuron j.

Neurons are organized in several layers. Here we only consider the feedforward neural network structure. In this structure, the layers are organized such that the outputs of one layer are the inputs for the next layer. The network does not contain interconnections in the same layer nor feedback between neurons of different layers (see Figure 14.4).

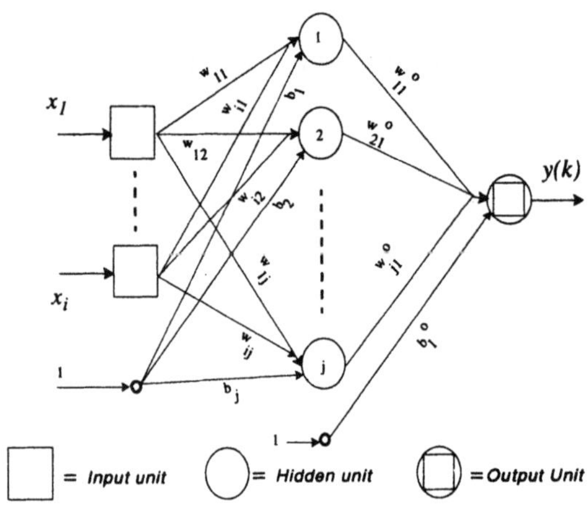

Figure 14.4: Feedforward neural network with N_i inputs (x_1,...,x_i) and one output (y). Input and output neurons do not include the non-linear function. For simplicity w^h is denoted by w.

Every input of a neuron in the hidden layer is multiplied by a weight W_{ij}^h. To these weighted inputs, a bias input is added:

$$z_j(k) = \sum_{i=1}^{N_i} \left[W_{ij}^h \cdot x_i(k) \right] + b_j^h \tag{14.3}$$

The output of the neurons in the hidden layer is then given by:

$$v_j(k) = f(z_j(k)) \tag{14.4}$$

The output of the neural network is given by:

$$y_l(k) = \sum_{j=1}^{N_h} \left[W^o_{jl} \cdot v_j(k)\right] + b^o_l \tag{14.5}$$

Subscript j denotes the j-th neuron in the hidden layer, i denotes the i-th input, l denotes the l-th output. The index k denotes the k-th event (i.e., sample). N_e is the number of events, N_i the number of inputs and N_o the number of outputs. N_h is the number of neurons in the hidden layer. The input bias is denoted as b^h and the output bias as b^o.

The input vectors and the bias vector can be arranged in a matrix, with one row for each event and one column for each input (including a unit column for the biases). Similarly, the output vectors can be grouped into a matrix. Then the neural net is expressed concisely as:

$$\begin{aligned} \mathbf{Z} &= \mathbf{X} \cdot \mathbf{W^h} \\ \mathbf{V} &= f(\mathbf{Z}) \\ \mathbf{Y} &= \mathbf{V}_b \cdot \mathbf{W^o} \end{aligned} \tag{14.6}$$

with input matrix $\mathbf{X} \in \mathcal{R}^{N_e \times (N_i+1)}$ and output matrix $\mathbf{Y} \in \mathcal{R}^{N_e \times N_o}$. Matrices $\mathbf{Z}$ and $\mathbf{V}$ contain intermediate results. Matrix $\mathbf{V}_b \in \mathcal{R}^{N_e \times (N_h+1)}$ is equal to $\mathbf{V} \in \mathcal{R}^{N_e \times N_h}$ augmented with a unit column for output bias b^o. The activation weights are arranged in matrix $\mathbf{W^h} \in \mathcal{R}^{(N_i+1) \times N_h}$, and the output weights in matrix $\mathbf{W^o} \in \mathcal{R}^{(N_h+1) \times N_o}$.

The neural network mapping from the inputs to the outputs then will be expressed conveniently by:

$$y = f_{nn}(x_1, \ldots, x_{N_i}) \tag{14.7}$$

In this equation $f_{nn}(\ldots)$ expresses the neural network, $x_1, \ldots, x_{N_i}$ are the inputs of the network and y is the network output.

14.3.2 Training of neural networks

Training of neural networks is, in fact, an estimation of the weights (the elements of $\mathbf{W^h}$ and $\mathbf{W^o}$). The weights are adjusted such that for a given input the neural network output is as close as possible to the desired output. This type of learning is called supervised learning. The weights can be estimated using a training set and a training algorithm. A training set is a data set which contains the inputs and desired (e.g., measured) outputs.

There are various training algorithms among which *backpropagation* is the most well-known (see for example: [8]). However, because it is a gradient descent optimization algorithm, it is rather slow and sensitive to local optima. Recently, faster algorithms have been proposed. te Braake and van Straten [3] for example, proposed the *RAWN* method based on fixing the randomly initialized weights between the input and the hidden layer, in such a way that a linear parameter estimation problem remains, which can be solved with standard least squares techniques. This results in an accurate neural model which is obtained in a fraction of required training time compared to backpropagation (e.g., seconds instead of hours). In the next section a more detailed explanation about the RAWN method will be given.

14.3.3 RAWN training method

The elements of the weight matrices $\mathbf{W^h}$ and $\mathbf{W^o}$ are parameters that must be found in order to obtain an acceptable fit to the available data. This process is called training. The idea of the RAWN method is to split the training problem into two subproblems which can be treated separately.

To obtain $\mathbf{W^o}$, first assume that the weights $\mathbf{W^h}$ are already known, and therefore $\mathbf{V}_b$ is known, too. Suppose that the true output can be modeled by:

$$\mathbf{Y} = \mathbf{V}_b\mathbf{W^o} + \mathbf{e} \tag{14.8}$$

The vector $\mathbf{e}$ denotes the modeling error. This parameter estimation problem is linear in the parameters. By minimizing the sum of squared modeling errors the well known least squares estimation of $\mathbf{W^o}$ then becomes [10]:

$$\widehat{\mathbf{W}}^{\mathbf{o}} = \left[\mathbf{V}_b^T\mathbf{V}_b\right]^{-1}\mathbf{V}_b^T\mathbf{Y} \tag{14.9}$$

The *hat* denotes the estimation of a variable. The matrix $\mathbf{V}_b^T\mathbf{V}b$ must be non-singular otherwise $[\mathbf{V}_b^T\mathbf{V}_b]^{-1}$ would not exist. This implies that $\mathbf{V}_b$ must have rank N_h.

Before Equation (14.9) can be applied, however, the matrix $\mathbf{W^h}$ must be obtained. The basic idea is that the input data multiplied with the activation weights (variable $z_j(k)$ in Equation (14.3)) must be *correctly* spread out over the relevant area of the non-linear activation function. If a *tanh*-activation function is used one can easily demonstrate that if $-c \leq z_j(k) \leq c$, $c \approx 3$, $z_j(k)$ is contained in the non-saturated part of this function (see Figure 14.5).

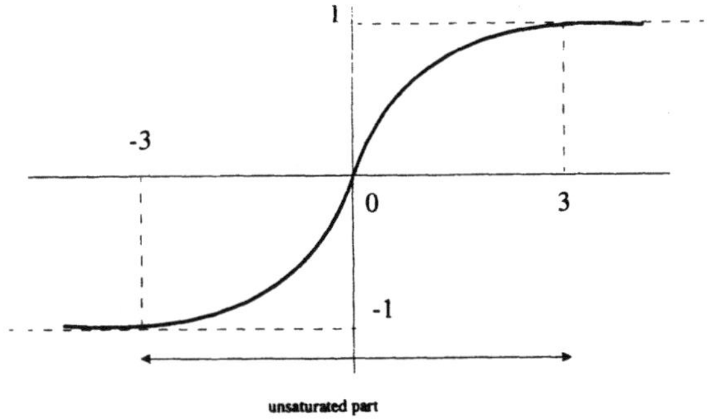

Figure 14.5: Graph of the tanh function.

If $|z_j(k)| \gg c$ for all j then $z_j(k)$ is in the saturated part of the activation function and then:

$$f\left(|z_j(k)|\right) \approx f\left(|z_j(k) + \delta|\right) \approx 1$$

with δ a small real number. The training therefore probably will fail to find a correct mapping between input and output.

If $|z_j(k)| \ll c$ then the mapping is concentrated around $-\epsilon < z_j(k) < \epsilon$, with ϵ a small real number. Then $f\left(z_j(k)\right)$ becomes a linear function of $z_j(k)$ which leads to an ill-conditioned matrix inversion $[\mathbf{V}_b^T \mathbf{V}_b]^{-1}$. To get a solution for the estimate of $\mathbf{W^o}$ (Equation (14.9)), matrix $[\mathbf{V}_b^T \mathbf{V}_b]^{-1}$ has to be non-singular. $\mathbf{V}_b$ therefore must be of full rank.

The matrix product $\mathbf{XW^h}$ causes that the rank of $\mathbf{Z}$ (see Equation (14.6)) equals to the rank of $\mathbf{X}$, provided that $\mathbf{W^h}$ is of full rank. If the function $f(\cdot)$ (as used in Equation (14.4) or (14.6)) is a linear function or a power function (e.g., $f(\cdot) = z^a$), the columns of $\mathbf{V}_b$ still will not be of full rank. On the contrary, if the function $f(\cdot)$ is a kind of exponential function (for example $f(\cdot) = e^z$), $\mathbf{V}_b$ will have full rank.

One way to obtain the elements of $\mathbf{W^h}$, such that $z_j(k)$ is neither only in the linear nor in the saturation part of the non-linear function, is to choose them randomly such that the maximum value of the variance of $z_j(k)$, for the whole data set, remains below a certain arbitrary value $0 < \alpha < c$. Or, in equation form:

$$\max_k \ \operatorname*{var}_j \left\{ \sum_{i=1}^{N_i+1} x_i(k) \cdot W_{ij}^h \right\} \leq \alpha \qquad (14.10)$$

This is equivalent to generating $\mathbf{W^h}$ according to:

$$W^h = \sqrt{\frac{\alpha}{\max_k \sum_{i=1}^{N_i+1} x_i(k)}} \cdot N(0,1) \tag{14.11}$$

where $N(0,1)$ denotes a generator of random numbers with zero mean normal distribution with standard deviation 1. After generating $\mathbf{W^h}$, the elements of $\mathbf{W^o}$ can be calculated with Equation (14.9).

As already stated, the RAWN approach divides the training problem into two subproblems. This does not imply that the overall problem can be solved optimally in a global sense. But it should be stressed that backpropagation also does not guarantee global minimization of the training error. Combined with the slow convergence properties of backpropagation, applying the RAWN method to a real-life process is attractive. In our case, better results have been achieved in a fraction of the time backpropagation needs.

14.3.4 Modeling non-linear dynamic systems with neural networks

The use of neural networks for the modeling of a dynamic process is explained with an example of a SISO (single input, single output) dynamic system. Assume that an input u is presented to an open loop non-linear system and that an output y is measured. Suppose it is known that the orders of a discrete-time model are 2, 1, i.e., two lagged outputs and one lagged input. Then the following neural model can be trained:

$$y_{k+1} = f_{nn}(u_k, y_k, y_{k-1}) \tag{14.12}$$

In this equation $f(\ldots)$ is a neural network with inputs u_k, y_k and y_{k-1} and output y_{k+1}. For comparison, a linear counterpart of this neural model would be $y_{k+1} = a_1 y_k + a_2 y_{k-1} + b u_k$, with parameters a_1, a_2 and b to be estimated.

To train the network, the inputs are arranged in a regression matrix $\mathbf{X}$ and the outputs in matrix $\mathbf{Y}$:

$$\mathbf{X} = \begin{bmatrix} u_2 & y_2 & y_1 & 1 \\ \vdots & \vdots & \vdots & \vdots \\ u_k & y_k & y_{k-1} & 1 \\ \vdots & \vdots & \vdots & \vdots \\ u_{N-1} & y_{N-1} & y_{N-2} & 1 \end{bmatrix} \qquad \mathbf{Y} = \begin{bmatrix} y_3 \\ \vdots \\ y_{k+1} \\ \vdots \\ y_N \end{bmatrix} \tag{14.13}$$

With these two matrices containing the measured data, the weights are found by applying the training algorithm (first generate $\mathbf{W^h}$ with Equation (14.11) and then calculate $\mathbf{W^o}$ with Equation (14.11)). The identified neural model can be used directly in the MBPC structure as shown in a later section.

The obtained model is validated in a simulation mode, which is explained in Figure 14.6.

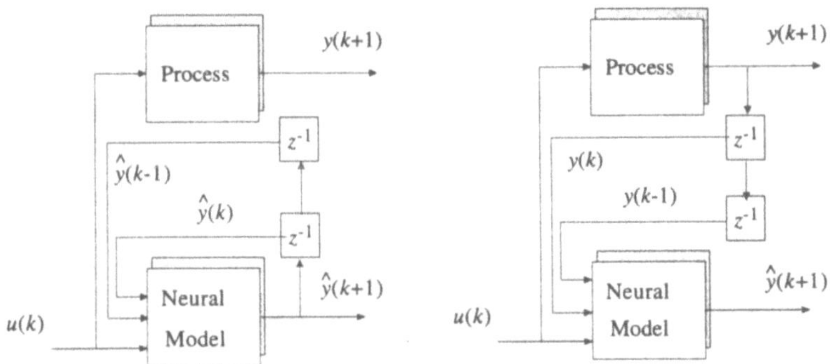

Figure 14.6: Left: Simulation mode. Right: One-step-ahead prediction. z^{-1} denotes a 1 sample lag.

The term *simulation* is used to stress the contrast with the so-called *one-step-ahead prediction*. In the latter case, the measured process output is fed back instead of the model output. Clearly, if the model is used for predicting the process output over a certain horizon (as is the case in MBPC), the simulation mode is a *heavier* test and better suited than validation of the one-step-ahead-prediction.

14.4 Modeling with Fuzzy Logic

In contrast to the *numerical nature* of neural networks, fuzzy sets theory and fuzzy logic are tools for handling imprecision and modeling linguistic semantics. The key idea of fuzzy modeling is to use a simple linguistic description of the process instead of precise mathematical relations between the variables. Fuzzy models are useful for processes where the underlying physical mechanisms are not completely known and when the understanding of the process behavior is mostly qualitative. Another important property of a fuzzy model is its capability to represent non-linear mappings. Therefore, fuzzy models also can be applied to systems that are well understood but are intractable with standard linear methods due to the non-linearities. Fuzzy sets are a powerful tool for formalizing heuristic knowledge expressed as rules and for blending this knowledge with information obtained from numerical data.

14.4.1 Fuzzy model structure

In literature, the term fuzzy modeling encompasses various kinds of models where the concept of a fuzzy set, i.e., a set with graded membership (in contrast to binary membership values (*crisp* values) in classical sets theory) is applied. Usually, fuzzy sets are used for partitioning the continuous domains of the model input and output variables into a finite number of mutually overlapping regions. A dynamic system is then represented as a static mapping between the fuzzy sets defined in the space of lagged model inputs and outputs. Note that in the same way the system dynamics are captured in other kinds of models like linear regression models or neural networks (see, e.g., Figure 14.7 and Equation (14.12)).

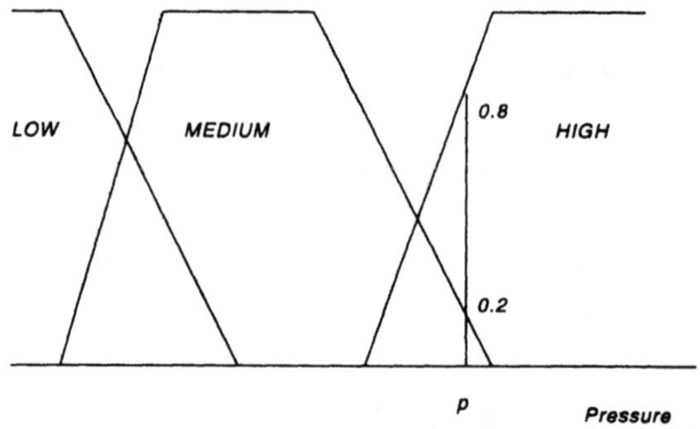

Figure 14.7: Fuzzy sets defined on the domain of pressure.

To illustrate the concept of graded membership, three fuzzy sets defined in the domain of a continuous process variable (pressure) are shown in Figure 14.7. For a particular pressure measurement p, a vector of membership degrees $\mu(p) = [\mu(p)_{LOW}, \mu(p)_{MEDIUM}, \mu(p)_{HIGH}]$ can be calculated. In our example $\mu(p) = [0, 0.2, 0.8]$. Note that as opposed to probabilities, the sum of membership degrees does not have to be one. The conversion of real numbers (measurements) into fuzzy representation in terms of membership grades is called fuzzification.

The mapping between model input and output fuzzy sets is usually defined as a set of *If/Then* rules, such as:

> *If pressure is LOW and outlet valve opening is MEDIUM,*
> *Then pressure will rise FAST.*

The *If* condition of the rule is called premise, the part following *Then* is called consequent. Linguistic terms (*LOW*, *MEDIUM*, *FAST*, ...) with a good semantic meaning are assigned to the fuzzy sets in order to make the model transparent and easily understandable. Membership functions provide the interface of the linguistic terms to the numerical variables, as shown in Figure 14.7. The complete fuzzy model consists of a set of rules (called the rule base) such that the combinations of premise fuzzy sets cover all the possible inputs. Because of its linguistic nature, such a rule-based model is also understandable to the users without any particular mathematical knowledge. However, it should be stressed that for complicated systems, a large number of rules may be needed in order to describe the system dynamics with a reasonable accuracy.

As an extension to the standard linguistic models described above, a class of so called fuzzy-linear models was introduced by Sugeno and his co-workers [21][19]. The rules in fuzzy-linear models combine linguistic premises with functional consequents, usually defined as linear functions of the input variables. A typical fuzzy-linear rule may look as follows:

> *If pressure $y(k)$ is SMALL and valve opening $u(k)$ is MEDIUM,*
> *Then pressure $y(k+1)$ is equal to $y(k+1) = ay(k) + bu(k) + c$.*

Such a fuzzy-linear model can be regarded as a collection of several linear models applied locally in the fuzzy regions defined by the rule premises. Each rule consequent defines different values for the parameters a, b, c. The idea behind fuzzy-linear modeling is similar to the well-known concept of gain scheduling and local linearization around operating points. The main advantage of fuzzy-linear models is their efficient way of representing non-linear functions by combining functional descriptions with a global rule-based character of the model. Fuzzy sets ensure smooth transition between individual local models. Also note that the model is linear with respect to the consequent parameters, which can be identified from numerical data using standard linear identification techniques (e.g., least mean squares).

14.4.2 Fuzzy inference mechanism

Once a numerical input to the fuzzy model is available, the corresponding numerical output of the model can be derived from the rules by means of so-called *fuzzy inference*. In standard mathematical terminology, fuzzy inference is nothing else than a multidimensional interpolation technique using base (membership) functions defined in the neighborhood of several fixed points (rules). The intuitive idea of fuzzy inference is very simple: the more the premise of a rule is satisfied (e.g., the higher the overall membership grade of the premise is), the more the rule consequent weighs. The overall membership grade of the rule premise is calculated as a combination of membership grades of the individual premise variables. If the premise variables are combined with a logical connective *and*, a conjunctive operator is used to combine the membership degrees. Typically, the minimum or product operators are applied. For example, with rules

R1.
If pressure $y(k)$ is SMALL and valve opening $u(k)$ is MEDIUM,
Then pressure $y_1(k+1)$ is equal to $y_1(k+1) = a_1 y(k) + b_1 u(k) + c_1$.

R2.
If pressure $y(k)$ is MEDIUM and valve opening $u(k)$ is MEDIUM,
Then pressure $y_2(k+1)$ is equal to $y_2(k+1) = a_2 y(k) + b_2 u(k) + c_2$.

...

the overall membership degree of the premise of rule R1 is calculated as

$$\mu_1 = \min\left(\mu_{SMALL}(y(k)), \mu_{MEDIUM}(u(k))\right) \tag{14.14}$$

and for rule R2 as

$$\mu_2 = \min\left(\mu_{MEDIUM}(y(k)), \mu_{MEDIUM}(u(k))\right) \tag{14.15}$$

The model output is computed as a weighted mean of the individual rule contributions:

$$y = \frac{\sum_{i=1}^{K} \mu_i y_i}{\sum_{i=1}^{K} \mu_i} \tag{14.16}$$

where μ_i is the overall membership degree of the i-th rule premise, y_i is the consequent value of the i-th rule and K is the total number of rules.

14.4.3 Identification of fuzzy models from data

If no process data is available, fuzzy models can be constructed using expert knowledge and heuristics. However, when measurements can be obtained, identification methods can be applied to fine-tune the model or to build it directly from the data. After choosing the model structure (i.e., finding the relevant inputs of the model and choosing the number of fuzzy sets per variable), the identification of a fuzzy model consists of the estimation of the membership function parameters (position, shape) and extraction of the rules (i.e., finding the mapping from the premise fuzzy sets to the consequent sets or functions). This is essentially a non-linear optimization problem, which can be more or less efficiently solved with a variety of methods, including neural network learning techniques, genetic algorithms or standard numerical methods (see, e.g., [22][21][19][14]). Also special techniques from the field of pattern recognition, such as fuzzy clustering have been successfully applied. The advantage of the fuzzy clustering approach is that the membership functions and the rules are identified simultaneously. Details can be found in [20][1][25][11].

14.4.4 Prediction with fuzzy models

The level of generality of fuzzy models places them between purely numerical models and qualitative models. There are several possibilities for predictive control schemes based on fuzzy models. First, a fuzzy model can be used for prediction on the numerical level, i.e., in the same way as, for instance, neural networks. In this case, the rule-based nature of the model does not yield any special advantages, except that the rules can give direct insight into the local behavior of the system, contrary to neural nets which are usually treated as completely black box models. For numerical predictions, fuzzy-linear models are more suitable than purely linguistic fuzzy models because of their higher degree of accuracy.

A fuzzy model also can serve as a predictor determining the future states in the form of linguistic terms. In this way, soft (fuzzy) constraints and goals can be defined in a linguistic way. This is suitable, e.g., for dealing with variables that humans define in a heuristic way, such as comfort, smell, colors, ... The degree of satisfaction of the fuzzy goals and constraints is based on similarity measures between the fuzzy sets describing the goals and constraints and the system states or outputs. In order to reduce the computations, only a finite set of discrete control inputs can be considered.

The optimization problem then boils down to a search in a discrete space of the control alternatives. Standard techniques known from operational research can be used, including dynamic programming, branch-and-bound method, ... [2]. Similar approach was used in the well-known application of the fuzzy predictive control concept to the Sendai subway system in Japan [24]. In this chapter, we illustrate the use of fuzzy-linear models for numerical predictions.

14.5 Application to Pressure Control in a Fermentor

The remainder of this chapter is devoted to an application of fuzzy and neural modeling to the predictive control of the pressure in a laboratory-scale fermentor. First, the modeling with a neural network and with fuzzy logic is addressed and next the use of these models in the MBPC structure is described.

It will be shown that the resulting controller performs better than a conventional linear MBPC or PI controller, but more importantly this controller structure is easier to design and to tune than conventional control schemes.

14.5.1 Experimental setup

The use of fuzzy and neural models in MBPC is demonstrated in the context of the real-time control of the pressure in a fermentor. Figure 14.8 shows the scheme of this setup.

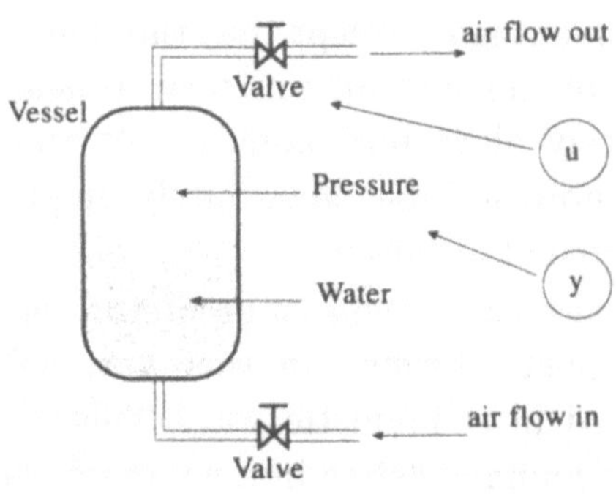

Figure 14.8: Experimental setup.

Air is supplied at the bottom of the vessel. The specified flow rate is kept constant by a local mass flow controller. The position of the outlet valve determines the pressure in the fermentor. By measuring various step responses, a time constant of about 45 seconds was found. The sample time therefore was chosen to 5 seconds. For controller implementation, no special hardware or software is needed. The controller was implemented in PASCAL and runs on a DOS based 486/50 MHz computer. The code for the on-line calculation of the fuzzy and neural model contains only a few lines. The optimizer (see Figure 14.2) uses standard optimization methods like the Nelder-Mead algorithm [12].

14.5.2 Building a neural network model for the pressure process

To find the parameters of a linear black box model, random input signals are often used, because these kinds of signals contain all frequencies. The superposition principle and the theory of persistent excitation can be applied. If a non-linear model must be found, the linear systems theory can no longer be applied. The choice of the training signals is then a *process-dependent* task and depends on the operating conditions of the process. In this study, 5-10 superimposed sinusoidal functions are used to build the training signal [6]. The following heuristic rule was used to determine which frequencies the superimposed wave functions should contain:

$$\frac{\tau}{10} < \frac{1}{f} < 10\tau \tag{14.17}$$

where f is the frequency of the wave function and τ is the (relevant) time constant of the process. One also should notice that not only the amplitude domain (i.e., the range of the process input) but also the amplitude range (i.e., the range of the process output) should cover the whole operating range of the model.

Training signals are obtained by measuring the pressure in the fermentor (in Figure 14.8 denoted y) while the designed input signal (i.e., the valve position u) is supplied at every time instant. With the training data, a training set can be constructed. The input matrix $\mathbf{X}$ and the output matrix $\mathbf{Y}$ are then given by (compare this with Equation (14.13)):

$$\mathbf{X} = \begin{bmatrix} u_2 & y_2 & y_1 & 1 \\ \vdots & \vdots & \vdots & \vdots \\ u_k & y_k & y_{k-1} & 1 \\ \vdots & \vdots & \vdots & \vdots \\ u_{N-1} & y_{N-1} & y_{N-2} & 1 \end{bmatrix} \qquad \mathbf{Y} = \begin{bmatrix} y_3 \\ \vdots \\ y_{k+1} \\ \vdots \\ y_N \end{bmatrix} \tag{14.18}$$

Thus, u_k, y_k and y_{k-1} are the inputs of the neural model and y_{k+1} is the output. Such a model is described by the following equation:

$$y_{k+1} = f_{nn}(u_k, y_k, y_{k-1}) \tag{14.19}$$

(for convenience the notation is changed from $y(k)$ to y_k, ...). The input of this network is given by a vector with elements u_k, y_k and y_{k-1}. $\mathbf{W^h} \in \mathcal{R}^{(3+1)\times N_h}$ and $\mathbf{W^o} \in \mathcal{R}^{(N_h+1)\times 1}$. The number of neurons in the hidden layer is 16 (i.e., $N_h = 16$). Note that when different network inputs are chosen, the matrix $\mathbf{X}$ in Equation (14.18) also must be reconstructed. The fast algorithm [3], described in Section 14.3.3, is used to obtain the weights. With this algorithm it takes only seconds to obtain a reliable neural model.

A test set is used to ascertain whether the model is correct. The selection of the test set is very important in deciding which model is the best. In theory, an infinite number of test signals must be used to decide whether the model is correct. In this study a test signal is used that differs from the training signals both in frequency content and amplitudes. Evaluation of the test results is necessary to check whether the model is accurate enough. If not, the model structure needs to be changed and tested again. This is a time consuming *trial-and-error* procedure, stressing the need for a fast training algorithm.

After testing various configurations, the one described by Equation (14.19) is found to be accurate enough. The result of this model on the test data is show in Figure 14.9.

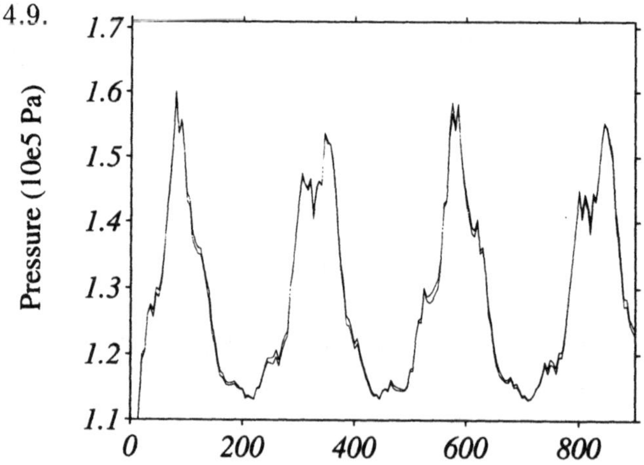

Figure 14.9: Validation of the neural model with 16 hidden neurons via simulation.

This model will be used in the MBPC controller mentioned later. In the next figure a scheme is given to summarize the necessary steps to obtain a correct neural network model.

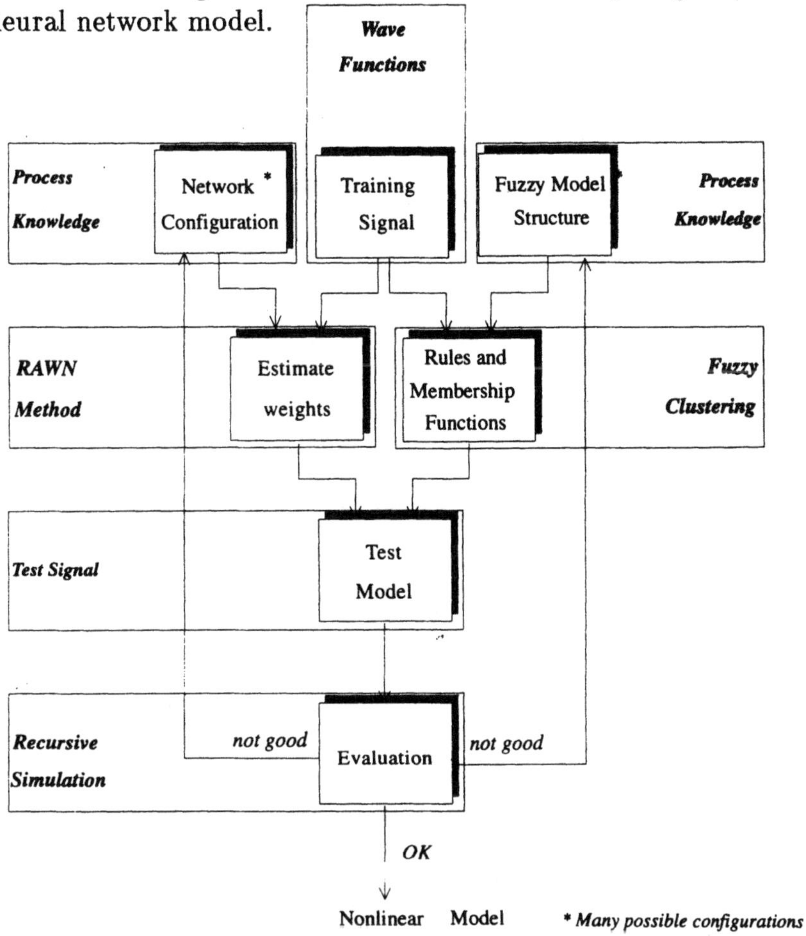

Figure 14.10: Modeling of non-linear processes by neural (left side of the diagram) or fuzzy model (right side of the diagram).

14.5.3 Fuzzy modeling of the pressure process

To demonstrate the fuzzy modeling methodology, a very simple fuzzy-linear model was identified for the pressure process, using a method based on fuzzy clustering [1]. The model rule base, containing only three simple rules, approximates a first-order non-linear regression model where the new pressure is a function of the current pressure and the valve position:

$$y(k+1) = f(y(k), u(k)) \tag{14.20}$$

The complete model was identified from a single data set of global process measurements. Figure 14.11 shows the three membership functions for the current pressure $y(k)$, labeled *Low*, *Medium* and *High*, respectively. Figure 14.12 shows the membership functions *Open*, *Half Closed* and *Closed* for the valve position $u(k)$. The membership functions were obtained as cluster projections.

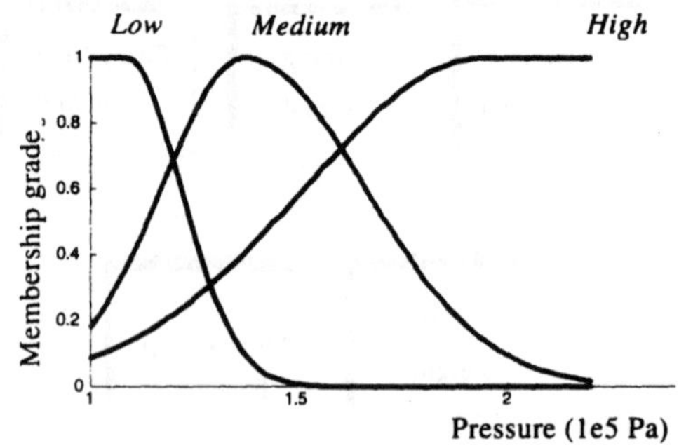

Figure 14.11: Membership functions for pressure.

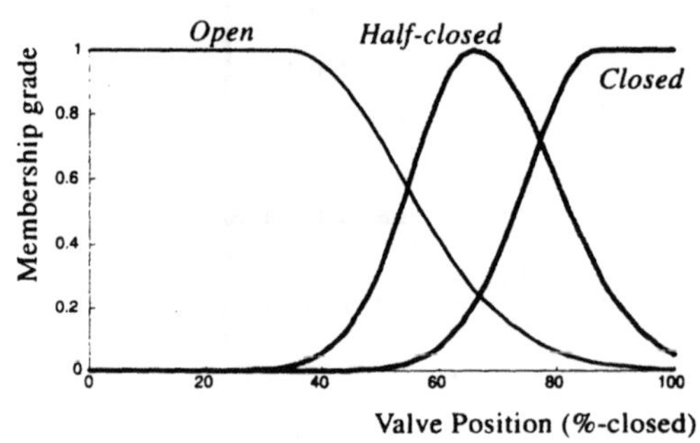

Figure 14.12: Membership functions for valve position.

In the following rule base, $y(k)$ and $y(k+1)$ denote the model output (pressure) at the discrete time instants k and $k+1$, respectively. $u(k)$ is the model input (valve position).

If pressure $y(k)$ is Low and valve $u(k)$ is Open,
then new pressure $y(k+1)$ is equal to
$y(k+1) = 0.67y(k) + 0.0007u(k) + 0.35.$

If pressure $y(k)$ is Medium and valve $u(k)$ is Half Closed,
then new pressure $y(k+1)$ is equal to
$y(k+1) = 0.80y(k) + 0.0028u(k) + 0.07.$

If pressure $y(k)$ is High and valve $u(k)$ is Closed,
then new pressure $y(k+1)$ is equal to
$y(k+1) = 0.90y(k) + 0.0071u(k) - 0.39.$

Note that some combinations of premise fuzzy sets are not included in the rule base. This is because the identification data set does not contain any information about the system behavior in these regions (e.g., valve *Open* and *High* pressure). Also note that such states even may not be physically reachable. For simulations, however, heuristic rules can be added to the initial fuzzy model to cover the situations not identified from the data. This simple fuzzy-linear model was compared in simulation with the real process. As shown in Figure 14.13 there is a small discrepancy between the model and process outputs, which can be removed almost completely by increasing the number of rules in the rule base. The clear advantage of the fuzzy model in this case is that the character of the process non-linearities can be easily understood from the rules. For instance, comparing the consequents of the three rules, one can see that both the process gain and the term corresponding to the time constant increase as the valve is more closed and the pressure increases. This is in a good agreement with the knowledge about the underlying physical mechanisms of the process.

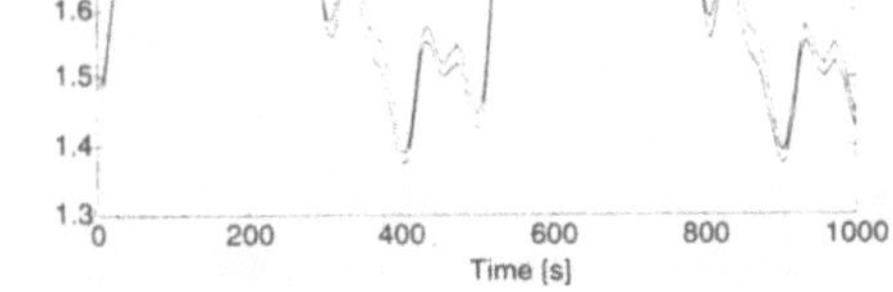

Figure 14.13: Validation of the fuzzy model via simulation.

14.5.4 Controlling the fermentor with non-linear MBPC

With the obtained models, a predictive controller is designed for the pressure process. Only the results of the neural net-based controller are shown. The controller based on the fuzzy model performs comparably.

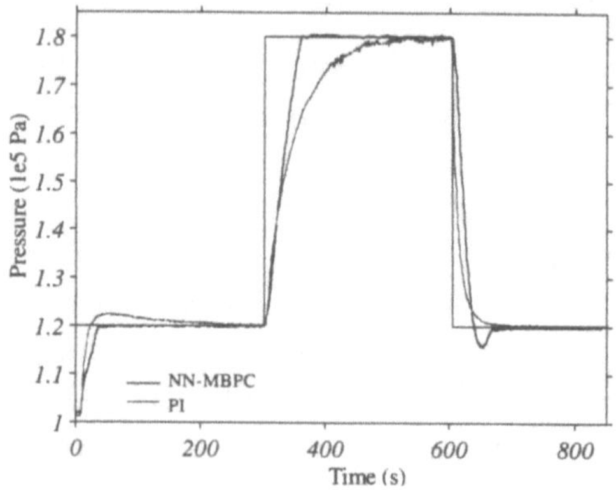

Figure 14.14: Performance of the neural model predictive controller for setpoint changes from 1.20 10^5 Pa to 1.80 10^5 Pa and back to 1.20 10^5 Pa.

As shown in Figure 14.14, Figure 14.15 and Figure 14.16, the controller is capable of tracking several different setpoint changes. In the experiment presented in Figure 14.16 the system dynamics are changed on purpose by applying a different input gas flow rate during the control experiment (15 L/min) than during the training (22.5 L/min). It can be seen that the system dynamics were slower due to the lower gas flow rate, but still the controller remains stable and is capable of tracking the setpoint.

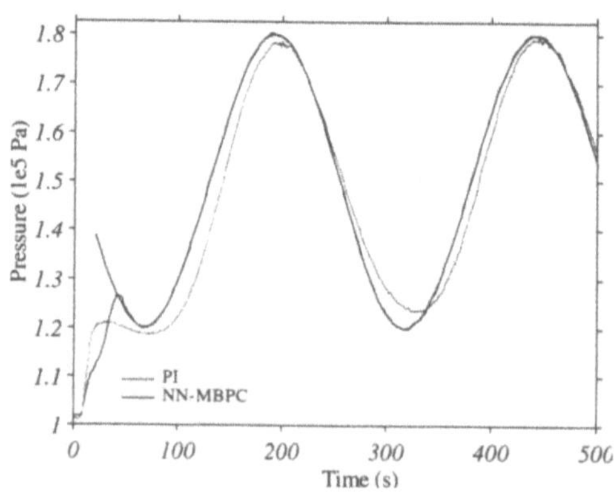

Figure 14.15: Performance of the neural MBPC controller on a sinusoidal setpoint change.

Figure 14.14, Figure 14.15 and Figure 14.16 also show the response of a carefully tuned digital PI controller. The sample periods of the PI and MBPC controller were the same (5 seconds). The response of the PI controller is slower than the response of the model-based predictive controller, especially for large setpoint changes.

We experienced that tuning the conventional controller was more cumbersome than designing a predictive controller based on a neural network or fuzzy model, especially because both types could easily deal with the non-linearities of the system.

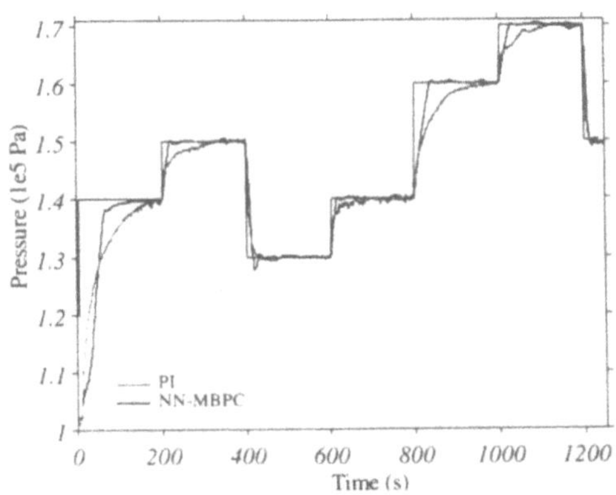

Figure 14.16: Performance of the neural model predictive controller on several setpoint changes (gas flow rate was 15.0 L/min).

14.6 Discussion and Conclusions

Neural and fuzzy models provide an efficient and convenient way for obtaining non-linear models valid over the whole operating range of the process. This makes the implementation and tuning of the predictive controller easier than with linear models, which are valid only around a certain operating point.

Compared to other non-linear models (such as white box models based on physical laws) they are easier to identify. Since the modeling part of the MBPC design is the most difficult one (according to Richalet [15] more than 80 % of the effort), significant savings of time and money can be achieved by applying neural network and fuzzy modeling techniques.

Both models can be used directly in a MBPC structure as numerical predictors forecasting the future process outputs and states. Due to their non-linear nature, analytical formulation of the control law is not straightforward. Therefore, the MBPC must rely on a good optimization method. In the case of complex, multivariable systems, computational limitations may become an issue. When comparing fuzzy and neural models several distinctive points should be mentioned.

- Neural networks are usually considered to be completely black box models which makes their analysis difficult. Good generalization properties are ensured by broad information contents of the training data and correct network structure.

- Fuzzy models partition the input space into a set of discrete regions. Rule base analysis can be used to find out the non-identified regions and additional rules can be supplied, using different strategies (heuristics, interpolation or extrapolation in the rule base, ...).

- One pays for the higher transparency of fuzzy models by more complicated identification methods, compared to the training of neural networks.

Fuzzy and neural net models can easily be implemented on standard computers without the explicit need for specialized hardware and/or software (even though both are nowadays commercially available).

14.7 Bibliography

1. R. Babuska and H.B. Verbruggen 1994. Applied fuzzy modeling. In: *IFAC Symposium on Artificial Intelligence in Real Time Control*, Valencia (Spain)

2. R.E. Bellman 1957. *Dynamic Programming*, Princeton University Press, Princeton, New Jersey

3. H.A.B. te Braake and G. van Straten 1995. Random activation weights neural net (RAWN) for fast noniterative training. *Engineering Applications of AI*, 8, 71-80

4. M.L. Brisk 1993. Process control: theories and profits. *Proceedings of the IFAC World Congress*, Sydney, 7, 241-250

5. J.J. Buckley and Y. Hayashi 1993. Fuzzy input-output controllers are universal approximators. *Fuzzy Sets and Systems*, 58, 273-278

6. H.J.L. van Can, H.A.B. te Braake, C. Hellinga, A.J. Krijgsman, H.B. Verbruggen, K.Ch.A.M. Luyben and J.J. Heijnen 1995. Design and real time testing of a neural model predictive controller for a nonlinear system. *Chemical Engineering Science*, 50, 2419-2430

7. G. Cybenko 1989. Approximation by superpositions of a sigmoidal function. *Mathematics of Control, Signals, and Systems*, 2, 303-314

8. K.J. Hunt, D. Sbarbaro, R. Zbikowski and P.J. Gawthrop 1992. Neural Networks for Control Systems - A survey. *Automatica*, 28, 1083-1112

9. H. Koivisto, V. Ruoppila and H.N. Koivo 1992. Properties of the neural network internal model controller. *Preprints 1992 IFAC/IFIP/IMACS Int. Sym. on A.I. in Real Time Control*, Delft University of Technology, Delft (The Netherlands), 221-226

10. L. Ljung 1987. *System Identification: Theory for the User*, Prentice Hall, Englewood Cliffs, New Jersey

11. S. Marsili-Libelli 1991. Fuzzy Clustering of Ecological Data. In: E. Feoli and L. Orloci (Eds.), *Computer Assisted Vegetation Analysis*, Keller Academic Publishers, The Netherlands

12. J.H. Mathews, 1992. *Numerical Methods for Mathematics, Science, and Engineering*, second edition, Prentice Hall, Englewood Cliffs, New Jersey

13. K.S. Narendra 1992. Adaptive control of dynamical systems using neural networks. In: *Handbook of Intelligent Control*, Multiscience Press Inc., 141-183

14. W. Pedrycz 1993. *Fuzzy Control and Fuzzy Systems*, second, extended edition, John Wiley, New York

15. J. Richalet 1993. Industrial Applications of Model Based Predictive Control. *Automatica*, 29, 1251-1274

16. J.A. Roels 1993. *Energetics and kinetics in biotechnology*, Elsevier, Amsterdam (The Netherlands)

17. M.L. Thompson and M.A. Kramer 1994. Modeling chemical processes using prior knowledge and neural networks. *AIChE Journal*, 40, 1328-1340

18. R. Soeterboek 1992. *Predictive Control; a Unified Approach*, first edition, Prentice Hall International, United Kingdom

19. M. Sugeno and K. Tanaka 1991. Successive identification of a fuzzy model and its application to prediction of a complex system. *Fuzzy Sets and Systems*, 42, 315-334

20. M. Sugeno and T. Yasukawa 1993. A Fuzzy-Logic-Based Approach to Qualitative Modeling. *IEEE Transactions on Fuzzy Systems*, 1, 7-31

21. T. Takagi and M. Sugeno 1985. Fuzzy identification of systems and its application to modeling and control. *IEEE Transactions on Systems, Man and Cybernetics*, 15(1), 116-132

22. R.M. Tong 1979. The construction and evaluation of fuzzy models. In: M.M. Gupta, *et al.* (Eds.), *Advances in Fuzzy Sets Theory And Its Applications*, North-Holland, 59-576

23. R.A.J. de Vries and T.J.J. van den Boom 1994. Constrained Predictive Control with guaranteed stability and convex optimization. In: *Proceedings of the American Control Conference*, Baltimore, Maryland (USA), 2842-2846

24. S. Yasonobu and S. Miymoto 1985. Automatic train operation system by predictive fuzzy control. In: M. Sugeno (Ed.), *Industrial Applications of Fuzzy Control*, North-Holland, 1-18

25. J. Zhao, V. Wertz and R. Gorez 1994. A fuzzy clustering method for the identification of fuzzy models for dynamical systems. In: *IEEE International Symposium on Intelligent Control*, Columbus, Ohio (USA)

MIX
Papier aus verantwortungsvollen Quellen
Paper from responsible sources
FSC® C105338

If you have any concerns about our products,
you can contact us on
ProductSafety@springernature.com

In case Publisher is established outside the EU,
the EU authorized representative is:
Springer Nature Customer Service Center GmbH
Europaplatz 3, 69115 Heidelberg, Germany

Printed by Libri Plureos GmbH
in Hamburg, Germany